KT-442-517

TRANSPUTER DEVELOPMENT SYSTEM

Second Edition

INMOS Limited
INMOS is a member of the
SGS-THOMSON Microelectronics Group

Prentice Hall
New York London Toronto Sydney Tokyo Singapore

First edition published 1988

This second edition published 1990 by
Prentice Hall International (UK) Ltd
66 Wood Lane End, Hemel Hempstead
Hertfordshire HP2 4RG
A division of
Simon & Schuster International Group

© INMOS Limited, 1988, 1990

INMOS reserves the right to make changes in
specifications at any time and without notice. The
information furnished by INMOS in this publication is
believed to be accurate, however no responsibility is
assumed for its use, nor for any infringement of patents
or other rights of third parties resulting from its use. No
licence is granted under any patents, trademarks or
other rights of INMOS.

●, inmos, IMS and occam are trademarks
of INMOS Limited. INMOS is a member of the
SGS-THOMSON Microelectronics Group.

INMOS document number: 72 TRN 011 01

All rights reserved. No part of this publication may be
reproduced, stored in a retrieval system, or transmitted
in any form, or by any means, electronic, mechanical,
photocopying, recording or otherwise, without prior
permission, in writing, from the publisher.
For permission within the United States of America
contact Prentice Hall Inc., Englewood Cliffs, NJ 07632.

Printed and bound in Great Britain at the
University Press, Cambridge

ISBN 0-13-929068-0

CIP data are available from the publisher

1 2 3 4 5   94 93 92 91 90

S. 223.51   571915   /45.75   CP.
£26.99

TRANSPUTER DEVELOPMENT SYSTEM

Second Edition

Other titles in this series

Transputer Reference Manual
Transputer Development System (Second Edition)
Communicating Process Architecture
Transputer Technical Notes
Transputer Instruction Set: A Compiler Writer's Guide
Digital Signal Processing

Contents

Contents overview

Preface

This manual describes the Transputer Development System, an integrated programming environment developed by INMOS to support the programming of transputer networks in occam. The Transputer Development System comprises an integrated editor, file manager, compiler and debugging system.

The Transputer Development System runs on a transputer board; for example it runs on an INMOS IMS B008 board with an IMS B404 TRAM (transputer module) containing an IMS T800 32-bit processor and 2 MBytes of memory. This board is installed inside an IBM PC/AT or similar computer, which provides a means of interfacing keyboard, screen and disks to the transputer.

The Transputer Development System allows occam programs to be written, compiled and then run from within the development system. Programs may also be configured to run on a target network of transputers; these may range from a single transputer on an evaluation board to networks of several hundred transputers. The code for a transputer network may be loaded directly from the Transputer Development System, through a link connecting the Transputer Development System transputer to the target network. Programs may also be placed into a file separate from the Transputer Development System, or into a ROM (Read-Only Memory), and used to load a network.

A post-mortem debugger allows programs running in the Transputer Development System environment or on a transputer network to be examined after they have been interrupted or have stopped as a result of an error. The line of source corresponding to a program error on one of the processors can be displayed, and the values of variables may be examined. The state of other currently active processes on this processor, and on other processors in the network, can also be examined.

The Transputer Development System software includes the interactive programming environment, the compilation utilities and other programming tools, a number of libraries to support program development (such as mathematical functions and I/O libraries), and an extensive set of examples in source form.

This manual is divided into two major parts: the *User Guide*, which introduces the system and takes the reader through the steps needed to write, compile and run programs, and the *Reference Manual*, which contains detailed reference information on the editor, utilities, tools, libraries and system interfaces.

The instructions on installing the software and a detailed list of the components of the release are contained in a separate *Delivery manual*, supplied with the software.

This manual corresponds to the IMS D700E (IBM PC) release of the Transputer Development System, which supports new transputer targets and is supported by, and can generate programs supported by, the server program `iserver`, used by all other INMOS hosted software products.

1 How to use the manual

The Transputer Development System Manual is broadly structured into four sections:

- Introduction
- User Guide
- Reference Manual
- Appendices

Each of the sections is briefly described below.

1.1 Introduction

This section gives a light, readable introduction to the transputer and the Transputer Development System (referred to as TDS in the rest of this manual). The rest of the manual does not require this to have been read and anyone reasonably familiar with the transputer can skip over this section. It does not require the reader to be sitting at a terminal, in fact it can be read anywhere: in an armchair or on a train for example.

1.2 User guide

The user guide provides the essential information for someone to start using the TDS. It provides an introduction to the facilities of the TDS and contains examples where appropriate. Most, but not all, is essential reading, depending upon one's individual interests.

Chapters 3 to 6, which introduce the development environment, should be carefully read by everyone.

Chapters 7 and 8 which deal with transputer networks and standalone programs need only to be read if they satisfy a user's interest.

Chapter 9 on debugging should be read by everyone, but not necessarily the sections relating to networks.

Chapters 10 and 11 which deal with EPROM programming and low level programming are not essential reading.

1.3 Reference manual

The reference manual gives the detailed, technical information that was not appropriate to the user guide. This part of the manual is not intended to be read as such, merely referred to.

1.4 Appendices

The appendices are there to provide rapid reference. As such certain of the information may duplicate that already found in the reference manual, but it is in a more accessible form.

1.5 Delivery manual

Additional information about installation and host-dependent aspects of the software will be found in a separate delivery manual shipped with the software.

2 Introduction

2.1 Overview

A transputer is a microcomputer with its own local memory and with links for connecting one transputer to another transputer.

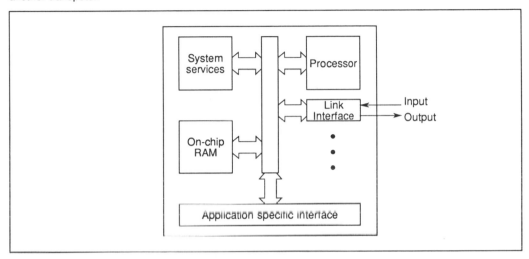

Figure 2.1 The transputer architecture

The transputer architecture defines a family of programmable VLSI components. A typical member of the transputer product family is a single chip containing processor, memory, and communication links which provide point to point connection between transputers. In addition, each transputer product contains special circuitry and interfaces adapting it to a particular use. For example, a peripheral control transputer, such as a graphics or disk controller, has interfaces tailored to the requirements of a specific device.

A transputer can be used in a single processor system or in networks to build high performance concurrent systems. A network of transputers and peripheral controllers is easily constructed using point-to-point communication.

Transputers and occam

Transputers can be programmed in most high level languages, and are designed to ensure that compiled programs will be efficient. Where it is required to exploit concurrency, but still to use standard languages, occam can be used as a harness to link modules written in the selected languages.

To gain most benefit from the transputer architecture, the whole system can be programmed in occam. This provides all the advantages of a high level language, the maximum program efficiency and the ability to use the special features of the transputer.

occam provides a framework for designing concurrent systems using transputers in just the same way that boolean algebra provides a framework for designing electronic systems from logic gates. The system designer's task is eased because of the architectural relationship between occam and the transputer. A program running in a transputer is formally equivalent to an occam process, so that a network of transputers can be described directly as an occam program.

The occam language used in this product is occam 2, this is a successor to the untyped language occam 1 or proto-occam.

2.2   System design rationale

The transputer architecture simplifies system design by the use of processes as standard software and hardware building blocks.

An entire system can be designed and programmed in occam, from system configuration down to low level I/O and real time interrupts.

2.2.1   Programming

The software building block is the process. A system is designed in terms of an interconnected set of processes. Each process can be regarded as an independent unit of design. It communicates with other processes along point-to-point channels. Its internal design is hidden, and it is completely specified by the messages it sends and receives. Communication between processes is synchronized, removing the need for any separate synchronisation mechanism.

Internally, each process can be designed as a set of communicating processes. The system design is therefore hierarchically structured. At any level of design, the designer is concerned only with a small and manageable set of processes.

2.2.2   Hardware

Processes can be implemented in hardware. A transputer, executing an occam program, is a hardware process. The process can be independently designed and compiled. Its internal structure is hidden and it communicates and synchronizes with other transputers via its links, which implement occam channels.

The ability to specify a hard-wired function as an occam process provides the architectural framework for transputers with specialized capabilities (e.g. graphics). The required function (e.g. a graphics drawing and display engine) is defined as an occam process, and implemented in hardware with a standard occam channel interface. It can be simulated by an occam implementation, which in turn can be used to test the application on a development system.

2.2.3   Programmable components

A transputer can be programmed to perform a specialized function, and be regarded as a 'black box' thereafter. Some processes can be hard-wired for enhanced performance.

A system, perhaps constructed on a single chip, can be built from a combination of software processes, pre-programmed transputers and hardware processes. Such a system can, itself, be regarded as a component in a larger system.

The architecture has been designed to permit a network of programmable components to have any desired topology, limited only by the number of links on each transputer. The architecture minimizes the constraints on the size of such a system, and the hierarchical structuring provided by occam simplifies the task of system design and programming.

The result is to provide new orders of magnitude of performance for any given application, which can now exploit the concurrency provided by a large number of programmable components.

2.3   occam model

The programming model for transputers is defined by occam. The purpose of this section is to describe how to access and control the resources of transputers using occam. A more detailed description is available in the occam reference manual.

Where it is required to exploit concurrency, but still to use standard sequential languages such as C or FORTRAN, occam can be used as a harness to link modules written in the selected languages.

In occam processes are connected to form concurrent systems. Each process can be regarded as a black box with internal state, which can communicate with other processes using point to point communication channels. Processes can be used to represent the behaviour of many things, for example, a logic gate, a microprocessor, a machine tool or an office.

The processes themselves are finite. Each process starts, performs a number of actions and then terminates. An action may be a set of sequential processes performed one after another, as in a conventional programming language, or a set of parallel processes to be performed at the same time as one another. Since a process is itself composed of processes, some of which may be executed in parallel, a process may contain any amount of internal concurrency, and this may change with time as processes start and terminate.

Ultimately, all processes are constructed from three primitive processes — assignment, input and output. An assignment computes the value of an expression and sets a variable to the value. Input and output are used for communicating between processes. A pair of concurrent processes communicate using a one way channel connecting the two processes. One process outputs a message to the channel and the other process inputs the message from the channel.

The key concept is that communication is synchronized and unbuffered. If a channel is used for input in one process, and output in another, communication takes place when both processes are ready. The value to be output is copied from the outputting process to the inputting process, and the inputting and outputting processes then proceed. Thus communication between processes is like the handshake method of communication used in hardware systems.

Since a process may have internal concurrency, it may have many input channels and output channels performing communication at the same time.

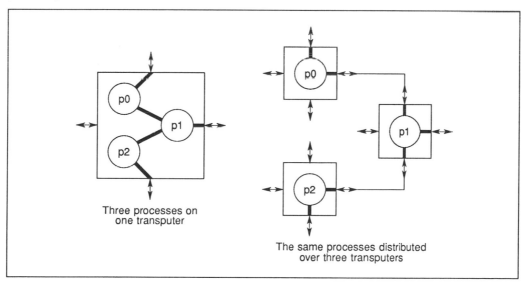

Three processes on
one transputer

The same processes distributed
over three transputers

Figure 2.2 Mapping processes onto one or several transputers

Every transputer implements the occam concepts of concurrency and communication. As a result, occam can be used to program an individual transputer or to program a network of transputers. When occam is used to program an individual transputer, the transputer shares its time between the concurrent processes and channel communication is implemented by moving data within the memory. When occam is used to program a network of transputers, each transputer executes the process allocated to it. Communication between occam processes on different transputers is implemented directly by transputer links. Thus the same occam program can be implemented on a variety of transputer configurations, with one configuration optimized for cost, another for performance, or another for an appropriate balance of cost and performance.

The transputer and occam were designed together. All transputers include special instructions and hardware to provide maximum performance and optimal implementations of the occam model of concurrency and communications.

All transputer instruction sets are designed to enable simple, direct and efficient compilation of occam. Programming of I/O, interrupts and timing is standard on all transputers and conforms to the occam model.

Different transputer variants may have different instruction sets, depending on the desired balance of cost, performance, internal concurrency and special hardware. The occam level interface will, however, remain standard across all products.

2.4   A programmer's introduction to the transputer

This description is intended to introduce the transputer to readers whose background is in programming rather than in hardware.

Most of the information here is taken from 'The Transputer Reference Manual'. Reference is also made to other INMOS publications where particular matters are discussed at greater depth.

Transputers are members of a family of VLSI components. Many properties and features of transputers are common to all members of the family, but some differ between members. This introduction concentrates on the things which are common to all transputers.

A transputer is a VLSI device (a chip) consisting principally of a processor, memory and communications links. Transputers can (but need not) have additional memory connected externally.

Transputers are designed to be connected to other transputers, through their communications links. Typically there are four links on a transputer, but there is nothing fundamental about this number.

2.4.1   Addresses and the memory

Addresses in a transputer are signed binary integers whose range is determined by the word length of the particular transputer. A word on the transputer may contain any integral number of 8-bit bytes. Existing transputers have either 16-bit (2-byte) words or 32-bit (4-byte) words. Although data transfer to and from memory is normally in multiples of words at the hardware level, the addressability is at the individual byte level.

Each byte in address space has a distinct address. The lowest is the most negative number expressible in the word size (-32768 on a 16-bit transputer). The highest address is the highest positive number so expressible. Because negative numbers in decimal representation are not particularly easy to think about, it is normal to use hexadecimal representation when talking about transputer addresses.

The address range of a 16-bit transputer is thus #8000 to #7FFF (64K bytes) and that of a 32-bit transputer is #80000000 to #7FFFFFFF (4 Gigabytes).

Bytes within words are addressed with increasing addresses from the least significant end of the word. The address of a word is always expressed as the byte address of its least significant byte. This 'little-endian' convention is applied uniformly to all data representation and addressing on the transputer. It is therefore conventional for programming language compilers to allocate arrays with the lowest subscripted element at the lowest address. This extends naturally to arrays of arrays.

Whether or not a particular address actually refers to a memory cell which can be read and/or written depends on the type of transputer and what connections have been made to its external memory interface pins.

For convenience we shall call the minimum address **Minint**. The addresses at and immediately above **Minint** are used to address the communications links, and for some special purposes which may be considered equivalent to additional special purpose hardware registers.

These addresses, and a block of addresses above these, refer to memory in the on-chip RAM of the transputer. The size of the on-chip RAM varies between transputer types (e.g. on the IMS T800 it is 4096 bytes). On-chip RAM has a faster access time than off-chip memory; the way program design can take advantage of this is mentioned below. The first address not used by the processor for special purposes is called **MemStart**.

Whether or not addresses further from **Minint** than the size of on-chip RAM identify actual memory locations depends on the external hardware design. External memory is normally connected in such a way that its addresses follow on directly from the on-chip RAM. Most compilers rely on this convention. Some of the highest end of address space may be allocated to ROM on self-starting systems. Other addresses above the rest of RAM may be allocated for special purposes such as memory-mapped peripherals, shared video-RAM or whatever a particular hardware design needs to provide in such a way that a program may access it as if it were memory.

2.4.2   Registers and instructions

For a more complete description of the instruction set of the transputer see the book 'Transputer instruction set: a compiler writer's guide'.

Like the majority of processors for which programmers write programs the transputer has a small number of registers and a repertoire of instructions, transferring bit patterns between registers and/or memory and performing operations on these bit patterns.

The registers of the transputer are special purpose, and are not explicitly referenced by instructions. it is not necessary for the beginning programmer to be aware of what registers there are, as high level programming languages hide these from the programmer. As mentioned above some of these registers are actually implemented in memory, the others are special purpose hardware.

Two registers, which are common to all transputers, are fundamental to the understanding of how programs on the transputer are constructed. These are the instruction pointer register **Iptr** and the workspace descriptor register **Wdesc**.

The instruction pointer register has the same size as the transputer word and so can hold a byte address to a location anywhere in address space. It holds the address of the next instruction to be executed by the processor.

The workspace descriptor register holds two items of information. Its least significant bit defines the processor priority. The rest of the register holds the address of a word in memory. As word addresses always have a zero in the least significant position, this word address is derived by forcing the priority bit to zero. The word address, known as **Wptr**, points to a place in memory conventionally used for the local variables of a procedure or other program unit. Any other information required to define the state of a process is addressed, directly or indirectly, relative to **Wptr**. Absolute addressing is reserved solely for the locations below **Memstart** and for addressing memory mapped peripherals, etc.

The design of the instruction set has two particularly important features. It is never necessary for absolute addresses to be embedded in code, and so code can always be made position independent. All instructions occupy one byte each and so the structure of the instruction set is independent of the word length of the transputer.

Each one-byte instruction consists of two 4-bit fields. The most significant 4 bits define the operation and the least significant 4 an operand. One of the 16 basic instructions is an 'operate code' whose operand extends the instruction set. Two of the operands out of the other 15 possible are used as special purpose prefix operations and allow either extended operation codes or operands up to the word length to be accumulated. The remaining 13 instructions use the operand as an immediate literal value or address offset. Address offsets are normally expressed in words, except for explicit byte accessing instructions and jumps and calls which access program code addresses.

The set of instructions provided is especially tailored to the support of programs compiled from high level languages, especially the occam language designed by INMOS for the transputer. occam is defined in the 'occam 2 reference manual'.

The instructions cover the usual range of operations including transfers between memory and registers, arithmetic and logical operations on values in registers and the usual unconditional and conditional relative jumps, etc. In addition there is a range of instructions which specifically support the occam process model. These instructions assume that the compiler has organised the run-time memory into workspace areas which can be pointed to by **Wptr**, and use the locations immediately surrounding the word identified by **Wptr** for special purposes connected with the management of processes and the switching of contexts.

Instructions are implemented by microcode in the processor, but the user need never be concerned with this level of processing. Some transputers have additional special purpose processing units for such operations as floating point arithmetic. From the programmer's point of view these are treated as an extension to the central processor with instructions encoded in a uniform manner.

2.4.3   Processes and communications

The transputer and occam model of computation is based on the concept of processes which communicate solely by transferring messages to each other along channels. A communication is strictly point to point with one sending process and one receiving process and is synchronised, insofar as either the sender or the receiver may be ready to communicate before the other, and is then unable to proceed until the other becomes ready and the transfer has taken place. A single communication may be any length from one byte upwards, restricted only by the fact that the count (of bytes) must fit in one word, interpreted as an unsigned integer.

The transputer instructions for communication and the hardware communication links have been designed so that it is possible to compile identical code for communications between processes sharing a single processor and for communications across hardware links. This makes it possible to design, and to compile components of, multi-process programs independently of the allocation of these components to distinct processors. This is one of the reasons for outlawing communication between processes by means of shared access to memory locations. The hardware does not protect against such shared access but the occam compiler does. Another important check that the compiler will perform if possible is that the sending process and the receiving process in a communication agree on the length of each message transmitted. When the lengths of messages are not known at compile time, it is conventional for them to be sent in a previous message along the same channel.

The channels used for communication are represented directly in the occam language. Any word in memory may be used as an internal (soft) channel. Such a word is named and allocated, in the same way as a program variable, in the workspace of a process. The channel word holds a pointer to the workspace of a process waiting for communication on that channel. This word will contain a unique null value except at the time when one process is ready to communicate on that channel and another is not yet ready. This pointer will in turn identify the instruction pointer value at which the other process may resume execution after the transfer is complete.

In order to ensure that the process at the other end of a communication will have an opportunity to become ready, the processes running concurrently on any one processor are time sliced. Time slicing can only occur after the execution of one of a limited set of instructions. These points are designed to be at positions in the execution sequence where the quantity of state information in registers that needs to be saved at the switch of context is minimised.

Processes awaiting communication are only identified in the channel word of that communication. Other processes that have been started, but which are not actually being executed, are held on queues represented by head and tail pointers in dedicated processor registers and the intermediate pointers in the workspaces of the processes themselves. A similar queuing mechanism is used to handle the transputer's real time clock, for which any process may wait using head and tail pointers in the reserved memory below **MemStart**. The representation of queues by links in the workspaces of the processes themselves does not impose a limit on the lengths of such queues, and so very large numbers of parallel processes, many of which may share common code, can coexist within a transputer.

The implementation of communication is such that when long messages are being transferred between external links and memory, the processor is not involved, and so other processes using other links or not using any links may proceed strictly in parallel. This feature, accompanied by careful program design, may be used to optimise the performance of distributed computations.

This brief introduction to the implementation of communications is intended only to give readers a general idea of the way this is done. In practice the behaviour is complicated by the existence of two priorities, and the need to be able to handle the occam concept of alternation, which allows a process to be ready to receive an input from any one of a number of channels at any one time.

2.4.4   Starting and stopping

Some of the pins on a transputer chip are provided to handle the special requirements of starting and stopping transputer programs. Starting is concerned with getting a transputer going when it is first switched on or when it must be restarted after stopping. Stopping is normally accompanied by the assertion of an error signal on an output pin.

The way these pins are made visible to the programmer is dependent on the design of the circuit board on which the transputer is mounted. In general it may be possible for external processors to set some of the input pins and read the state of some of the output pins, or their state may depend on microswitch settings or other hardware devices.

The normal method of starting a transputer is to cause it to execute a bootstrap program, which may in turn load and/or execute any other program in the memory.

The **BootFromRom** pin determines whether the bootstrap is assumed already to exist in ROM or must be received as a communication on a link before being executed. These two ways of starting are called *boot from ROM* and *boot from link*, respectively. A **Reset** signal initiates the bootstrap process.

When booting from ROM the necessary boot program must have been written into the ROM chip, and the chip wired into address space, in such a way that the two bytes at the extreme top of address space are a jump instruction to the bootstrap program itself. According to the system design the rest of the program may then be read from a link or another peripheral, or may already be resident in the ROM. Code for other processors connected by links may also reside in the ROM of a transputer acting as network master and may be transmitted to these processors as part of the bootstrapping process.

When booting from link the transputer starts in a state in which it can respond to any input on any one of its links. The first byte received determines the next action. If this byte has any value greater than 1, it defines the length of the bootstrap program which follows. Such a bootstrap program is read directly from the link into memory starting at the address **MemStart**, and the **Iptr** and **Wdesc** are initialised so that the instruction at **MemStart** is the first to be executed, and the **Wdesc** is set for low priority with a workspace starting at the first word above the bootstrap program. The identity of the host link and the previous values of **Iptr** and **Wdesc** are preserved in registers.

If the first byte received when a transputer is set to boot from link is a 0 or a 1, one of two special actions occurs. These special actions consist of writing or reading the contents of any one word in address space. The read command (0) is expected to be followed by a single word which is interpreted as an address; the transputer then sends the contents of that address out on the output link corresponding to the input link on which the command was received. The write command (1) is expected to be followed by two words, the first is interpreted as an address and the second as a value to be written there.

These commands, alternatively known as peek and poke, may be used by a program at the other end of such a link to determine the state of a transputer which has been reset but not booted, or to perform any other diagnostic actions. Any number of peek or poke commands may be received and obeyed before a bootstrap program is received.

The transputer has an error flag which can be set on the occurrence of arithmetic overflow and similar conditions. It can also be set explicitly. The state of this flag is represented by the **Error** pin whose state can be detected externally. INMOS evaluation boards are designed so that after an error is detected an **Analyse** signal may be sent to the network of transputers including the one which has set error.

The **Analyse** signal forces a clean close down on each transputer and leaves it in a state in which its memory and register contents at the time of error may be recovered. This feature is used by the debugger program.

INMOS have established a set of conventions for the use of these signals for the loading and post-mortem analysis of transputer programs. These are discussed in Technical Notes 33 and 34.

2.4.5   Programs

A transputer program, like any other, must be designed, written, compiled, linked, stored somewhere, loaded, entered and run. A transputer program, unlike most others, may be a multi-processor program.

A transputer program may be constructed to be totally self-contained, handling all its communications with its peripheral devices itself, or it may be designed to require support at run time from some other program (a run time system). Communication between a program and its run time system should be designed to use occam channels. If the run time system is on a processor also used by the program then it is conventional to build the program as an occam procedure (**PROC**) and to pass these channels to it as procedure parameters. Initial values of any kind may also be passed into such a program as parameters.

Otherwise the run time system may be any process connected to the program only by hard channels (links). Such a run time system may be another transputer program, or any suitably designed program on other kinds of processor connected in such a way that it can communicate using one or more INMOS links. INMOS evaluation boards designed for mounting inside an IBM PC or compatible computer achieve this by using an INMOS link adaptor which appears to the transputer as a pair of hard channels, but which appears to the 8086 (or similar) processor in the PC as a collection of addressable locations in memory.

2.4.6   Multi-processor programs

As it is possible for arbitrarily long messages to be communicated in a single transfer, the process of loading a program into a transputer can be made very simple. A multi-processor program may be loaded into a network of transputers from any point in the network, using any graph of link connections which spans the network. Each processor in the network first receives a bootstrap program, which in turn performs its part in the loading of more remote processors and finally itself. The code on each processor is then entered at its entry point, and can perform an arbitrary amount of local initialisation or other computation before interacting with one or more connected processors by attempting a link communication.

The stored form of a multi-processor program must be constructed using knowledge of the network connectivity to be used for loading. The individual code blocks transmitted to the network have to have their destinations identified. These destinations are defined solely by the path, in terms of link numbers at each processor on the path, to be taken by the code. There is no concept of processor naming in the stored form of a program.

The software tool which constructs the stored form of a multi-processor program from its components is called a configurer.

2.4.7   Conventions for the code on each processor

Conventions adopted by INMOS software tools define the structure of the code area and the data area on each transputer processor. The hardware architecture does not impose any such structure, beyond requiring each process to have, at any time, a block of words in memory to which the workspace pointer **Wptr** points. As addressing relative to **Wptr** is the most efficient method of memory access, and offsets up to 15 words from **Wptr** are the most efficiently accessed of these, compilers generate code which allocates these words in blocks as stack frames to local variables, procedure parameters, etc.

It is a feature of occam 2 that the sizes of stack frames can be determined at compile time. This enables a simple approach to be adopted to the allocation of stack frames for called procedures at addresses immediately below the frame for the calling procedure. Parameter locations are so positioned that they can be equally readily accessed by the calling code and by the called code after **Wptr** has been adjusted appropriately. The size of each stack frame is called the workspace size of the procedure. This is handled internally by the compiler for locally declared procedures, but must be kept in a compiled code descriptor for separately compiled modules, such as those in a library of commonly used procedures.

Parallel processes, including those executing common code, are allocated their own disjoint workspaces within which all locally declared variables will be allocated. In occam the sizes of these are also determinable at compile time.

Languages with recursion and/or arrays whose sizes are determined at run time do not allow such a simple approach to the allocation of stack frames. There will inevitably be more overhead at run time associated with procedure calling in such languages.

The variables of a program may be classified into the scalars, occupying up to 8 bytes each, and the arrays or vectors, which are larger blocks normally accessed by indexing. Compilers may decide whether or not to include the vectors in the workspace stack frame for a procedure. If a compiler allows vectors to be allocated in a separate area, performance will probably be enhanced because of the more efficient access to scalars with smaller offsets, but there will be added complexity at compile time as both scalar and vector workspace sizes will need to be managed in compiled code descriptors.

The separate allocation of vectors has an additional benefit on the transputer where the lowest memory addresses are those of the on-chip RAM which has the fastest access speed. By allocating scalar workspace at the lowest addresses, followed by the code, followed in turn by the vector workspace, an optimum use of on-chip RAM is likely to be achieved for the majority of programs.

When the user's program shares a single transputer with a run time system such as is provided by the INMOS TDS, it is conventional for the run time system to be allocated at the high end of available memory. When loading a user's program, the workspace requirement of that program, and the run time system's use of memory for its own purposes can be taken into account before the called program is entered, and the called program can be informed of the size of otherwise unused memory as a procedure parameter. This technique is used by software tools within the TDS to create working arrays with sizes proportional to the size of available memory.

2.5 Program development

The development of programs for multiple processor systems can involve experimentation. In some cases, the most effective configuration is not always clear until a substantial amount of work has been done. For this reason, it is desirable that most of the design and programming can be completed before hardware construction is started.

2.5.1 Logical behaviour

An important property of occam in this context is that it provides a clear notion of 'logical behaviour'; this relates to those aspects of a program not affected by real time effects.

It is guaranteed that the logical behaviour of a program is not altered by the way in which the processes are mapped onto processors, or by the speed of processing and communication. Consequently a program ultimately intended for a network of transputers can be compiled, executed and tested on a single computer used for program development.

Even if the application uses only a single transputer, the program can be designed as a set of concurrent processes which could run on a number of transputers. This design style follows the best traditions of structured programming; the processes operate completely independently on their own variables except where they explicitly interact, via channels. The set of concurrent processes can run on a single transputer or, for a higher performance product, the processes can be partitioned amongst a number of transputers.

It is necessary to ensure, on the development system, that the logical behaviour satisfies the application requirements. The only ways in which one execution of a program can differ from another in functional terms result from dependencies upon input data and the selection of components of an **ALT**. Thus a simple method of ensuring that the application can be distributed to achieve any desired performance is to design the program to behave 'correctly' regardless of input data and **ALT** selection.

2.5.2   Performance measurement

Performance information is useful to gauge overall throughput of an application, and has to be considered carefully in applications with real time constraints.

Prior to running in the target environment, an occam program should be relatively mature, and indeed should be correct except for interactions which do not obey the occam synchronization rules. These are precisely the external interactions of the program where the world will not wait to communicate with an occam process which is not ready. Thus the set of interactions that need to be tested within the target environment are well identified.

Because, in occam, every program is a process, it is extremely easy to add monitor processes or simulation processes to represent parts of the real time environment, and then to simulate and monitor the anticipated real time interactions. The occam concept of time and its implementation in the transputer is important. Every process can have an independent timer enabling, for example, all the real time interactions to be modelled by separate processes and any time dependent features to be simulated.

2.5.3   The transputer development system

The transputer development system is an integrated development system which can be used to develop occam programs for a transputer network. It consists of a plug in board for an IBM PC with a transputer module, such as an IMS B404, and all the appropriate development software, see figure 2.3.

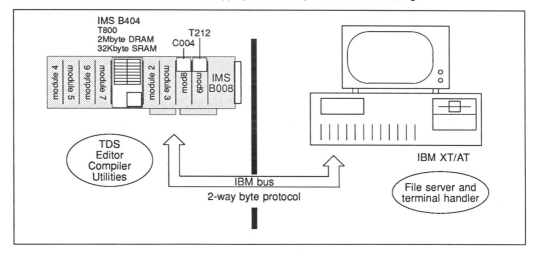

Figure 2.3 Transputer development system

Most of the development system runs on the transputer board; there is a program on the IBM PC called a 'server', which provides the development system with access to the terminal and filing system of the IBM PC.

Using the TDS a programmer can edit, compile and run occam programs entirely within the development system. occam programs can be developed on the TDS and configured to run on a network of transputers, with the code being loaded onto the network from the TDS. Alternatively an operating system file can be created which will boot a single transputer or network of transputers. As a final variation, the TDS can be used to create programs for single transputer or networks of transputers that operate completely independently of the TDS; such code could be placed in EPROM for example. Programs that work independently of the TDS are known as 'standalone' programs.

The TDS comes with all the necessary software tools and utilities to support this kind of development. There is a variety of libraries to support mathematical functions and input/output for example. There is a sophisticated debugging tool and software to analyse the state of a network.

The user guide

3 Directories

The software components of the Transputer Development System are supplied as compressed files. As part of the installation procedure a directory **\ARCD700E** is created, where the compressed files are placed. The directories required by the system are created and the appropriate files extracted from the compressed ones and placed in the correct directories. The compressed files may then be deleted.

The main directory created is called **\TDS3** and all files are stored in subdirectories

The actual subdirectories used are (some in turn have subsubdirectories):

\TDS3\SYSTEM   System and utility files.
These files must be accessible from any working directory.
If the operating system on the computer is DOS
then a path must be set to this directory.

\TDS3\COMPLIBS   Compiler libraries.

\TDS3\TOOLS   Software tools supplied with the system e.g. debugger.

\TDS3\IOLIBS   Libraries of input/output procedures.

\TDS3\HOSTLIBS   Libraries of procedures supporting the **iserver** interface.

\TDS3\MATHLIBS   Libraries of additional mathematical functions.

\TDS3\TUTOR   Tutorial material described in the user guide.

\TDS3\EXAMPLES   Additional example programs that are supplied with the system.

\TDS3\SERVER   Server files.

\TDS3\INMOS   Libraries used by TDS tools and examples only.

Some of the terms used above will not be familiar to many people at this stage. They will become clearer by carefully reading and working through the user guide. There is a glossary at the end of the book and the reference manual describes the more technical aspects of the system in greater detail.

4 The editing environment

4.1   Introduction

The Transputer Development System (TDS) consists of a plug-in transputer board and development software which runs on the transputer board. This combination provides a complete, self-contained development environment in which programs can be developed, compiled and run. Programs can also be developed and compiled on the TDS to run on a network of transputers, the code being loaded on to the network from the TDS. In this case the combination of transputer board and PC is referred to as the 'host computer', and the transputer network is known as the 'target system'. Finally, as is probably more realistic for most applications, programs can be developed to run on transputers completely independently of the TDS; these are known as 'standalone' programs.

The principal interface to the system is an editor; as soon as the system starts up the user is placed in an editing environment, and all program editing, compilation and running can be carried out within that environment, by the use of a set of function keys. Instead of having a special command language to the operating system to manage the filing system, file operations occur automatically as a result of certain editor operations. There is also a set of 'utility' function keys which may be assigned to different functions during a session. Throughout this manual the convention of referring to function keys (including utility function keys) by name will be followed; for example: $\boxed{\text{CURSOR UP}}$ or $\boxed{\text{COMPILE}}$. In fact these logical names may correspond to a combination of physical keypresses at the terminal. The actual keys associated with these function key names are given in the keyboard layout diagrams in appendix A.

The editor interface is based on a concept called 'folding'. The folding operations allow the text currently being entered to be given a hierarchical structure ('fold structure') which reflects the structure of the program under development.

Because of the importance of folding within the TDS, this chapter starts by explaining folding. It then describes how to boot up the TDS. As with many systems, the best way to start learning about the TDS is to start using it. For this reason a tutorial file is provided; this does not assume any knowledge about the TDS so it can be worked through before reading the rest of the chapter. Section 4.3 describes how to find the tutorial file. The rest of the chapter describes the editor interface in some detail, and then describes the facilities for loading and running code within the editing environment.

4.1.1   Folding

Just as a sheet of paper may be folded so that portions of the sheet are hidden from view, the folding editor provides the ability to hide blocks of lines in a document. A fold contains a block of lines which may be displayed in two ways: open, in which case the lines of the fold are displayed between two marker lines (called creases), or closed, in which case the lines are replaced by a single marker line called a fold line.

To create a fold the user inserts creases around the text to be folded; the fold is closed automatically when the second crease is made. Any text may be placed on the fold line to indicate what the fold contains; this text is called the 'fold header'.

A fold may be removed, so that its contents are once again placed in sequence with the surrounding lines.

Folds may contain text lines and also fold lines; therefore folds can be nested. Folds can be nested to a maximum depth of 50.

An example of how folds are displayed by the editor follows. The fold line is marked with three dots (...). A top crease is marked with the symbol {{{. A bottom crease is marked with }}}. There are two folds in this program: one marked **Declarations**, and one marked `initialise`. In the second example the fold `initialise` has been opened.

Example: program with closed folds

```
... Declarations
SEQ
  ... initialise
  WHILE going
    process(ch, going)
```

Example: program with open fold

```
... Declarations
SEQ
  {{{  initialise
  going := TRUE
  input ? ch
  }}}
  WHILE going
    process(ch, going)
```

A fold has an indentation associated with it; the fold and crease line markers begin at this indentation level. No text may be inserted within the fold to the left of this indentation. In occam the indentation of a line is significant; the folding features of the editor make it relatively easy to change the indentation of part of an occam program.

Folding, in conjunction with the ability to nest folds, provides a way of organising a large document or program as a hierarchy. The editor has functions to 'enter' a fold, which opens the fold and moves down into it, and also to 'exit' the fold, which closes the fold and returns to the level from which the fold was entered. For example, entering the fold marked **Declarations** in the example above would make the following lines the only visible lines on the screen.

Example: entering a fold

```
{{{  Declarations
INT ch:
BOOL going:
PROC process()
  ... body of process
  :
}}}
```

Here the line marked **body of process** is a fold nested inside the fold **Declarations**.

Any document can be folded in such a way that most of the folds are shorter than the length of the screen. Fold operations then become the principal method of traversing a document, with screen scrolling operations used only for small local movement.

Because a closed fold is represented by a single line on the screen, some editor line operations may act on fold lines as well as text lines. When such an operation is applied to a fold line it also applies to the fold contents. For example, deleting a fold line deletes all its contents as well. This means that operations to transform the fold structure, (such as moving, copying, and deleting folds) appear identical to the line operations which are familiar to any user of a screen-oriented editor.

So far folds have been described as sequences of text lines; however, not all folds are text folds. There are also data folds, which are created by certain utilities in the system to store data. For example, when the occam compiler compiles a section of source code it places the resulting code in a data fold. Data folds appear as a single line on the screen, but cannot be opened and displayed by the editor.

In order to allow the system to distinguish the different types of folds, each fold has attributes to indicate the nature of its contents.

There are two attributes of interest:

1 The 'fold type' attribute which indicates to the editor the general nature of the contents of the fold (e.g. text, data)

2 The 'fold contents' attribute which indicates in more detail the nature of the contents of the fold (e.g. program text, comment text, compiled code, compiled and linked code).

The possible values of these fold attributes are listed in appendix F. Attributes remain with a fold until it is removed.

4.1.2 Files as folds

The folding editor allows a fold to be designated a 'filed fold'. The effect of this is to indicate that the fold contents are to be stored in a separate file. When the fold is first opened, the contents of the file are read in, and the fold may then be edited. When a filed fold is closed the system will write out the contents of all the files which have changed since they were last written out.

Many of the data folds produced by the utilities are in fact filed folds. The attributes of a filed fold are stored with the fold header, in the enclosing file, not in the file containing the fold's contents.

A large document or program consists of many files, organised in a nested structure. For example, consider the following program:

Example: use of filed folds

```
{{{F "filename" Example program        -- top crease of a filed fold

...F "filename" Declaration of PROC p1  -- filed fold
...F "filename" Declaration of PROC p2
PAR
  p1()
  p2()

}}}
```

The filed fold marked **Example program** contains filed folds which contain the declarations of **PROC p1** and **PROC p2**. Opening the filed fold marked **Declaration of PROC p1** causes the appropriate file to be read in and inserted at that point in the text.

The file containing the declaration of **p1** might also contain other filed folds. This shows how nested filed folds can be used to make up a large document. The document can be navigated in the same way as a small document, with only the explicitly opened sections of the document being read in by the editor. Most operations which can be carried out on fold lines may also be applied to filed fold lines, including those that contain nested files. So, for example, copying a filed fold line will make a copy of the file and all its nested files.

A directory used by the TDS contains a small number of root or 'top level' files, within which all other files are contained.

4.2   Starting and finishing the system for the first time

4.2.1   Starting the system

This section describes how to start the transputer development system from DOS command level, and how to start using the system.

To start the system for the first time move to an empty directory, ensure that there is a DOS path to the directory **\TDS3\SYSTEM** and type:

    tds3

In response to the **tds3** command the system will display a welcome message followed by:

    TDS system file   : *file path name*
    Board memory size : *x* Bytes

Once the TDS is loaded from disk, the system clears the screen and displays the top level view, which consists of all the files in the current directory with the extension **.TOP**. In the case of a new directory, there will be only one top level file: **TOPLEVEL.TOP**. The screen will appear as follows:

    Press [ENTER FOLD] to start session
    ...F "TOPLEVEL.TOP"

The principal operations available on these top level filed fold lines are ENTER FOLD, which enables a fold to be entered and edited, and FINISH, which ends the session. Most of the normal editing operations and utilities are disallowed here.

To enter one of the folds the cursor should be placed on the appropriate fold and the ENTER FOLD key pressed. The contents of the fold will be read in and displayed.

4.2.2   The TDS3 command

The TDS3 command has the form:

    tds3

This calls a command file **TDS3.BAT** in the directory **\TDS3\SYSTEM**. This normally contains the following command:

    \tds3\system\iserver /sb \tds3\system\tdsload.b4
                     -f \tds3\system\tds3.xsc

This runs the server program **ISERVER.EXE** on the host, and the TDS loader program **TDSLOAD.B4** on the transputer. The TDS loader program loads the file given by the **-f** parameter. See chapter 16 for full details of the **iserver** command line.

4.2.3   Problems starting the system

If no transputer board is connected the system may hang, or it may display one of the messages:

    Unable to access a tranputer

    or another message as listed in section 16.4.3

This will also happen if the transputer board does not have its reset link connected correctly or if the system is being run from an IBM PC which is neither a PC-XT nor a PC-AT. See the appropriate board manual for details on these matters.

The system may hang if the wrong link adaptor addresses are used. This may occur if a TDS configured for a different machine is used, or if the **/sl** command line option is incorrectly specified. See chapter 16 on System interfaces, which describes the server.

4.2.4   Keyboard layout

To display a map of the keyboard layout, press the HELP function key, which is assigned to the F1 key on the keyboard in the standard IBM PC layout. A keyboard map will appear; you can return to the normal editor display by pressing any key. Keyboard layouts are also shown in appendix A.

4.2.5   Repainting the screen

The function key REFRESH repaints the entire screen. This may be useful to check that the editor is driving the screen correctly, or if the terminal is accidentally switched off.

4.2.6   Ending the session

It is only possible to end the session from the outermost level (i.e. where the top level filed folds appear). Pressing FINISH here returns to the operating system. If any of the folds have been entered they must be exited back to this level before FINISH can be used.

4.2.7   Interrupting and rebooting the TDS

The server supporting the TDS can be interrupted by pressing the interrupt key. This is 'control-break' on IBM machines, but is 'control-c' on some others. The system will offer the option of exiting, calling an operating system shell or continuing. The user can press 'x' which returns to DOS command level, after giving the user an option to reboot the TDS in analyse mode for debugging.

The only time it should be necessary to press the interrupt key is when a user program fails to terminate and the system needs to be restarted. Subscript range and similar errors in a user program will cause the server to terminate automatically.

The interrupt key can also be used to prevent the TDS from writing out any more files, if a catastrophic edit has been done. This should not be done, however, if the system is actually in the process of writing out files.

If the TDS is called from a suitable command file, it may be re-entered automatically after the server is interrupted so that the debugger may be used to determine the cause of failure or state of the program at interruption.

4.2.8   Suspending the TDS

The key SUSPEND TDS can be used to suspend the TDS temporarily and return the user to the host operating system, so that operating system commands can be issued (for example, getting directory listings, or formatting floppy disks). In DOS typing the command **exit** returns to the TDS, which is in the same state as it was when SUSPEND TDS was pressed.

Before resuming the TDS, the current directory *must* be the same as it was when the TDS was suspended.

This facility works, in DOS, by making the server call the command file associated with the environment variable **COMSPEC**. The file associated with this variable can be changed by putting a **set** command of the following form into the **AUTOEXEC.BAT**:

> **set COMSPEC=***filename*

DOS commands which reset the transputer board (for example, running a server with another transputer boot file) will cause the state of the suspended session to be lost, and typing **exit** will then cause the system to hang up. The interrupt key can be used to release the system from this state.

4.3   Tutorial file

There is a file included with the system which provides an introduction for those starting to use the system. The file is in **\TDS3\TUTOR** and is called **TUTORIAL.TOP**. This file contains a detailed practical example on using the TDS and anyone new to this system is strongly advised to work through it.

To use the tutorial move to the directory **\TDS3\TUTOR**, then type:

    tds3

to start the system.

It is advisable to have nearby the appropriate keyboard layout. Keyboard layouts appear in appendix A.

When the system starts up ensure the cursor is on the line **TUTORIAL.TOP**, and then press ENTER FOLD to read in and display the file.

The contents of the file will then give you detailed instructions on how to proceed.

4.4   The editor interface

This section defines some terms which are used to describe the behaviour of the editor keys. Figure 4.1 shows a graphical representation of these terms.

4.4.1   Editor's view of a document

At any time during the session, the editor has a view of the document, consisting of a sequence of text lines, closed folds and open folds. This is called the current view.

The current view of the document at any time is principally determined by the fold operations which have been carried out. At the start of the session the current view contains a sequence of lines which correspond to the set of toplevel files in the current directory. When ENTER FOLD is pressed on one of these lines, the contents of the filed fold, surrounded by top crease and bottom crease lines, become the current view.

Whenever ENTER FOLD is pressed on a fold line, the current view is stacked up, and the contents of the fold become the current view. After editing the contents of the fold it is possible to return to the previous view using EXIT FOLD.

4.4.2   The screen display

The screen is divided into two parts. The top line of the screen is used to display messages. The rest of the screen displays a 'window' into the current view (that is, it displays as many lines of the current view as will fit on to the screen).

The editor provides functions to move the screen window up and down the current view, thus providing a scrolling facility . These functions do not change the editor's view of the document, merely what is visible in the screen window.

The cursor is used to point to a position in the screen window; functions are provided to move the cursor around the screen. The cursor cannot be moved below the end of the current view.

The current column is the column which the cursor is on. The current line is the line which the cursor is on. The current enclosing fold is the fold which contains the current line, or, if the current line is a crease line, the fold formed by that crease and its partner.

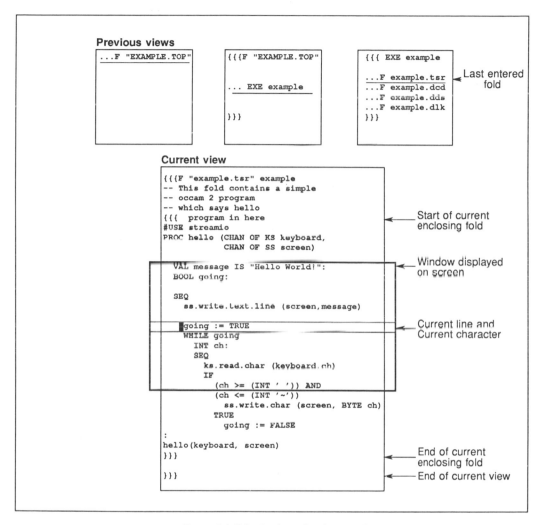

Figure 4.1 Editor's view of a document

4.4.3   Line types

Four general types of line may be displayed; they are text lines, top creases, bottom creases, and fold lines. Top creases and fold lines also have filed fold versions.

Fold lines and crease lines start with a marker symbol. The different types of marker symbols are:

Fold line	**. . .**
Filed fold line	**. . .F** *"filename"*
Top crease	**{ { {**
Filed fold top crease	**{ { {F** *"filename"*
Bottom crease	**} } }**

All marker symbols consist of the textual symbol above, plus one or two following spaces to give the symbol a width of five characters plus any filename. The marker, including the filename when present, is protected from change by the editor.

4.5   Editor functions

This section introduces and describes the functions provided by the editor. A detailed listing of the keys used and messages given by the editor is available in chapter 12. The mapping of key names to keys on the keyboard is given in appendix A.

4.5.1   Overview of editor functions

The editor accepts and acts on sequences of keystrokes from the user. If any of the sequences are not recognised the terminal bell rings. The table below provides an overview of the available editor functions, which are described in detail in the following sections. In addition there are function keys for loading and running code within the TDS, which are described in section 4.6.

Moving the cursor	CURSOR UP CURSOR LEFT START OF LINE	CURSOR DOWN CURSOR RIGHT END OF LINE	WORD LEFT WORD RIGHT	TOP OF FOLD BOTTOM OF FOLD
Scrolling the screen	LINE UP PAGE UP	LINE DOWN PAGE DOWN		
Fold browsing	ENTER FOLD OPEN FOLD	EXIT FOLD CLOSE FOLD	BROWSE FOLD INFO	
Inserting and deleting characters	Character keys DELETE	RETURN DELETE RIGHT	DELETE WORD LEFT DELETE WORD RIGHT DELETE TO END OF LINE	
Fold creation and removal	CREATE FOLD	REMOVE FOLD	MAKE COMMENT	
Storing text in files	FILE/UNFILE FOLD			
Deleting lines	DELETE LINE	RESTORE LINE		
Moving and copying lines	MOVE LINE	COPY LINE	COPY PICK PUT PICK LINE	
Defining and using a keystroke macro	DEFINE MACRO	CALL MACRO	SAVE MACRO	GET MACRO

4.5.2 Editor modes

At certain times when using the editor, only a limited subset of the editor functions may be available. For example, a fold is created by two presses of a key called CREATE FOLD; one to mark the top of the fold and one to mark the bottom of the fold. Between these two presses normal editing operations are not allowed; the only keys which the editor will accept are those needed to change the cursor position and the help key. All other keys cause the terminal bell to ring. When the editor is only accepting a restricted subset of keys, this is known as an editor 'mode'. It is indicated by a message on the top line of the screen which persists until the operation requiring the mode has been completed.

In the rest of this chapter, where a function results in an editor mode, this is indicated in the appropriate section.

4.5.3 Moving the cursor

The normal cursor positioning functions are used to move the cursor around the screen window. The cursor may be moved into any part of the screen, except the message line. In addition there are functions to move the cursor to the start or the end of the current line, and one word to the right or left on the line.

The cursor keys cause the screen to scroll when used at the top and bottom of the screen. Separate screen scrolling functions can be used to scroll the screen up and down the current view; these are described in the next section.

CURSOR UP moves the cursor up one line.

CURSOR DOWN moves the cursor down one line.

CURSOR LEFT moves the cursor left one column.

CURSOR RIGHT moves the cursor right one column.

END OF LINE places the cursor after the last significant character on the current line (which is normally the last non-blank character).

START OF LINE places the cursor on the first significant character of the current line (normally the first non-blank character).

Two keys, WORD LEFT and WORD RIGHT, are provided to move the cursor one word at a time. A word consists of a sequence of alphanumeric characters or a single non-alphanumeric character. More precise definitions of the word move operations are given in chapter 12 under the definitions of the relevant keys.

WORD LEFT moves the cursor one word to the left of the current cursor position.

WORD RIGHT moves the cursor one word to the right of the current cursor position.

TOP OF FOLD moves the cursor to the top crease line of the current enclosing fold. If the top crease line is not within the screen window the screen will be scrolled.

BOTTOM OF FOLD moves the cursor to the bottom crease line of the current enclosing fold. If the bottom crease line is not within the screen window the screen will be scrolled.

4.5.4 Scrolling and panning the screen

These functions scroll the screen up and down the current view by a line or a page at a time. A page is the number of lines in the screen window.

LINE UP moves the screen one line up the current view, if there are lines in the current view above the screen.

LINE DOWN moves the screen one line down the current view, if there are lines in the current view below the screen.

PAGE UP moves the screen one page up the current view, or to the top of the current view, whichever is the nearest.

PAGE DOWN moves the screen one page down the current view, or to the bottom of the current view, whichever is the nearest.

None of the above functions affect the position of the cursor on the screen

There are no keys which explicitly control panning, but if actions are taken which drive the cursor off the sides of the screen the whole screen will be repainted with the cursor position on the screen and an indication of how many columns are missing at the left in the message line at the top of the screen. There is an absolute maximum line length of 255 characters which can never be exceeded and it is recommended that the need to pan is minimised by keeping lines within the capacity of the screen.

4.5.5   Fold browsing operations

Opening and closing folds

This section describes the keys which are used, along with the cursor positioning keys, to move around a document. There are two pairs of fold browsing operations, one pair being ENTER FOLD and EXIT FOLD, and the other pair being OPEN FOLD and CLOSE FOLD.

The folding features of the editor give a document a hierarchical structure. The keys ENTER FOLD and EXIT FOLD are used to move around the hierarchy. When ENTER FOLD is pressed on a fold the screen is cleared and the contents of the fold become the current view. The previous view is stacked up, and can be returned to using EXIT FOLD

ENTER FOLD is appropriate when the fold contains a reasonably self-contained piece of text. However, it may often be more desirable to view a piece of text in its surroundings; for example the body of a **WHILE** loop may be folded up, and it may be best viewed with the **WHILE** condition displayed above it. OPEN FOLD and CLOSE FOLD are provided for this purpose.

OPEN FOLD inserts the contents of a fold between the surrounding lines, bracketted with top and bottom creases. CLOSE FOLD may be used to close an opened fold, and replace the displayed contents with a single fold line.

ENTER FOLD is useful where a quick return up to a particular position is required; doing an ENTER FOLD at that position will allow, at some future time, an EXIT FOLD to cause a return back up to that position.

At the outermost level, only ENTER FOLD may be used. Once the outermost level has been left by entering a fold, this starts the normal editing mode. All the editor functions are available in this editor mode, as well as the utilities.

Fold information

The key FOLD INFO, used on a fold or a crease line, displays the attributes of the fold on the message line. On a filed fold line, the message also includes the name of the file in which the contents of the fold are stored.

Browsing mode

Sometimes when viewing an existing document it is useful to set the editor into a mode so that you can not accidentally change the document. The key BROWSE can be used to get into and out of this mode. While in this mode a message is displayed continually on the message line of the screen, and all editor functions which could change the document are disallowed.

4.5.6   Inserting and deleting characters

In general characters may be inserted or deleted at the cursor position, but there are some exceptions, as follows:

1   Text may not be be inserted when the cursor is on the top crease of the view, a fold or crease marker, or when the cursor is to the left of the leftmost column of an open fold.

2   The indentation of a closed fold may be changed by inserting or deleting spaces to the left of a fold marker symbol. No other text may be inserted there.

Insertion

A character or space can be inserted in the current column position and the cursor, the character underneath the cursor and all subsequent characters on the line are moved right by one place.

$\boxed{\text{RETURN}}$ is used to split lines and insert blank lines. It has no effect on a fold line if used between the first and last significant characters of the line.

Deletion

$\boxed{\text{DELETE}}$ is used to delete the character to the left of the cursor. This causes the character underneath the cursor and the rest of the line to the right to be moved one place to the left. If $\boxed{\text{DELETE}}$ is used at the extreme left of a line it concatenates the line with the preceding line, if that line is not a long line. It has no effect if used at the extreme left of a fold or crease line.

$\boxed{\text{DELETE RIGHT}}$ deletes the character under the cursor. All the characters to the right of the cursor are moved left by one place. The cursor remains in the same position.

Character deletion has no effect on the top crease of a view, when the cursor is on part of a marker symbol, or is to the left of the leftmost column of an open fold.

Spaces may be deleted to the left of a closed fold to change the indentation of the fold.

$\boxed{\text{DELETE TO END OF LINE}}$ deletes all text from the character under the cursor, to the last significant character on the line, inclusive. The cursor remains in the same position.

Deletion can take place a word at a time. A word can be considered to be a sequence of alphanumeric characters or a single non-alphanumeric character, as for cursor movement.

$\boxed{\text{DELETE WORD LEFT}}$ deletes the word to the left of the cursor.

$\boxed{\text{DELETE WORD RIGHT}}$ deletes the word to the right of the cursor.

4.5.7   Fold creation and removal

Fold creation is achieved by marking the top and bottom of the sequence of lines required to form the contents of a fold. Two presses of $\boxed{\text{CREATE FOLD}}$ are needed to do this. Firstly the cursor should be placed at the start of the top line and $\boxed{\text{CREATE FOLD}}$ pressed. The column of the cursor at this point (i.e. how far it is from the left hand side of the screen) determines the indentation of the created fold. The cursor should then be moved to the line below the bottom line to be folded, and $\boxed{\text{CREATE FOLD}}$ pressed again. For this to work, all lines between the top and bottom lines must be indented at least as far as the indentation of the fold to be created; if this is not the case an error message is displayed.

After $\boxed{\text{CREATE FOLD}}$ has been pressed once the editor changes its mode and all normal editing functions are suspended until this key has been pressed again to complete the process of fold creation.

Once a fold has been created, it is good practice to add a comment by inserting text after the fold marker. This text is known as the fold header.

The created fold has an indentation associated with it, given by the indentation of the fold line marker when it is closed, and the indentation of the creases when It is open. It is not possible to insert text to the left of this indentation.

When a fold is newly created it is given default attributes: 'text' for the fold type attribute, and 'source' for the fold contents attribute. Other fold types are made using appropriate utilities.

An empty fold can be created above the current line by pressing CREATE FOLD twice in succession.

A fold may be 'commented out' by pressing MAKE COMMENT. This encloses the fold in another fold with the attribute 'comment text'. The word **COMMENT** is inserted on the new fold line before a copy of the previous fold header.

A fold can be removed by placing the cursor on a fold line and pressing REMOVE FOLD. The fold contents are inserted between the lines above and below the fold. This key should also be used for removing a comment fold.

If CREATE FOLD is pressed accidentally the fold must be completed by pressing CREATE FOLD again. The resulting empty fold may then be removed.

4.5.8   Filed folds

The editor provides the ability to store sections of a document in separate files. This can be done by creating a fold around the text and making the fold a filed fold. As described earlier, a filed fold is similar to an ordinary fold, but it has its contents stored in a separate file. When a filed fold is first opened, the contents of the file are read in and displayed. When a filed fold is closed, a new version of the file is automatically written out if the contents have changed since they were last written out.

When a fold is made into a filed fold the file must be given a name. In order to supply a name to the editor, the fold header may be edited before the fold is filed. The name given to the file will then be derived from the sequence of characters at the start of the fold header. It is not necessary to include an extension with the name; that is provided by the editor. The system checks that the file name generated is different from that of any existing files; if necessary, it adds numeric characters to the name to ensure this.

Filed folds may be treated in exactly the same way as ordinary folds, and most of the same operations apply. Filed folds may be copied, in which case a copy is made of all of the contents of the fold, including any nested files. New unique names are generated for copies of files, in the same manner as described above. Filed folds may also be deleted, in which case the corresponding file, and any nested files, are deleted.

The use of filed folds in the system allows the user to make up a document consisting of multiple files and browse through it in exactly the same way as browsing of a single folded file is done. No explicit commands to read or write files need be given. A file is always written back if it has changed. The new version of the file replaces the old version.

If it is necessary to back up a copy of a file before editing it, this may be done by duplicating the filed fold using the COPY LINE function key (see section 4.5.10). The duplicate may then be moved elsewhere in the fold structure.

One function key, FILE/UNFILE FOLD, is provided to convert ordinary folds to filed folds, and vice versa. Before a fold is filed the fold header should be edited so that the name intended for the file is written at the start of the fold header, as previously discussed in this section. The name of the file associated with a filed fold will be displayed in double quotes on the fold line.

Storage of files in memory

The editor reclaims storage room from data copied out to files. This reclamation is done when extra room is needed, and so a filed fold may be repeatedly opened and closed without constant re-reading of the file.

The following message may appear:

     Warning : running out of room

This indicates that the editor's storage room is getting low. Any open filed folds not currently in use should be closed to make some more room; alternatively some new filed folds may have to be made. If this message is persistently ignored then the editor may run out of room and refuse to allow any more insertions until more room has been made.

The use of a TDS loader option to change the amount of room available in the fold manager buffer is discussed in section 16.2

File extensions

When a filed fold is made the file name is given an extension corresponding to the attributes of the fold. The first character corresponds to the fold type attribute, and the second and third characters correspond to the fold contents attribute. The most common extension for files made using the [FILE/UNFILE FOLD] function key is `.TSR`, which is a 'text' and 'source' fold; the standard type of fold made by [CREATE FOLD].

See appendix F for a description of the fold attributes and their correspondence with file extensions.

The following extensions are generated for files created by the system:

    `.TOP`   top level files
    `.TCM`   comment text files (ignored by compilers)
    `.TSR`   program source
    `.TCI`   configuration information
    `.DDS`   compiler descriptor
    `.DLK`   compiler linkage information
    `.DDB`   debugging information
    `.DCD`   object code
    `.DMP`   coredump file
    `.CUT`   utility packages
    `.CEX`   user programs
    `.CPR`   program code fold
    `.CSC`   code SC file

In addition, the system creates the files **TOPLEVEL.MOV**, **TOPLEVEL.TKT** and **TOPLEVEL.PCK**. These are used to store, respectively, the current contents of the buffer associated with [MOVE LINE], the contents of the toolkit fold and the contents of the pick buffer (all of which are discussed later). All these are preserved between sessions. There is also a **TOPLEVEL.DEL** file which is not preserved between sessions; this contains the line associated with the last line deletion (see later).

Writing back files

The system takes some trouble to ensure that the versions of the files on disk at any one time are consistent. Operations on the filing system cause the system to write back all files which are open and which have changed since they were last written back to the disk. This operation is called a 'flush'.

A flush is performed whenever one of the following filing system operations occurs:

    1 Closing a filed fold

2 Creating a filed fold

3 Copying a filed fold

A flush is also performed before a running utility or user program reads or writes a file, before suspending the TDS, and when entering the toolkit or code information folds (these are discussed in section 4.6).

An error can occur on writing back a filed fold if the file with the name given on the filed fold cannot be opened, or if a filing system error occurs in the process of writing. If this happens, the editor converts the filed fold into an ordinary fold. The main circumstance under which this can occur is when attempting to write back to a file or directory which is write-protected.

If it is possible to write the file back to another directory, then the name on the fold header can be edited to give a suitable file name, and the FILE FOLD function applied to the fold line. The file can then be retrieved from the other directory later.

Several nested filed folds may be written in a single flush, but normally no information will be lost as a result of one or more write failures; all open filed folds which fail to write are simply converted into ordinary folds. However, an outermost level filed fold will not be automatically converted into an ordinary fold in this way. Instead, the fold is closed, and all changes to the information in the fold since the last flush will be lost.

4.5.9   Deleting lines

DELETE LINE deletes the current line from the document. If this is a fold line, the fold and all its contents are deleted. If it is a filed fold line, or contains a filed fold line, the associated file (or files, if there are nested files) will be deleted from the directory. Since this makes DELETE LINE a very powerful operation, it should be used with care.

On a filed fold line, or a fold line containing a filed fold, the editor asks for the DELETE LINE key to be repeated before deleting the line, as a precaution against accidental deletion.

There is a function RESTORE LINE to undo a deletion, restoring the last deleted line at the current position in the document. However, RESTORE LINE only works until the next flush takes place (as described in the previous section). At the next flush, any required file deletions are carried out, and the delete buffer is cleared.

Only one deletion can be restored, so a deletion cannot be recovered if another subsequent deletion has been done. If a catastrophic deletion has been done, then the TDS interrupt key (see section 4.2.7) can be used to leave the TDS immediately without writing any more files.

4.5.10   Moving and copying lines

Often when using an editor it is necessary to make structural changes to the text, moving lines and blocks of lines around. In the TDS editor, the representation of folds as lines on the screen means that substantial structural changes can be made to a document in the same manner as reorganisation of lines. An individual line can be picked up, or a block of lines can be folded and then picked up.

The functions COPY LINE and MOVE LINE are used to copy and move sections of the document from one place to another. A text line or fold can be duplicated with the COPY LINE function, or moved to another position in the document using MOVE LINE.

COPY LINE duplicates the current line, inserting the copy in the text. If the current line is a filed fold, or a fold containing a filed fold, then copies of the files contained within the fold are made. Before starting the copy operation on a filed fold, the editor asks for the COPY LINE key to be repeated to confirm the operation, as the copying may take some time. The new files are given names derived from the names of the files in the original fold.

Two presses of MOVE LINE are needed to move a line from one part of the document to another; one to pick up the line, and one to put it down. If a sequence of lines is to be moved, the lines should be folded up first.

A buffer (the 'move buffer') is used to store the line between the two operations. There is no need to go and put the line down immediately; the buffer will be retained until the next press of MOVE LINE, even if that is not done until a later session using the TDS.

Using the above keys, it is difficult to collect a number of different parts of a document before putting them down together. Here PICK LINE and COPY PICK are more appropriate. These make use of a different buffer (the 'pick buffer') that is accumulative. This enables the user to gather together, in the buffer, various pieces of text that can be put down in one place. PUT is used to put down the text in the buffer, which is emptied at the same time. As with the move buffer, the pick buffer is also preserved between sessions using the TDS.

PICK LINE is used to pick up a line, which may be a fold line, so that it may be moved to another place in the document. It removes the current line from the document and appends it to the end of the pick buffer.

COPY PICK is used to copy a line, which may be a fold line, so that it may be moved to another place in the document. It makes a copy of the current line and appends it to the end of the pick buffer. If the line is a filed fold, or is a fold containing a filed fold, COPY PICK must be pressed again for confirmation, as the copying may take some time.

PUT puts down the contents of the pick buffer at the current position in the document. It inserts a fold line at the current line, containing the sequence of lines placed in the pick buffer using PICK LINE and COPY PICK. The pick buffer is cleared. If there are no lines in the pick buffer PUT has no effect on the document.

Lines and folds moved by these operations may come from, or be put into the toolkit fold or other display folds created by the actions of utilities.

4.5.11 Defining keystroke macros

The key DEFINE MACRO can be used to define a sequence of keys (which are commonly going to be used together during a session) and assign the sequence to a single keystroke. Two presses of DEFINE MACRO are needed to define a key sequence; the required keys (which may not include DEFINE MACRO or CALL MACRO) should be pressed between the two presses of DEFINE MACRO. The sequence may contain up to 64 keys. Any previously defined macro is forgotten. The defined macro sequence may be invoked using the CALL MACRO key. The currently defined macro may be saved in a new fold above the current line by pressing SAVE MACRO. A saved macro may be recovered from a fold by pressing GET MACRO.

4.6 Utilities and programs

In order to be an integrated programming environment rather than just an editor, the TDS needs two things: the ability to load and run a programming utility, such as an occam compiler, and the ability to load and run programs written by the user. This section describes the aspects of the TDS which concern loading and running code.

The TDS provides the facility to read a transputer code file into memory, where it may be run without leaving the TDS environment. The code file, which appears as a filed fold within the TDS fold structure, may be a file provided with the system, or it may result from the compilation of a user's program. A function key called GET CODE is used to load a code file into memory.

There are two kinds of code files suitable for loading and running within the TDS: utility sets, which are usually marked with the text **UTIL**, and executable programs, which are usually marked with the text **EXE**. These two kinds of code files are introduced below; their use is described in more detail in the following sections.

A utility set provides a number of different functions (up to 10) within a single code file. When a utility set is loaded, the functions it provides are mapped onto a set of ten function keys on the terminal. The utility function keys are shown in the keyboard maps in appendix A, or can be found by using the CODE INFO key at the terminal. Pressing one of these function keys will invoke one of the functions in the utility set.

Utility sets provide a group of commonly needed functions for developing programs within the editing environment. There are two sets of utilities supplied with the standard release of the TDS: the compiler package and the file handling package. The compiler package is introduced in the next chapter, while the file handling package is described later in this chapter. Other utility sets may be provided by INMOS from time to time to extend the functionality of the development system.

An executable program is a single unit of code. Once loaded it can be run by pressing a function key called RUN EXE. It is also known as a 'user program', as it is normally a program being developed by the user of the TDS, although a number of the tools supplied by INMOS with the system (such as the debugger) are executable programs.

More than one set of utilities can be resident in memory. Of these sets only one, known as the current utility set, is immediately accessible by means of the utility function keys, but it is possible to switch between the resident utility sets, using a function key called NEXT UTIL. Similarly, more than one executable program can be in memory at the same time. It is possible to select any of the resident programs as the current one, using a function key called NEXT EXE.

The available memory within the TDS is shared between the code for the currently loaded utilities, the code for the currently loaded user program, and the data space needed for the utilities and program to run. Special function keys, called CLEAR EXE, CLEAR UTIL, and CLEAR ALL can be used to clear the memory associated with loaded code, in order to make more space available.

4.6.1   The toolkit fold

In addition to the normal fold structure, which contains user data and programs under development, there is an additional fold called the toolkit fold. This may be accessed from the editing environment, except when browsing. The contents of this fold (which may include nested filed folds) are stored in between sessions using the TDS.

At any point in a session, the toolkit fold can be entered using the ENTER TOOLKIT function and then edited. Once ENTER TOOLKIT has been pressed, the fold can be viewed and edited until a complementary EXIT FOLD causes a return to the place where ENTER TOOLKIT was pressed. While in the toolkit fold, most editor functions are allowed; the principal exceptions are those which have to make use of the toolkit fold, for example it is not possible to run utilities or programs.

The toolkit fold contains a sequence of folds, each fold normally containing one of the following:

- Utilities and programs for loading.

- A selection of utilities and programs contained in a fold marked **Autoload**.

- Default values of parameters for use by utilities.

- Logical names for libraries.

The **Autoload** fold and the parameters for utilities are described later in this chapter. The use of libraries and logical names is described in chapter 5.

The existence of utilities and programs in the toolkit fold means that at any time it is possible to switch to the toolkit fold and load any code needed to carry on with the task at hand.

Since text and fold lines can be moved in or out of the toolkit fold, the toolkit fold can also be used to store data temporarily while moving around the fold structure. In addition, since the toolkit fold may be entered and viewed at any time, it may be useful for storing information which has to be referred to frequently while working.

The standard toolkit fold supplied with the TDS release appears as follows:

```
{{{F "\tds3\system\toplevel.tkt" Toolkit fold
...  Autoload
...  Tools
...  Library logical names
}}}
```

The **Autoload** fold contains the two standard utility sets, plus the source-level debugger. The **Tools** fold contains a number of useful tools, supplied as **EXE** programs.

The contents of the toolkit fold are stored in a file called **TOPLEVEL.TKT**. The TDS searches for this file first in the current directory and then in any directories specified in the path variables **TDSSEARCH** (see section 16.1). The standard toolkit fold is shipped in the directory **\TDS3\SYSTEM**. This may be used from any directory or may be copied from the TDS system directory to any other directory where it is needed. The filename of the current toolkit fold is included in its top crease line.

4.6.2   Loading utilities and programs

When the TDS starts up, there are no utilities associated with the function keys, and no resident programs. The code for these may be loaded by using the $\boxed{\text{GET CODE}}$ key applied to a filed fold containing the code. More than one code item may be resident in memory at the same time; in fact up to 32 code items may be resident. The workspace available for running the current code is the memory remaining within the system.

One utility set is 'current', which means that the utility function keys are bound to the functions of that particular utility. The identity of the current set is indicated on the message line; the text on the filed fold line from which the code was loaded is remembered and displayed when that set is current.

When $\boxed{\text{GET CODE}}$ is used to get a utility set, that set is made the current one. Any of the resident utility sets can be made the current one; a key called $\boxed{\text{NEXT UTIL}}$ can be used to cycle through the available utility sets. There is also a $\boxed{\text{CLEAR UTIL}}$ key which clears the current utility out of memory and makes the next resident utility set current (if there is one).

There is also a current **EXE**, which is the program which is run when $\boxed{\text{RUN EXE}}$ is pressed. When $\boxed{\text{GET CODE}}$ is used to get a program, that program is made the current one. Any of the resident **EXEs** can be made the current one; a key called $\boxed{\text{NEXT EXE}}$ can be used to cycle through the available programs. There is also a $\boxed{\text{CLEAR EXE}}$ key which clears the current program out of the memory and makes the next resident program current (if there is one).

The key $\boxed{\text{CLEAR ALL}}$ clears all loaded code items, both **UTILs** and **EXEs**.

It is often necessary to know which utility sets and programs are currently loaded, and which functions are currently bound to the utility keys. For this purpose, there is a function $\boxed{\text{CODE INFORMATION}}$, which creates and displays the resident code information fold. The user can view this fold, but cannot edit it; pressing $\boxed{\text{EXIT FOLD}}$ returns to the normal editing environment.

At the top of the code information display fold is some help information for the current utility set; this lists the utilities in the set and indicates which function key they are mapped onto. The rest of the fold lists the utility sets and executable programs currently loaded and, for each utility set, there is a fold containing the utility help information. The current utility set and program are marked with >. The help information for the other utility sets can be obtained by moving to and opening the appropriate fold. In addition to the help information, the code size and data space requirement for each of the code items is given.

4.6.3   Loading code from the toolkit fold

The toolkit fold contains references to the standard utilities and programs provided with the system. These references are in the form of filed fold lines referring to the files in the appropriate TDS directory. So the toolkit file contains only references to the utilities and programs, not the actual code.

The fold marked **Autoload** in the toolkit fold contains the selection of utilities and programs for normal use. After the TDS has been started, pressing the key AUTOLOAD loads all of the code contained in the autoload fold, as if GET CODE had been applied to each line individually. The last utility set loaded is made the current set, and similarly for the last program. Therefore the most frequently required utility set should be the last one in the autoload fold.

Thus, to set up the standard utilities, the user can either:

- enter the toolkit fold, using ENTER TOOLKIT and then load the required set of utilities from the autoload fold using GET CODE, or

- use AUTOLOAD to load the two sets of utilities, subsequently using NEXT UTIL, if appropriate, to change the current set.

The fold marked **Tools** in the toolkit fold contains a number of useful programs, described in this chapter or later in the manual. The tools can be selected when required, by going into the toolkit fold, and using GET CODE on the appropriate line. If a particular tool is needed frequently it can be moved into the autoload fold.

The space available for loading utilities and tools is limited by the size of memory on the transputer board on which the TDS is running. If running on a small (e.g. 1Mbyte) board, it will not be possible to load other utilities or tools at the same time as the compilation utilities, and to be able to compile non-trivial programs. In such circumstances AUTOLOAD is of limited value and the contents of the Autoload fold should be carefully chosen.

4.6.4   Running a utility

A utility in the current utility set is run by pressing the appropriate utility function key.

Before invoking a utility, it is often necessary to place the cursor in a particular position to indicate an object on which the utility is to operate. For example, when the occam syntax checker is invoked it needs a sequence of text lines containing the program to be checked. In a normal command environment this would be done by storing the text in a file and then giving the name of the file as a parameter. In the TDS the cursor is placed on a fold line and the CHECK key is pressed. This indicates that the checker should take the contents of the fold as the text to be processed. This operation is normally termed 'applying' the utility to the fold. Because of the representation of files as folds, the same utility can sometimes be applied to a few lines of text, a complete program in a file, or a large program made up of many files.

When a utility is running, it may read and write data in the fold to which it is applied. In addition it may display messages on the message line to indicate what it is doing; the rest of the screen appears as it was when the function key was pressed.

Certain utilities need to be supplied with parameters to determine selected options. This is done by making use of a 'parameter fold', and is described in the next section.

The key SET ABORT FLAG can be used to abort a utility when it is running. This sets a flag to indicate that the abort key has been pressed. Utilities and programs can periodically test the value of this flag and terminate when it is found to have been set.

Before finishing, a utility may clear the screen and display a fold of information to the user. The user can browse this fold and edit it. Items in this fold may be picked or moved to the normal editing enironment if they are required for permanent use. Pressing EXIT FOLD leaves the fold and returns control to the utility.

When both a utility package and a user program have been loaded, it is possible that there is not enough memory available for the data space of the utilities. If this condition occurs, when a utility function key is pressed the following message appears:

        Data requirement too large

The condition is also indicated by the removal of the utility package help lines from the code information fold. They are replaced by the text

> **Utility workspace is larger than free storage**

If this occurs it is necessary to use the code clearing function keys to make more memory available.

4.6.5   Supplying paramotoro to utilitio

A utility which requires a set of parameters in order to run obtains them from a 'parameter fold'. When the utility is first run it creates a parameter fold containing the default values for the utility's parameters and displays it to the user, as if the user had chosen to enter the fold. The fold will contain a sequence of lines of text appearing as occam constant definitions.

For example, the parameter fold for the search utility (containing two string parameters and four boolean parameters) appears as follows:

```
{{{   Search and replace
VAL search.string    IS "" :   -- ""
VAL replace.string   IS "" :   -- ""
VAL case.sensitive   IS TRUE :
VAL global.replace   IS FALSE ·
VAL forward.search   IS TRUE :
VAL forward.replace  IS TRUE :
}}}
```

The displayed fold may be edited to set the parameter values, before the utility is allowed to continue. Pressing EXIT FOLD supplies the parameters to the utility and allows it to continue.

A function SELECT PARAMETER is provided to facilitate the editing of parameters. It moves the cursor to the parameter value section, and allows the user to toggle between a number of possible values of the parameter. For example, it could be used on a boolean parameter to toggle between **TRUE** and **FALSE**. On a line of the form:

> **VAL parameter IS value1 : -- value1 | value2 | value3**

the SELECT PARAMETER key will cycle the parameter between the three allowable values. The empty strings after the -- in the search and replace example above are provided to simplify the changing of these parameter values.

The SET ABORT FLAG key can be used when a utility has popped up its parameter to cancel the utility before it runs. EXIT FOLD still needs to be pressed to exit from the parameter fold.

Once a parameter fold has been used by a utility, it is stored in the toolkit fold. To change the parameters before the next run of the utility, ENTER TOOLKIT can be used. Once inside the toolkit fold the individual parameter folds can be entered and the parameters edited as required. The next run of a utility will take its default values from the toolkit; if this is not required the parameter fold should be deleted from the toolkit. Each parameter fold is recognised by the text of its fold comment. A fold may be made invisible to a utility by changing this text. A parameter fold must contain definitions of all the parameters required by a utility that uses it. It may also contain additional definitions not so used.

4.6.6   When a utility finishes

When a utility finishes running, it normally outputs a message indicating either successful completion or a condition which it wishes to bring to the user's attention.

On successful completion of a utility the current editing position normally remains as it was when the utility was started. Sometimes the utility will need to identify a line in the fold structure (e.g. where a syntax error was found, or the occurrence of a string being searched for). It does this by 'locating' the line; that is, moving the current editor position to that line, opening folds as necessary to reach it, and positioning the screen so that the line appears in the middle of the screen (or as near it as possible). The utility then finishes, and control is returned to the user.

If the utility is of the type which is 'applied' to a fold (i.e. the fold line on which it is placed determines the portion of the fold structure upon which it operates) then, before locating, the fold is entered. This means that the user may easily return to the position before the location was done by using $\boxed{\text{EXIT FOLD}}$.

4.6.7   Running executable programs

The current executable program may be run by using the $\boxed{\text{RUN EXE}}$ function key.

In a similar manner to utilities, an executable program may be given a portion of the fold structure on which to operate by means of cursor positioning before the program is run. Unlike utilities, executable programs do not make use of the message line or parameter folds, but can access the whole screen for interactive communication with the user. The screen is cleared when the program starts, and the fold structure is repainted when the program terminates.

The preparation of user programs for running within the TDS is covered in the next two chapters of this manual. A program suitable for loading and running within the TDS must be an occam process, the environment to which is supplied by a number of channels. The environment allows the program to read from the keyboard, write to the screen, and read and write data within the fold structure. The channels available, and the protocols which should be used on these channels, are introduced in chapter 6.

An executable program may either terminate naturally, or may set the transputer's error flag to denote a run-time exception, or may deadlock. The way to proceed in either of the latter cases is described in chapter 9 on the debugger.

4.7   File handling utilities

One of the two standard utility sets provided with the system is the file handling package. This is a set of utilities for the manipulation of TDS files. The help information for the set appears as follows:

```
1   [ATTACH/DETACH]       - attaches or detaches a file
2   [COPY ATTACH]         - copies files and attaches copy to current fold
3   [COMPACT LIBRARIES]-  copies files out, compacting libraries
4   [RENAME FILE]         - rename a filed fold
7   [COPY IN]             - copy files from another directory
8   [COPY OUT]            - copy files to another directory
9   [READ HOST]           - read host file into fold structure
0   [WRITE HOST]          - write TDS file to host file
```

All of these utilities are introduced below and discussed in more detail in chapter 13.

The utilities fall into four groups:

- Attaching to and detaching from existing TDS format files.

- Changing file characteristics, such as DOS file names.

- Copying TDS format files between directories (including between devices) from within the TDS.

- Reading and writing host operating system format files to and from TDS format files from within the TDS.

Attaching and detaching files

Occasionally it is necessary to take an existing file, which could be in a separate directory or on a separate drive, and make it part of a larger document. This operation is done by 'attaching' a file to a fold.

To attach a file an empty fold should be made and the header edited to include the name of a file which already exists. The name should include the extension, which must be one of the standard extensions supported by the system (see appendix F). The attributes of the fold will be set to reflect the attributes associated with that extension. When the utility ATTACH/DETACH is applied to this empty fold, the file specified is then 'attached' to that fold so that future opens of the fold will cause the file to be read in at that position.

Correspondingly, files can be detached from the fold structure using ATTACH/DETACH. Applying the utility to a filed fold converts it into an empty fold, losing the reference to the file. The detached file is not deleted from the directory.

When the file is attached using the ATTACH/DETACH utility a reference is set up to the file. This means that a file could be attached to more than one place in more than one document. A consequence of this is that any editing of the file in one document will be reflected in all other places to which the file is attached. It may be more appropriate to make a copy of the file before it is attached; this maintains the integrity of any existing document structures. The utility COPY ATTACH should be used when a copy of the file is to be made. The COPY ATTACH function will cause the file whose name is given on the fold line, along with any nested files it may contain, to be copied, and the resulting file attached to the fold structure. Unique names are generated for any new files created, as described previously.

Changing file characteristics

The name of a file associated with a filed fold can be changed. The name is altered by placing the cursor on the filed fold and editing the fold header so that the contiguous sequence of characters after the current filename in quotes up to the first space or dot is the required new name of the file. Then the RENAME FILE utility key should be pressed. The underlying file will be renamed and the local filename reference updated to the new name.

In order to prevent files from being altered or deleted they can be write protected. To do this suspend the TDS, using SUSPEND TDS, and use the appropriate host operating system command. Protection is particularly useful for files that are multiply attached, such as the standard utility sets. However, care should be taken, if write-protecting ordinary TDS text files, not subsequently to attempt to edit the files while they are write-protected, as the TDS will be unable to write back the changes.

Copying TDS files

TDS files in other directories, including any nested files, can be copied into a filed fold in the current directory by using COPY IN. The full name of the source file must be given, including the directory name and drive if necessary. The name of the new file will be the same as that of the source file unless there is a name clash with an existing file in the local directory, in which case the TDS will modify the name to make it unique. In a similar manner the contents of a filed fold, including all nested files, can be copied to another directory by using COPY OUT.

The COMPACT LIBRARIES utility also copies files between directories, but is only intended for use with library files. Library compaction is discussed in chapter 5.

Reading and writing host files

Although the TDS is a self-contained development environment, there are times when it can be useful to read and write host operating system files. READ HOST copies a host operating system file into a filed fold, converting the format of the file to TDS format. The reverse process is performed by WRITE HOST which copies a filed fold, including all nested files, to a host operating system file, converting from TDS format to the host operating system file format, stripping off the folded structure.

4.8   Searching and replacing

SEARCH and REPLACE are two of the utilities in the compiler utility set. They are used in conjunction; SEARCH searches for a text string specified by the user and REPLACE replaces one text string with another that has been specified by the user. They are introduced here, but described in more detail in chapter 13.

The string to be searched for, and the string to replace it, are contained in a parameter fold (mentioned in section 4.6.5). If the parameter fold does not exist when SEARCH and REPLACE are invoked, a new one is created and is popped up onto the screen so that the strings can be entered. Thereafter the strings remain in the parameter fold and are used whenever these utilities are invoked. The values of the strings are maintained in the toolkit fold between sessions. To change the strings it is necessary to enter the toolkit fold and edit the parameters.

According to a boolean parameter the search takes place in a forward or backward direction from the current cursor position and continues to the end (or start) of the current view or until a match has been found, whichever comes first. All nested folds are searched. When a match has been found the utility may be invoked again to continue searching.

It is also possible to set a parameter that causes repeated replacements of all matching strings to the end (or start) of the current view.

4.9   Listing programs

There are three ways to list programs and other folded text: the LIST utility in the compiler utilities, the WRITE HOST utility in the filer utilities, and the lister program which can be found in the **Tools** fold in the toolkit.

The LIST utility has one particularly valuable feature which enables open fold structures with some nested folds left closed to be sent to a printer or file. This may be used for producing documentation fragments including program text with selectively opened folds.

The WRITE HOST utility may be used for producing hard copy listings of programs. This can be used to write a program as a DOS file, which can be printed out later, or printed immediately if DOS supports a predefined filename for the printer or by using SUSPEND TDS to temporarily leave the TDS, print the file and then return to the TDS.

Both these utilities are fully described in chapter 13

4.9.1   The lister and unlister programs

Alternative listing facilities are given by the lister and unlister programs, which are contained in the **Tools** fold. This pair of programs facilitates the conversion of occam source from TDS files to and from DOS text files. The lister gives its user the opportunity to select lines from the source file on a variety of criteria and so is useful for many program documentation and maintenance tasks.

The lister is an **EXE** which may be applied to any fold containing occam source. If the fold is a bundle of folds then the input is the first fold in the bundle and the user has the option of storing the output as a new last fold in the bundle or to a DOS file (which may be a printer). Otherwise the whole fold is processed and the output is always to a DOS file. If the input contains nested filed folds any of these which contain occam source are included in the output in a single large file, including sufficient information to enable the unlister to reproduce the original folded file structure.

The following options are presented to the user in an interactive menu:

- Output to screen and/or printer, DOS file, or filed fold.

- Option to exclude folds of one or more of these kinds: comment folds, foldsets, folds whose comment includes the words 'NO LIST'.

- Representation of fold creases as braces ({{{ and }}}), occam comments or as commented braces.

- Option to include full analysis of fold attributes.

- Selection of: all lines or file names only, file headers, fold headers, procedure and function headers, procedure and function calls, lines containing strings matching a search string provided (including alternative strings, wild cards, etc.).

The unlister can take any DOS text file (or occam fold) and convert it into a folded file structure. If the input includes creases and other fold information in the form generated by the lister then the fold structure will be regenerated. Alternatively a large 'flat' file may be split into chunks small enough to be handled by the TDS fold manager by creating folds each containing a number of text lines requested by the user.

The source code for the programs described above is provided with the system, in the directory \TDS3\TOOLS\SRC. As the use of these programs is straightforward, they are not documented in the reference section of this manual.

4.10   Transferring TDS files between computers

TDS files may be transferred between host computers using the operating system facilities available on the host. In addition the TDS provides a program to send or receive a folded file structure on a transputer link. This is often the most convenient way to transfer files between TDS systems running on different machines (particularly if the disk formats are incompatible).

The link transfer program can be loaded from the **Tools** fold in the toolkit. The cursor should then be placed on the fold to be sent, or on an empty fold to receive the data. When the program is run it will prompt for the link number to be used, and whether data is to be sent or received. If it is the sending program it also offers the option of sending text folds only, text and descriptor folds (see chapter 5 for a definition of descriptor folds), or all folds.

Once a sending program has been run on one TDS, and a receiving program on another TDS, with the appropriate links connected, the programs will make contact and start to transfer the data. The link transfer can be interrupted using the [SET ABORT FLAG] key, on either the sending TDS or the receiving TDS.

The source code for the link transfer program is provided with the system, in the directory \TDS3\TOOLS\SRC. As with the lister program above, it is not documented in the reference section of this manual.

5 Compiling and linking occam programs

5.1 Introduction

Throughout this chapter and the rest of the manual frequent use is made of occam concepts and example occam program text. Any reader not familiar with occam at this stage should read the occam tutorial to gain an introduction to the language. It will also be useful for all readers to have a copy of the occam 2 Reference Manual available.

This chapter discusses in some detail how to compile and link occam programs using the Transputer Development System. Early on, a simple example is introduced, which is compiled, linked and then shown running within the TDS. Later in the chapter a larger example is introduced and discussed. This example is used to show how large programs might be structured and developed. The following three chapters make use of the same example where it is shown:

- Running within the TDS

- Running on a network

- Running as a standalone program

At the end of the chapter there are some technically more detailed sections, describing the implementation of occam by the compiler, which may be omitted when first reading the manual.

5.2 The compiler utility set

As discussed in chapter 4 the TDS editing environment is not just an editor but a complete development environment. occam programs can be compiled, linked and run without leaving this environment. To do this, the compiler utility set must be loaded into the development system. This is one of the standard utility sets in the toolkit fold and it provides the facilities to compile and link occam programs. It also enables programs to be configured to run on transputer networks, and loaded onto a target network from the TDS.

The loading of utility sets from the toolkit fold was discussed in chapter 4. Pressing the [AUTOLOAD] key loads the standard utility sets from the autoload fold; since the compiler utility set is normally the last in the fold, this leaves the compiler utilities as the current set after autoloading is complete.

The utilities in the set are as follows:

1	CHECK	Syntax check an occam program
2	COMPILE	Compile an occam program
3	EXTRACT	Link and extract code
4	LOAD NETWORK	Load a compiled program onto network
5	RECOMPILE	Recompile a program with old parameters
6	COMPILATION INFO	Display information about the compiled program
7	MAKE FOLDSET	Make a 'foldset' suitable for compilation
8	SEARCH	Search for a string
9	REPLACE	Replace the string at current cursor postion
0	LIST FOLD	List current fold on printer

The [SEARCH] and [REPLACE] utilities were discussed in chapter 4. The other utilities are discussed in this chapter, except the use of the utilities to prepare a program for a transputer network, which is discussed in chapter 7.

All of the utilities are described in more detail in chapter 13.

5.3   Preparing a program for compilation

5.3.1   Creating a compilation fold

Before an occam program can be compiled two conditions must have been met. Firstly, the fold containing the source must be filed, and secondly, this source fold must be enclosed by a 'compilation fold', to which the compiler will be applied. The type of the 'compilation fold' indicates what type of compilation unit the fold contains. There are five types of compilation unit as described below:

EXE — an 'executable' program designed to run within the TDS. It is an occam process that can access channels which communicate with the screen, keyboard, fold system and server. Most programs written to run within the TDS are **EXE** programs. A full description of **EXE** programs is given in chapter 6.

UTIL — a program to be run as a utility set within the TDS. A utility program consists of a process which has a more complex environment than an **EXE**. The utility interfaces are currently not available to normal TDS users.

PROGRAM — a program intended to run on a network of one or more transputers. The **PROGRAM** contains configuration information that enables the development system to load the program into a transputer network. A **PROGRAM** cannot run within the TDS. Chapter 7 describes **PROGRAM** creation and compilation in detail.

SC — a 'Separate Compilation' unit. This is not a complete program in itself and is normally contained within another compilation unit or library. An **SC** unit contains one or more occam procedure or function declarations. Separate compilation is described later in this chapter, in section 5.6.

LIB — a library compilation unit. It contains a number of constant, protocol, procedure and function declarations that may be shared between parts of a program or between different programs. Libraries are described later in this chapter, in section 5.6.

To create a compilation fold, the cursor is placed on the filed fold containing the source of an occam program, and the MAKE FOLDSET utility invoked. This will prompt for a parameter of the form:

```
VAL make.foldset.type IS SC: -- SC | EXE | UTIL | PROGRAM | LIB
```

The value of the parameter selected by the user determines the type of fold created by the utility. For example, to make a program to run as an executable program within the TDS, the user selects the value **EXE**. The MAKE FOLDSET utility creates a compilation fold of the selected type around the source fold. The new fold has its attributes set to indicate that it is suitable for compilation and the fold header is marked with some text to indicate the type of compilation unit that is enclosed within the fold.

For example, in order to compile a section of code as a program to be run within the TDS, the following two folds might be created around it:

```
{{{   EXE myprog -- compilation fold
{{{F "prog.tsr"   -- filed fold
...   Program text
}}}
}}}
```

The **EXE** fold is the compilation fold produced by MAKE FOLDSET. The compilation fold, together with the filed fold or folds inside it, is known as a 'foldset'. A foldset is a compilation fold with one or more subsidiary filed folds. When a compiler is applied to the compilation fold, it takes the first subsidiary filed fold as the source text to be compiled, and creates other subsidiary filed folds containing (for example) code produced as a result of the compilation.

5.3.2   Comment folds

When developing programs it is often desirable to comment out part of a program so that it is ignored by the compiler. This can be done by placing the program text in a fold and then applying $\boxed{\text{MAKE COMMENT}}$ to the fold. This produces a new fold which encloses the original fold. The header of the new fold contains the original fold header prefixed by the letters **COMMENT**.

A comment fold can be removed by applying the $\boxed{\text{REMOVE FOLD}}$ key to it.

The contents of a comment fold will be ignored by the $\boxed{\text{CHECK}}$ and $\boxed{\text{COMPILE}}$ utilities, but not by $\boxed{\text{SEARCH}}$ or the file handling utilities. The lister program described in chapter 4 includes an option to include or omit comment folds from a listing.

5.4   Using the compiler utilities

Once a program has been placed within a compilation fold, the compilation utilities can be used to compile the program. This section gives a simple introduction to using the compiler utilities, and provides enough information to allow the reader to work through the example program in the next section.

The steps to compile an occam program are as follows:

1 Check the syntax of the program.

2 Compile the program, producing some data folds as a result of the compilation.

3 Link the program together with any libraries it uses, creating a self contained code file.

These are described in more detail below.

5.4.1   Compilation for different transputers

The compiler produces code targetted at a particular transputer type or class. All compilations for a single processor must be for the same or a compatible transputer type.

Transputer classes

The compiler can produce code that will run on different transputers by taking advantage of commonality in their instruction sets. Provided that no code is written which compiles into instructions which are not shared between different processors, the code will run normally.

The commonalities that exist between different processors are as follows:

- All T2 and M2 series 16-bit transputers have compatible instruction sets; this class is called T2.

- T414 and T425 transputers share the same instruction set except for CRC and 2D block move operations; this class is called TB.

- T425 and T800 transputers share the same set except for floating point operations; this class is called TC.

- T414, T425, and T800 transputers share the same set except for CRC, 2D block move, and floating point operations. Programs which use none of these instructions may be compiled for class TA.

- T801 and T805 transputers are instruction compatible with T800 transputers; this class is called T8.

Code compiled for a transputer class must be able to run on any member of that class. If the source code would compile into transputer code that is not common for all members of the class then an error is reported.

For example, code compiled for class TC cannot contain floating point and extended arithmetic because operations on **REAL** numbers are implemented differently on the two machines. On the T425 the implementation is in software whereas on the T800 it uses the on-chip floating point processor. Similarly, code compiled for class TB can contain no CRC or 2D block move operations because the respective transputer instructions are not implemented on the T414. Code compiled for class TA can contain no floating-point, CRC, or 2D block move operations. Code compiled for TA cannot call real arithmetic procedures in the compiler libraries.

The restrictions on floating point arithmetic apply only to *operations* on the variables, or the returning of a **REAL** result from a function, because these cause dissimilar instructions to be used. The declaration of **REAL** variables and the passing of **REAL** parameters into procedures or functions is not prohibited. Library procedures for conversion between **REAL** values and strings do not use floating point arithmetic.

5.4.2   Mixing code for different transputers

By using transputer classes for compilations it is possible to produce code that may be mixed with code for other transputer types and classes. It should be noted that this not possible for all compilations, but only where instruction sets overlap.

The rule for mixing code is as follows:

> Code may be called provided it is compiled for a class which is the same or is a superset of the calling code.

The code that can be run on different processor types is listed below.

Processor	Compatible code
T2 series	T212
T800	T800, TC, TA
T425	T425, TC, TB, TA
T414	T414, TB, TA
TC	TC, TA
TB	TB, TA
TA	TA

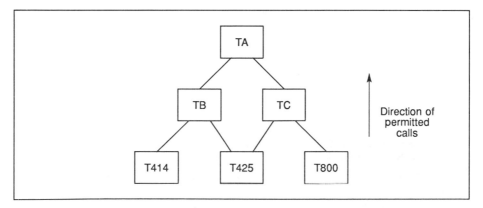

When compiling for a transputer class the compiler will report an error if the source is such that it cannot be compiled for that class. This will often take the form of an undeclared procedure or function from the compiler libraries.

Wherever possible libraries supplied with the TDS are compiled for T2 and TA and so may be called from code compiled for any target. If a program uses any of the instructions which are not common to a class of transputers it will be necessary to target the compilation to a particular transputer type.

5.4.3   Error modes of compilation

For systems that require maximum security and reliability, the error behaviour is of great concern. occam 2 specifies that run-time errors are to be handled in one of three ways, each suitable for different programs. The error mode to be used is supplied as a parameter to the occam 2 compiler.

The first mode, called HALT system mode, causes all run-time errors to bring the whole system to a halt promptly, ensuring that any errant part of the system is prevented from corrupting any other part of the system. This mode is extremely useful for program debugging and is suitable for any system where an error is to be handled externally. HALT system mode is the default for the compiler, and you should use this mode when you may want to use the debugger.

The second mode, called STOP mode, allows more control and containment of errors than HALT mode. This maps all errant processes into the process STOP, again ensuring that no errant process corrupts any other part of the system. This has the effect of gradually propagating the STOP process throughout the system. This makes it possible for parts of the system to detect that another part has failed, for example, by the use of 'watchdog' timers. It allows multiply-redundant, or gracefully degrading systems, to be constructed.

The third mode, called REDUCED or UNDEFINED mode, is to ignore all run-time errors. This is potentially dangerous, but there are occasions when it is useful to avoid the run-time overhead of error checking, for example, where a program has already been proven correct. A second example is where results are being checked elsewhere.

See also section 5.7.2. If there is doubt as to which mode to use, HALT mode should be chosen.

5.4.4   Mixing code with different error modes

In some circumstances it may be desirable to omit the run time error checking in one part of a program for example, in a time-critical section of code, while retaining error checks in other parts of a program, for debugging purposes. The compiler allows the mixing of unchecked code (REDUCED) with code of other error modes, in a restricted manner.

To prevent accidental mixing of UNDEFINED code, an extra mode has been added, called UNIVERSAL mode. UNIVERSAL code is the same as REDUCED, but has the property that it may be called from any other error mode. Code compiled in UNIVERSAL mode can only call code which is also in UNIVERSAL mode.

Note: UNIVERSAL mode is not intended as a general purpose facility and should be used with great caution, because it disables the security associated with error checking. It should only be used when error checking is not required and would be undesirable, such as with time critical code that is already proven.

Although the code produced in UNIVERSAL mode is the same as REDUCED mode it is important to distinguish between them. The behaviour of a system when an error occurs is not necessarily the same for both modes. This is because UNIVERSAL mode may be mixed with other modes, so an error occurring in UNIVERSAL code could halt the processor if it is mixed with HALT code. However, for the same code in REDUCED mode the behaviour at an error is not predictable.

Because the compiler libraries are only available in the HALT, STOP and REDUCED error modes you cannot use UNIVERSAL mode for any occam which requires the compiler libraries. When you compile any source in UNIVERSAL mode you should always disable the compiler libraries by setting the parameter **"use.standard.libs"** to **FALSE**.

5.4.5   Checking occam programs

The ⌑CHECK⌑ utility can be used to check the syntax of occam programs. When a program is compiled, the program syntax is checked, so the use of ⌑CHECK⌑ is optional; however, it is often faster to use the checker to eliminate syntax errors before running the compiler. The checker can be applied to any filed fold, or to a compilation unit fold.

When it is first run, the checker creates a parameter fold and puts it up on the screen for editing. The occam checker shares the same parameter fold as the occam compiler, but only uses a few of the parameters in the fold. The compiler parameters are described below.

If an error in the occam source is discovered, a message is displayed, and the editor moves to the line in the fold containing the error. The located line is placed as near to the centre of the screen as possible and the cursor is positioned on the located line. The effect is the same as entering the checked fold, followed by screen moves and OPEN FOLD operations to find the correct line. Thus it is possible to return to the fold line on which the checker was started by typing EXIT FOLD.

5.4.6   Compiling occam programs

To compile a compilation unit, the cursor should be placed on the compilation fold and the COMPILE utility key pressed.

When it is first run, the compiler creates a parameter fold and puts it up on the screen for editing (unless such a fold already exists in the toolkit fold). The parameters are required to set a number of compiler options, such as the checks done on the program source, some characteristics of the compiled code, and whether a debug data fold is produced. See section 5.7 later in this chapter for a full description of these parameters.

A collection of compilation units may be compiled by a single use of the COMPILE utility if they are enclosed within an outer text fold.

If the compiler detects an error it reports it in the same way as the checker. Compilation is not continued after an error has been found.

If the compilation succeeds, the compiler creates several new folds within the compilation fold to contain the results of the compilation (code, debug information and so on). The new folds created as part of the compilation process are automatically filed. The filenames for these folds are derived from the name of the source fold's file with the appropriate extensions added.

An example of a compiled foldset is given below:

```
{{{   SC mysc
...F "prog.tsr" mysc
...F "mysc.dcd" code HT8
...F "mysc.dds" descriptor
...F "mysc.dlk" link
...F "mysc.ddb" debug
}}}
```

The data folds, marked **code**, **descriptor**, **link** and **debug**, are subsidiary data folds produced by the compiler. The **code** fold contains the compiled code for this compilation unit, but does not include the code for any libraries used. The error mode and compilation target are shown on the fold line. The **descriptor** fold contains some information about the compiled code. The contents of the descriptor fold depend upon the type of compilation unit, but give details of things such as data space size, code size, libraries used etc. The **link** fold contains relocation information. The **debug** fold contains information to allow the debugger to relate the state of a stopped program to the original source code.

There is a utility COMPILATION INFO which reads the information in a descriptor fold and displays it as user-readable text. It is possible to move this text into the fold structure so that it may be printed or kept for future reference.

5.4.7   Linking occam programs

A compiled occam program needs to be linked before it can be run, In the case of an **EXE** the program is automatically linked when it has been successfully compiled. The linking process involves including with the code any library routines that are required. Libraries may be those known about by the compiler, or may be product libraries or user-written libraries referenced explicitly by the user.

An example of a compiled and linked foldset is given below:

```
{{{   EXE prog
...F "prog.tsr" prog
...F "prog.dcd" code HT8
...F "prog.dds" descriptor
...F "prog.dlk" link
...F "prog.ddb" debug
...F "prog.cex" CODE EXE myprog time and date of compilation
}}}
```

The linked compilation unit has an extra filed fold created at the end of the foldset, here marked **CODE EXE myprog**. This fold, referred to as a 'CODE EXE', contains the linked code in a format suitable for loading into memory using GET CODE. This fold can be left within the foldset, or it can be moved to another part of the fold structure, and used on its own with GET CODE.

The CODE EXE file normally has the same name as the source file, but with the extension **.CEX**. The only exception to this is if a previous version of the CODE EXE file has been kept, in which case the file name will be modified to avoid a clash with the existing file.

An **SC** compilation unit will be linked when a PROGRAM using it is configured. The CODE SC file so generated will have the same name as the source file but with the extension **.CSC**

The EXTRACT utility may be used to link an **SC** compilation unit which is to be dynamically loaded (see section 11.3).

5.5   Compiling a simple example program

This section is a tutorial section, giving explicit instructions to compile, link and run a simple example program. It requires some program text to be entered. The interactive tutorial, described in section 4.3, is an alternative way of learning how to compile and run a simple program.

It is a good idea to have a keyboard layout with you when you are working through this tutorial. Keyboard layouts are given in appendix A. Start by running the TDS in the directory **\TDS3\TUTOR**. In the top level file **EXAMPLES.TOP** is a fold called 'Simple example'. Enter this fold, which is empty. Make a new fold (put the cursor on the bottom line and press CREATE FOLD twice) and label it **hello**. Enter this fold and type in the following program, adhering strictly to the indentation. The first line of the program should start at the left hand side of the screen. Indentations are two character spaces.

```
#USE streamio
VAL message IS "Hello World !" :
INT key.char :

SEQ
  ss.write.string (screen, message)
  ks.read.char (keyboard, key.char)
```

Exit the fold **hello** and file it by placing the cursor on the fold and pressing FILE/UNFILE FOLD. The following message will appear:

Filed OK as hello.tsr

When it is run this program writes the simple message **Hello World !** to the screen.

The first line of the program references the general purpose I/O library **streamio**; see chapter 14 for more details of this and other libraries. **streamio** contains the two procedures **ss.write.string** and **ks.read.char** which are used to write to the screen and read from the keyboard.

At the start of the program, the constant **message** is declared, along with an integer variable **key.char**.

The executable code begins with a **SEQ**, indicating that the statements following are to be executed sequentially. The first statement outputs the **Hello World !** message to the screen handler.

The last statement inputs a value from the keyboard to the variable **key.char**.The program waits at this point until the input can proceed; i.e. until a key is pressed. This allows the **Hello World !** message to be read before returning to the TDS.

5.5.1   Getting the compiler utilities

Before the example program can be compiled it is necessary to load the compiler utilities.

Press AUTOLOAD to load the standard utilities. The AUTOLOAD function loads the following code items

- The file handling utility set.
- The compiler utility set.
- The debugger.

As it loads each of these a message is displayed on the message line:

Getting *text* **. . .**

Once all the loading has been done, the message line of the editor displays:

CODE UTIL occam 2 compiler utilities.

This indicates that the current utility set is the compiler utility set described at the beginning of this chapter. The utilities are called by pressing the utility function keys. If necessary, use the HELP key to find out which these are.

Once the utilities have been loaded, pressing CODE INFORMATION shows the following on the screen:

```
1   [CHECK]             - check current fold
2   [COMPILE]           - compile current and nested foldsets
3   [EXTRACT]           - extract code and put into foldset
4   [LOAD NETWORK]      - export code to transputer network
5   [RECOMPILE]         - use descriptor fold for parameters
6   [COMPILATION INFO]  - display compilation information
7   [MAKE FOLDSET]      - make compilation fold
8   [SEARCH]            - search for text string
9   [REPLACE]           - replace found text string
0   [LIST FOLD]         - list current fold
```

There will also be other text on the screen relating to memory usage and loaded utilities. The above display states which utilities are associated with which utility keys. To return to normal editing, press EXIT FOLD.

5.5.2   Making an EXE fold

Before the hello program can be compiled an **EXE** fold must be created around **hello.tsr**. Place the cursor on the fold **hello.tsr** and press MAKE FOLDSET . The parameter fold for the utility MAKE FOLDSET is then displayed. It contains the following text:

VAL make.foldset.type IS SC : -- SC | EXE | UTIL | PROGRAM | LIB

The text shows the current value of the parameter, which is **SC**, and on the right the five possible values that the parameter can take. As the current value of the parameter is **SC**, the fold needs to be edited. The editing can be carried out using SELECT PARAMETER. Press SELECT PARAMETER once; this moves the cursor to the first occurrence of **SC**. Press it again; the second press replaces **SC** with the next parameter in the list on the right, which is **EXE**, the required one. Now leave the parameter fold by pressing EXIT FOLD. An **EXE** fold, labelled **EXE hello** is created around **hello.tsr**.

5.5.3   Checking and compiling the example program

The program can now be checked and compiled. Place the cursor on the fold **EXE hello** and press CHECK; the parameter fold for the checker and compiler is displayed, as shown below:

```
VAL    error.checking        IS HALT : -- REDUCED | STOP | HALT
VAL    alias.checking        IS TRUE :
VAL    usage.checking        IS TRUE :
VAL    separate.vector.space IS TRUE :
VAL    create.debugging.info IS TRUE :
VAL    range.checking        IS TRUE :
VAL    compile.all           IS FALSE :
VAL    force.pop.up          IS FALSE :
VAL    use.standard.libs     IS TRUE :
VAL    target.processor      IS T8 :   -- T2|T4|T8|T425|TA|TB|TC
VAL    code.inserts          IS NONE :  -- NONE | RESTRICTED | ALL
VAL    ring.bell             IS NEVER :  -- NEVER | ERROR | ALWAYS
VAL    tds2.style.exe        IS FALSE :
```

None of the parameters need to be changed to compile this example. The parameters are described in section 5.7.

Press EXIT FOLD; the parameter fold will disappear and the checker will run. If it finds no errors it will respond with the message:

Checked (T8 - HALT) EXE hello OK

If an error is found, the checker indicates it by displaying an error message and placing the cursor on the line in error. If you have mis-typed part of the program, this will happen. Correct the error and then press EXIT FOLD; this will return you to the compilation fold so that the checker can be run again.

If no errors are found the program can be compiled. Place the cursor on **EXE hello** and press COMPILE. The compiler creates new folds within **EXE hello.tsr** to hold the compiled code, and then links the program. When compilation and linking is complete, the compiler responds with the message:

Linked EXE hello OK

5.5.4   Running the example program

The program is now ready to be run within the TDS. Place the cursor on **EXE hello** and press GET CODE to place the code in the user program buffer. The TDS responds with:

Got code ok

Pressing RUN EXE runs the program, which displays the message:

Hello World !

and then waits for a key to be pressed before returning to the TDS.

5.5.5   Compilation information

It is sometimes necessary to check how much code has been generated by a compilation, and how much workspace (data space) will be required to run the code. This information is stored in the descriptor fold, and can be displayed using the utility COMPILATION INFO utility.

This can be used on the example that has just been compiled and run. With the cursor on the line **EXE hello** press the key COMPILATION INFO. The following information is displayed:

- Target processor (e.g. **T4** for IMS T414 or **TA** for any 32-bit transputer).

- occam compatibility (i.e. which versions of the compiler this compilation is compatible with).

- Compiler version (i.e. which particular version of the compiler was used for this compilation).

- Compiler options used in this compilation.

- Whether the program contains any nested **SC** (separate compilation) units or alien language programs (programs written in a language other than occam).

- Code size of this compilation unit.

- Entry points (just one in the case of an **EXE**) and channel usage information.

- Data space required to run the program. Workspace Slots are words holding scalar data and Vector Slots are words holding vector or array data.

- Library usage (i.e. names and version numbers of libraries which may be needed by the program).

- Total linked code size.

To view all of the information provided it is necessary to scroll the screen. Once you have finished viewing this, press EXIT FOLD to return to the normal editing environment.

This concludes the tutorial section.

5.6   Separate compilation and libraries

. The TDS supplies two mechanisms to support the development of large programs:

- Separate compilation (**SC**) units

- Libraries

These are introduced in this section, along with a description of how to compile and link programs made up of more than one compilation unit. Section 5.8.1 describes how to make user-defined libraries.

5.6.1   Separate compilation

Separate compilation allows a program to be split up into parts which may be compiled individually. Using separately compilable units reduces the time taken to recompile a complete occam program because only those units that have been changed since the last compilation need to be recompiled. The separate compilation system is useful for 'top-down' decomposition of programs into major sequential and parallel sections.

A program compilation unit, such as an **EXE**, can contain one or more **SC** compilation units. Separate compilation units may be nested, in a hierarchical fashion, so a large program may consist of a nested structure of separately compilable units.

An **SC** unit consists of one or more occam procedure or function declarations. A procedure or function can call any other procedure or function defined before it in the **SC** unit. The **SC** may also contain constant and protocol declarations, and library usage directives, before the procedures; these may be used by the procedures, but are not visible outside the **SC**. The text of an **SC** must be self-contained; it must not refer to anything declared outside of it (except declarations imported by library directives).

To make a section of source text into an **SC** unit put it into a filed fold and apply the MAKE FOLDSET utility with its parameter set to **SC**. This creates an **SC** foldset around the filed fold. The results of compiling this **SC** will be stored in the foldset, as for an **EXE**.

For example, a program might have the form:

```
{{{  source
...  SC PROCs P1 and p2
...  SC PROC P3
CHAN OF INT c1, c2, c3:
PAR
  P1(c1, c2)
  P2(c2, c3)
  P3(c3, c1)
}}}
```

In the example above, the folds marked with the letters **SC** are compilation folds including separately compiled procedures. The first fold contains the procedures **P1** and **P2**; the second contains the procedure **P3**.

5.6.2   Libraries

Libraries provide a means of sharing common declarations and code between separately compiled parts of a program, and between different programs. They are used by the compiler for pre-compiled procedures such as those which implement some of the extended types in occam.

There are two types of libraries in normal use:

- 'Header libraries' containing declarations of constants and **PROTOCOLs**.

- 'Code libraries' containing collections of compiled procedures and functions.

In fact there is no real distinction between these — libraries may be made containing both header text and code — but in practice it is useful to separate them out.

A header library comprises a sequence of text folds, containing constant and **PROTOCOL** definitions.

A code library comprises a sequence of **SC** folds, each containing compiled procedures and/or functions. The **SC** folds in the library must not include any nested **SCs**.

A combined library includes both text folds containing constant and **PROTOCOL** definitions, and **SC** folds.

Libraries may not contain text lines or blank lines, but may contain **COMMENT** folds.

A library is used within a compilation unit by means of a **#USE** directive. When the unit is compiled any constant and **PROTOCOL** definitions in the library come into scope as do any appropriate procedures or functions declared in **SCs** within the library. When the program is linked the linker will include the code for those **SCs** containing procedures which have been used by the program.

There are a variety of libraries provided with the TDS to perform, for example, many of the mathematical functions and the input/output facilities that a programmer might require.

An example of the use of these libraries has been shown in the example in the previous section of this chapter; the directive **#USE streamio** in the program caused the compiler to use the library **streamio** when compiling the program.

The code libraries provided with the TDS are described in detail in chapter 14 of the manual. The headers for use with these libraries are listed in appendix D.

5.6.3   Compiling and linking large programs

Compiling a program which includes separate compilation units and library references is very straightforward. Separate compilation units in the program can be compiled individually by applying the compiler to them. Alternatively, the compiler can be applied to the whole program, and it will search within the program for any separate compilation units requiring compilation. These nested compilation units are compiled, in a bottom-up order, and then the top level of the program is compiled; finally the whole program is linked together. This can all be done with a single press of the COMPILE utility.

For an **SC** unit the descriptor fold contains all the information about the procedures in that unit (names, formal parameters, workspace and code size etc.) needed to compile calls to the procedures.

When the program is linked the code folds for all the separate compilation units in the program are copied into a linked code file. In addition, code for any libraries used is included in the file. Where libraries contain more than one compilation unit, only those compilation units containing routines actually required in a program are linked into the final code. This helps to minimise the size of the linked code.

When using the CHECK utility on a program containing nested separate compilation units, it should be noted that this utility also needs the information in the descriptor fold to check the calls to procedures in an **SC**. So when using CHECK, all nested compilation units within the text being checked must already have been compiled.

5.6.4   Changing and recompiling programs

When a change is made to part of a compiled program, it is necessary to recompile the program to create a new code file reflecting the change. The purpose of the separate compilation system is to split up a program so that only those parts of the program which have changed need to be recompiled, rather than needing to recompile the whole program. However, it would be tedious for the user to have to remember which portions of a program had been edited in an editing session. For this reason, the TDS remembers which compilation units have been edited since they were last compiled. This ensures that **SC** folds will always be recompiled where necessary, and the compiler is able to tell automatically which **SC** units require compilation.

When editing a program, if a change is made to the source of a compilation unit, then an attribute on its compilation fold is set to indicate that it is now invalid. As folds are closed, the invalid attribute is propagated up to any compilation units above it in the fold structure. All of these invalid compilation units will be recompiled when the compiler is next applied to this program. The actual implementation details of this are described in the next section.

An **SC** to be shared between more than one section of code should be placed in a library. Libraries have a version number associated with them, as described later in section 5.8. When a program is recompiled, the compiler will ensure that all compilation units have been compiled with the latest versions of the libraries; any compilation units previously compiled, but with an old version of a library, are automatically recompiled. This ensures that when the latest version of the library is linked in to a program, all compilation units requiring the library have been compiled with that version.

The compiler also recompiles any compilation units it finds which are not compiled in a manner compatible with the current program being compiled.

To summarise, the compilation system within the TDS ensures that when the compiler is applied to a program:

- If a compilation unit has been changed, it is recompiled.

- If a new version of a library has been made, then any parts of the program dependent on the library are recompiled.

- Any units compiled for a different processor type, or with a different error mode (see section 5.7.2) are recompiled.

- Any units which have been compiled with an old, incompatible, version of the compiler are recompiled. This is the purpose of the 'compiler compatibility string' stored in a descriptor.

5.6.5   The implementation of change control

The change control of compilation units is implemented using the 'fold type' attribute of a compilation fold. The type is set to **ft.foldset** when the program is compiled. Following any change to the contents of the fold, the type is set to **ft.voidset** when the compilation fold is closed, or when it is next written to the filing system.

The fold attribute value can be found by using the FOLD INFO key. This displays **compiled fold set** after the program has been compiled. Following a change to the contents of the fold, it will display **uncompiled fold set**.

When a compilation unit is compiled, the names and version numbers of any libraries it uses are recorded in the descriptor. When the compiler next examines the compilation unit to see whether recompilation is needed, it compares the current version number of each of the libraries against the values recorded in the descriptor. If any of these differ, the compilation unit is recompiled.

5.7   Compiler parameters

This section explains the meaning of the compiler parameters and how they are used during compilation. The parameters are found in a fold labelled **Occam 2 compilation parameters** which is created in the toolkit fold when the first compilation is attempted. The values set up by the user will remain unchanged for future use unless explicitly edited.

5.7.1   The parameter fold

The compiler makes use of the following parameters:

error.checking Default is **HALT**. This selects the type of error checking. The options are **REDUCED, STOP, HALT** and **UNIVERSAL**. See section 5.7.2.

alias.checking Default is **TRUE**. When this parameter is **TRUE**, the compiler does full alias checking. See section 5.7.3.

usage.checking Default is **TRUE**. When this parameter and the **alias.checking** parameter are **TRUE**, the compiler does full usage checking. See section 5.7.3.

separate.vector.space Default is **TRUE**. When this parameter is **TRUE**, the compiler creates separate workspaces for scalars and vectors within the programs being compiled. See section 5.7.4.

create.debugging.info Default is **TRUE**. This allows the debugger to be used with a program when it is run. See chapter 9 for information on the debugger.

range.checking Default is **TRUE**. Setting this to **FALSE** causes the compiler to omit certain checking code (e.g. array bounds checking). It has no effect when the **error.checking** parameter is set to **REDUCED**, as no checks at all will be inserted in **REDUCED** mode. See section 5.7.2.

compile.all Default is **FALSE**. This parameter forces the compiler to recompile all nested compilation units encountered. This is useful if it is necessary to ensure that a program has been compiled uniformly: for example, to ensure that a whole program or set of programs is compiled with the latest version of the compiler, or if changing a program compiled with vector space off to be compiled with vector space on.

force.pop.up Default is **FALSE**. This parameter forces the parameter fold to be displayed whenever the checker or compiler is invoked. This is useful if it is necessary for the user to check and alter the compiler parameters each time the compiler is run.

use.standard.libs Default is **TRUE**. This parameter causes the compiler to use its standard arithmetic libraries within this compilation. For normal compilation the value should be **TRUE**. Setting it to **FALSE** will prevent the compiler from compiling any programs with extended arithmetic, and the compiler will not recognise certain implicit library procedures. Must be **FALSE** when compiling general purpose tools for **TA** transputer class.

target.processor Default is **T8**. This parameter is used to set the target processor when compiling for transputer networks. The following target processors and processor classes are supported:

  T8 the IMS T800, IMS T801 and IMS T805 transputers.

  T425 the IMS T425 transputer

  T4 the IMS T414 transputer.

  T2 the IMS T212, IMS M212, IMS T222 and IMS T225 transputers.

  TA all 32-bit transputers.

  TB all 32-bit transputers without hardware floating point.

  TC all 32-bit transputers with 2D block move and CRC instructions.

code.inserts This parameter determines whether assembly-code insertions are allowed within the program. Values are **NONE**, **RESTRICTED** or **ALL**. The default is **NONE**. See chapter 10 for a description of code insertion.

ring.bell Default is **NEVER**. Determines whether the bell is sounded on completion of compilation. The options are **ALWAYS**, **ERROR** or **NEVER**.

tds2.style.exe Default is **FALSE**. Forces the compiler to restrict the parameters of an **EXE** or **UTIL** to those provided in the obsolete TDS2.

5.7.2   Error modes and range checking

The occam compiler in the TDS implements all four error modes; the mode is specified by the **error.checking** parameter to the compiler. All **SCs** for a single processor must be compiled in the same error mode except that any code can call code compiled with **UNIVERSAL** error checking. Where a library reference is used, the **SCs** of the appropriate error mode will be selected from the library.

On the IMS T414, **HALT** mode does not work for processes running at high priority, as the *HaltOnError* flag is cleared when going to high priority.

In some circumstances it may be desirable to omit the runtime error checking in one part of a program (e.g. in a time-critical section of code), while retaining error checks in other parts of a program, for debugging purposes. The **range.checking** parameter to the compiler has been included to control these checks. Normally when compiling in **HALT** or **STOP** mode, the **range.checking** parameter should be set to **TRUE**. Setting it to **FALSE** allows part of a program to be compiled with certain error checking code omitted. This should be done with great caution; it loses the security associated with error checking. It should only be done if the program is believed to be correct, and there are good reasons for wanting that part of the program to omit error checks.

5.7.3   Alias and usage checking

The compiler implements the alias and usage checking rules described in the occam 2 reference manual. Alias checking prevents an element from being referred to by more than one name within a section of code. Usage checking ensures that channels are used correctly for unidirectional point-to-point communication, and that variables are not altered while being shared between parallel processes. For a further discussion of the motivation behind these rules, see INMOS technical note 32 'Security aspects of occam 2'.

The checking of the alias and usage rules during a compilation is controlled by the **alias.check** and **usage.check** parameters to the compiler. It is possible to turn off alias and usage checking by setting these parameters to **FALSE**. It is also possible to carry out alias checking without usage checking. However, it is not possible to do usage checking without alias checking, as the usage checker relies on the absence of aliasing in the program.

If a program is compiled with **alias.checking** on, the compiler may insert extra code for checking array accesses which cannot be checked until runtime. However, alias checking can also improve the quality of code produced, since the compiler may be able to make some extra optimisations if it knows that names in the program are not aliased.

The usage checker detects illegal usage of variables and channels, for example, assigning to the same variable in parallel. The checker performs most of its checks correctly, but with certain limitations. Normally, if the checker is unable to implement a check exactly, it will perform a stricter check. For example, if an array element is assigned to, and its subscript cannot be evaluated at compile time, then the usage checker will assume that all elements of the array are assigned to. No illegal programs, other than certain programs which use subscripted arrays with replicated PARs will be accepted by the checker. If a correct program is rejected because the usage checker is imposing too strict a rule, it is possible to switch off the checker. A more detailed discussion of the implementation of usage checking is given in section 5.11.3.

5.7.4   Using the separate vector space

The compiler has a parameter called **separate.vector.space**. With this option set to **TRUE** the vectors (arrays) declared within a compilation unit are allocated into a separate 'vector space' area of memory, rather than into workspace. This decreases the amount of stack required, which has two benefits: firstly, the offsets of variables are smaller (therefore access to them is faster), and secondly, the total amount of stack used is smaller, allowing better use to be made of on-chip RAM. If this parameter is **FALSE**, the implementation places vectors in the workspace.

When a program is run within the TDS or loaded onto a transputer in a network, memory is allocated in the following order:

- workspace

- code

- separate vector space

This allows the workspace (and possibly some of the code) to be given priority usage of the on-chip RAM. Generally, the best performance will be obtained with the separate vector space switched on.

The default allocation of a vector can be overridden by an allocation immediately after the declaration of an array. This allocation has one of the forms:

> **PLACE** *name* **IN VECSPACE :**

> or

> **PLACE** *name* **IN WORKSPACE :**

For example, in a program which is normally using the separate vector space, it may be advantageous to put a crucial buffer into internal RAM. The program would be compiled with **separate.vector.space** set to **TRUE**, but would include something like:

```
[buff.size]BYTE crucial.buffer :
PLACE crucial.buffer IN WORKSPACE :
```

For a program where it is required to put all of the data into the workspace, apart from one large array, the program would be compiled with separate vector space off, but with a **PLACE IN VECSPACE** allocation after the declaration of the large array.

Within a program it is possible to mix code compiled with separate vector space on and code compiled with separate vector space off. The parts of the program which have been compiled with separate vector space on will be given use of the vector space.

5.8   Creating and using libraries

Libraries were introduced in section 5.6. This section describes how to create a library, and gives more information on the use of libraries. In particular, it describes how to use 'library logical names' which map library names onto file names in the directory.

Normally, when developing a library, the code will be developed and tested as a set of separately compiled procedures within a test program. The library system has been designed to make it easy to move a set of procedures developed in a test program into a library which can be shared between programs. The work involves collecting together the set of compiled procedures, putting them into a particular kind of fold, and storing them in a file from which they can be accessed. These steps are described in more detail in section 5.8.1.

A library created as above can be shared between parts of a program or between different programs in a single directory. This may be sufficient for a single user. However, if a number of users are working together on a project, or if a user is working on a number of different projects, it will be necessary to share libraries by placing them in a directory shared between users. To make a set of libraries to be shared between directories, an operation called 'library compaction' is required. This collects all the data in the library into a single file. Normally this does not include the source code, but it can include a copy of the source code if required (e.g. for debugging). Library compaction is useful for producing a staged 'release' of a library while development work continues on the sources. In addition, it improves the speed of access to a library as only one file has to be read. Library compaction is described in section 5.8.6.

5.8.1   Creating libraries

This section describes how to create a library, as a series of steps. It assumes that a number of separately compiled procedures have been developed and tested, and it is now required to make these into a library. These separately compiled procedures should not include nested **SCs**; this is a restriction of the library system.

Step 1

Create an empty fold by pressing the $\boxed{\text{CREATE FOLD}}$ key twice. Type something on the fold line, if required. For example:

```
...   mylib
```

Step 2

Apply the $\boxed{\text{MAKE FOLDSET}}$ utility to the empty fold, with the parameter set to **LIB**. The utility will pop up its parameter fold; cycle through the available values using $\boxed{\text{SELECT PARAMETER}}$ until **LIB** is selected, and then press $\boxed{\text{EXIT FOLD}}$. This creates a library fold (marked **LIB**) containing a single fold marked as **Library version**. The library version fold is used to ensure that when a library changes, any programs subsequently linked which refer to that library are first recompiled; see section 5.9.

The fold will now look like:

```
...   LIB mylib
```

with the contents:

```
{{{   LIB mylib
...   Library version date of validation
}}}
```

Step 3

A sequence of text folds containing constant and **PROTOCOL** definitions, and compiled **SC** folds may be placed after the **Library version** fold. For example:

```
{{{   LIB mylib
...   Library version date of validation
...   text fold containing constant definitions
...   text fold containing PROTOCOL definitions
...   SC PROC p1
...   SC PROC p2
}}}
```

The text folds may not be filed (but may contain filed folds). The **SC** folds in a library must appear directly under the version fold; they may not be contained in another fold. There should be no text or blank lines in the fold.

The **SC** folds in a library need not all be compiled for the same target processor type or in the same error mode; they may be 'mixed'. When using the library, the compiler will select the procedures compiled with parameters suitable for the program using the library.

Step 4

Now the library fold should be closed and the COMPILE utility applied to the closed library fold. If the library contains mixed SCs, use RECOMPILE. All compilation units in the library will be inspected, and, if necessary, recompiled. Then the library is made valid; the fold attribute will be set to **compiled fold set**. An error message will occur at this stage if any of the items in the fold is not correct. The syntax of constant and **PROTOCOL** definitions is not checked.

Step 5

To be able to use the library it is necessary to place the library within another, filed, fold. The host filename of that fold is used to identify the library. For example, make a fold which when open looks like this:

```
{{{F "mylib.tst" mylib
...   LIB mylib
}}}
```

Then the library may be used, by quoting its filename, or by using a logical name, as described in the next section.

5.8.2   Using libraries

A library is normally referenced from a compilation unit by a 'logical library name'. A reference to a library in a **#USE** directive takes the form:

   #USE logical.name

The logical name is associated with a real host file name by means of a line in a **Library logical names** fold used by the compilation utilities. This fold is stored in the toolkit fold. A standard version of this is supplied with the TDS system, containing the logical names for the compiler libraries and the libraries supplied with the TDS. This can be added to for user-defined libraries. Libraries that are never used may be deleted.

The form of the lines in the logical names fold is described in detail in section 5.8.5. For now it is only necessary to know that there is one line in the fold for each library. The information on the line includes the directory in which the library file is placed, the error modes and transputer target types supported in the library, and the logical name by which the library will be referenced.

As an alternative to using a logical name, a directive of the form:

```
#USE "host.file.name"
```

may be placed in a compilation unit which wishes to use the library. Using logical names is recommended as it makes it easier to move libraries around the directory structure, or to replace one version of a library with another one.

If the directory in which the library is filed is **\libdir**, then to use the library created in the previous section, either of the following lines may be used:

```
#USE "\libdir\mylib.tsr"

#USE mylib
```

For the second directive to be valid, the programmer must include a line in the logical names fold as follows:

```
{\libdir\} "mylib.tsr" HT8 mylib
```

The first item on the line is the directory, surrounded by curly braces. The second item is the name of the file containing the library, surrounded by double quotes. The third item **HT8**, implies that the library contains code compiled in **HALT** mode for **T8** transputers. The final item is the logical name for the library.

5.8.3   Using protocols with separate compilation

A **PROTOCOL** may be declared and used within a compilation unit according to the rules of the language. Where a protocol is to be used across separate compilation boundaries, the protocol should always be placed in a library; the library should be referenced in any **SC** where the protocol is needed, and in any enclosing compilation unit. For example, suppose we have a protocol **p** defined in a library **my.protocols**. We might then use it as follows:

```
PROC main()
  #USE my.protocols

  {{{  SC do.it
  {{{F "doit.tsr" PROC do.it
  #USE my.protocols
  PROC do.it(CHAN OF p channel)
    ...
  :
  }}}
  }}}
  CHAN OF p actual.channel :
  PAR
    do.it(actual.channel)
    ...
:
```

Since the protocol name **p** occurs in the parameter list of the separately compiled procedure **do.it**, the enclosing compilation unit must include a **#USE** statement, above the declaration of **do.it**, to introduce the name **p**.

5.8.4   How the library system works

It may be useful to know something of how the library system works, in order to resolve problems that may occur when using libraries.

A library contains program text and compiled compilation units. When the compiler encounters a usage of a library it reads in the text and the descriptors of the compilation units, as if they had appeared in the program text. However, if the compiler finds an error in the library header text it cannot report the line in error. So all declarations in a library should be checked before being placed in the library.

The compiler selects compilation units from a library on the basis of their error mode and transputer target type. Only those units with the same error mode and target type, or a transputer class including the target type, as the current compilation will be selected. This may lead to unexpected effects; for example, if a library only contains procedures compiled in **HALT** mode, and the current program is being compiled in **STOP** mode, then the use of the library will not bring any of the procedures into scope and error messages will merely report the use of an undeclared name.

Having made use of a library to compile a compilation unit, the compiler records in the descriptor which libraries have been used. When the program is linked, the linker reads these libraries, and extracts from them any code which is required to link into the program. The **SCs** within a library may themselves refer to other libraries, in which case these are also read. If an **SC** in a library contains one or more procedures which have been used in the program, then the code for that **SC** is linked in. The linker only includes code from those **SCs** containing procedures which have been actually been used in the program. So the only extra code linked into the program, beyond the code actually needed, is the code of unused procedures in library **SC** folds containing at least one used procedure.

The list of libraries used by a compilation unit can be found by using [COMPILATION INFO]. Note that the list also includes libraries used by nested **SC** units within the compilation unit being viewed. The list shows all the libraries used (by means of **#USE** directives) within this compilation unit; if none of the procedures in a library are actually called, then the code for the library will not be included.

Some restrictions of the library system which the user needs to be be aware of are as follows:

- **SCs** within libraries may not contain nested **SCs**.

- A procedure or function name must be unique within a program; the linker will complain if, when linking a system, it finds two library entries of the same name.

The latter condition is flagged at link time by a message of the form:

> **Symbol** *name* **multiply defined in library**

It is up to the user to identify the libraries involved. This message can also appear unexpectedly when using host file names instead of logical names to identify libraries. This can usually be traced to two uses of the same library, using different strings to identify the file (e.g. upper case in one, and lower case in another).

5.8.5   The library logical names fold

The **Library logical names fold** should be provided in the toolkit fold. It defines the mapping from library file names to library logical names. It includes the mappings for the libraries used by the compiler (such as those to support long arithmetic), and the mappings for the library files provided with the TDS.

The library logical names fold is a text fold. It may contain nested text folds. Any comment folds, occam comments on text lines in the fold, or blank lines in the fold, are ignored. Each non-comment text line in the fold either introduces a directory path name, or describes a particular library, and is known as a 'library text line'.

Each library text line corresponds to one library file and describes for that library the error mode and target processor types supported in the library, and the logical name by which the library is to be known. The same logical name may appear in a number of library text lines. When compiling a program, the compiler reads a logical name from a **#USE** statement in the program and uses the logical names fold to find the library file corresponding to the particular combination of the logical name with the target processor and error mode values being used in the compilation.

A library text line consists of a sequence of items. The items are:

- a *directory.name* in braces (e.g. **{c:\tds2\complib\}**) — this may appear on a separate line,
- a *file.name* in quotes (e.g. **"userio.tsr"**),
- one or more *keywords* defining error mode and target processor,
- a *logical.name*.

A *logical.name* may be any contiguous sequence of characters. Filenames or logical names may not contain quotes or spaces. Directory names may not contain braces or spaces.

The keywords defining stopping mode and target in the present implementation are: **HT2 HT4 HT5 HT8 HTA HTB HTC ST2 ST4 ST5 ST8 STA STB STC RT2 RT4 RT5 RT8 RTA RTB RTC XT2 XT4 XT5 XT8 XTA XTB** and **XTC**. The first character is an upper case letter defining the error mode (HALT, STOP, REDUCED or X for UNDEFINED) and the other characters define the target processor (5 for the IMS T425) or processor class.

When a group of libraries in one directory is being specified, it is not necessary to repeat the directory name on every line. The first library text line for that directory must include the directory name. The following lines need not include the directory name; the directory name from the line above is used. A directory name may appear on a line of its own, in which case it applies to the following lines, up to the next line including a directory name. When opening a file the directory name text is simply concatenated with the filename text; note that for this reason the directory name must include the closing backslash (****).

A typical library text line is thus:

```
{c:\userlibs\}"mylib.tsr" HTA RTA STA XTA HT2 RT2 ST2 XT2 mylib
```

indicating that the library file **c:\userlibs\mylib.tsr** contains library code for all modes and targets and may be referenced by the logical name **mylib**.

The complete library logical names fold is read by the compiler when it is started; it is checked for validity and to ensure that any particular combination of logical name, error mode and target processor only maps onto one possible filename.

5.8.6   Library compaction

Library compaction is required whenever a library is to be used in a directory other than the one it was developed in. The compiler is unable to read filed folds nested within a file in another directory, so to make a library available from another directory, all the information in the library has to be placed into a single file. When doing this, it is normal practice to remove the source text from the compacted copy of the library, as it is not needed to use the library, and including it increases the size of the library and the time taken to read it. The source text can be included in the compacted version if required; the main reason for this is to allow the source-level debugger to be able to display the source of a library in which an error occurs.

A utility to compact libraries, COMPACT LIBRARIES, is provided in the file handling utility set. It behaves in a similar manner to the COPY OUT utility in the same set, but compacts any library files encountered on the way. It can thus be used to compact a group of libraries. It is suggested that users adopt the practice of compacting libraries to a different directory from that containing the source, to avoid file name clashes between the original and the compacted versions.

COMPACT LIBRARIES copies the contents of a filed fold, including nested files, to another directory. Any valid library folds encountered are compacted, that is, all information in the library is written into a single file. A parameter **DeleteSource** allows source to be removed from the library as it is compacted. The name of a file being written is normally the same as that of the file being read. A parameter **OverwriteFiles** determines whether existing files are overwritten.

To compact a single library place the cursor on the filed fold containing the library fold, and run the utility COMPACT LIBRARIES. Supply as **DestinationFileName** the intended name of the compacted library. This must include a directory name (which may be the current directory).

To compact a group of libraries, make a fold around the libraries, place the cursor on the fold, and run COMPACT LIBRARIES. Supply as **DestinationFileName** a file name to contain the compacted libraries. This may be a **.TOP** file to allow the files in the other directory to be accessed using the TDS. The destination file name should normally include a directory name; this should be a different directory than the current one, as compacting a set of libraries to the current directory will either overwrite existing library files, or will produce unprodictable file names (depending on the value of the **OverwriteFiles** parameter).

5.9 Changing and recompiling libraries

A previous section (section 5.6) described how the compiler behaves when recompiling a program after a new version of a library has been made. This section discusses the topic in more detail, and describes how to ensure that libraries are recompiled correctly.

5.9.1 Change control

When the text of a library fold is edited, its compilation fold is made invalid, just as for other kinds of compilation unit. The library may not be used again until it has been recompiled.

When all or part of a library is recompiled, its version number is incremented. This is the purpose of the **Library version** fold contained within a library. This fold, which is not openable by the editor, contains a version number for the library, which is incremented every time the library is made valid.

The compiler will refuse to use any libraries which are invalid, and will stop compilation to report an error. If this happens the user should go to the source of the library and recompile it before attempting compilation again.

The compiler records, within the descriptor, the version numbers at compile-time of all the libraries used by a compilation unit. These are checked by the linker against the versions available to it at link time.

The version number of a library can be found by using COMPILATION INFO on the library fold. The version numbers of libraries used at compilation time can be found for any compilation unit using COMPILATION INFO on the foldset.

5.9.2 Library dependencies

As discussed in section 5.6, when compiling a program, the TDS will automatically cope with changes in libraries used by the program. However, there are still some problems of library dependency which may occur. Suppose that library **a** is used by a separately compiled procedure **p** in a program, and that library **a** in turn uses library **b**. If library **a** has changed, then, when compiling the program, this will be noticed and **p** will be recompiled. However, if library **b** is changed, but library **a** has not been recompiled, then the program will compile, but an error will be reported when the whole system is linked together.

Tracking library dependencies of this kind can be aided by a suitable organisation of libraries in a fold structure. The COMPILE utility can not only be applied to a compilation fold, but in fact can be applied to *any* source fold, in which case all compilation units within the fold will be examined, and recompiled if necessary. A collection of libraries involving compilation dependencies may be placed together in a fold, with the 'lowest level' libraries earliest in the fold. If one of the low-level libraries changes, then the COMPILE function can be applied to the collection, to ensure that any dependent libraries are also recompiled.

5.9.3   Recompiling mixed libraries

When using the COMPILE function, parameters are supplied at the time of starting up the compiler, and these parameters apply to all compilation units compiled during that run. For example, the error mode to be used, and the target transputer type are normally the same for all compilation units in a program. To ensure this the compiler also recompiles any units it finds which have been compiled with a different error mode or target type.

This is inconvenient when compiling libraries containing **SCs** compiled for different targets, or in different error modes. For this reason the RECOMPILE function has been provided. When applied to a previously-compiled program, RECOMPILE does the same job as COMPILE, recompiling compilation units as necessary, but for each unit compiled it uses the parameters from the descriptor fold left from the last compilation. So when building a library containing **SCs** for a range of targets or error modes, the compiler parameters for each **SC** unit need only be supplied once, at the time of first compilation, and thereafter the RECOMPILE utility can be used.

The RECOMPILE utility can also be used for recompiling transputer network programs which include code for more than one processor type. It is also useful for other compiler parameters; for example, if one compilation unit in a system needs to be compiled with usage checking off, while the rest are compiled with usage checking on.

5.9.4   Compacting recompiled libraries

Since the compiler can compile all units in any fold it is run on, and the library compacter can compact multiple libraries, it is possible to recompile a set of libraries and then compact them to another ('release') directory using two keystrokes. Note that if libraries in this set depend on libraries earlier in the compilation sequence, then the logical name system should be used as follows:

- Within the building directory, the logical names should refer to the local files, so that the latest compiled version of the library is picked up.

- Within any other directory from which the libraries are being used, the logical names should refer to the files in the release directory.

5.10   The pipeline sorter example

This section introduces a more substantial example which serves to show how a larger program might be structured, in terms of **SC** units and libraries.

Although introduced in this section, the example is also used in the following three chapters, where it is shown running:

- within the TDS

- on a transputer network

- as a standalone program

The application used for this example sorts a sequence of characters into alphabetical order. The basic algorithm, which is discussed in the occam tutorial, uses a number of similar parallel processes. The code for one of these processes is listed opposite.

Since this example does not use real arithmetic or any other transputer operations implemented by different instructions on different target processor types all example compilations are done for the **TA** transputer class in **HALT** error mode.

```
PROC element (CHAN OF letters input, output)
  INT highest, next:
  BOOL going, inline:
  SEQ
    going := TRUE
    WHILE going
      input ? CASE
        terminate
          going := FALSE
        letter; highest
          SEQ
            inline := TRUE
            WHILE inline
              input ? CASE
                letter; next
                  IF
                    next > highest
                      SEQ
                        output ! letter; highest
                        highest := next
                    TRUE
                      output ! letter; next
                end.of.letters
                  SEQ
                    inline := FALSE
                    output ! letter; highest
            output ! end.of.letters
    output ! terminate
:
```

The occam tutorial example has been adapted to have a **WHILE** loop instead of a replicated sequence in order to sort variable length strings of characters. An outer **WHILE** loop separates global program termination from terminating the end of character sequences. Other differences involve using a variant protocol for communicating letters between sorting elements. The **PROTOCOL** is as follows:

```
PROTOCOL letters
  CASE
    letter; INT
    end.of.letters
    terminate
:
```

The example is contained in the directory **\TDS3\TUTOR**, in the top level file **EXAMPLES.TOP**, and in the fold marked **Pipeline sorter example**.

It uses three user-defined libraries:

"header.tsr" contains all the constants and protocol definitions for procedure declarations to come.

"problem.tsr" contains the three separately compiled procedures that make up the body of the application itself.

"monitor.tsr" encloses a procedure used to interface between the application program and the TDS.

The code in these libraries has been put into libraries to facilitate the development of a group of related programs. For any one version of the example a simpler program would result by incorporating this code in-line in a single source file, optionally using one or more separate compilation units.

The three libraries are contained in separate folds in a fold called **Libraries**. Entering this fold shows the three user defined libraries as filed folds:

```
{{{   Libraries

...F "header.tsr" header

...F "problem.tsr" problem

...F "monitor.tsr" monitor

}}}
```

In order to allow these libraries to be used, the **Library logical names fold** in the toolkit fold contains a fold setting up logical names for these libraries. This appears as follows:

```
{{{   pipeline sorter
{\tds3\tutor\}
"header.tsr" HTA header
"problem.tsr" HTA problem
"monitor.tsr" HTA monitor
}}}
```

The following subsections look at the contents of these folds and their structures.

5.10.1   The 'header.tsr' library fold

The **header.tsr** filed fold contains constants and protocols used in the rest of this example. The most important parts of this library are the protocol definitions for **string** and **letter**.

```
{{{F "header.tsr" header
{{{   LIB
...   Library version date of validation
{{{   protocols
PROTOCOL string IS INT::[]BYTE:

PROTOCOL letters
  CASE
    letter; INT
    end.of.letters
    terminate
:
}}}
...   program constants
}}}
}}}
```

The **string** protocol is used for communications between the monitor interface and the application program. As these programs are running in parallel with each other they will be referred to as parallel processes.

The application is made up of many parallel **element** processes, all of which communicate using the **letters** protocol. The **letters** protocol is a variant protocol. This is the method by which differing types of data may be communicated using the same occam channel. With a variant protocol every communication is preceded by a tag to identify the type of the data to follow. These tag names are defined by the programmer. When the tag name itself conveys the desired message no further communication is required. The application reads a stream of letters followed by an **end.of.letters** tag. This is followed by either another stream of letters or a **terminate** tag.

The program constants are selected values from those available for interfacing with the TDS.

5.10.2   The 'problem.tsr' library fold

There are two procedures called **inputter** and **outputter**, which have been put together into an **SC** fold and then placed inside a library fold.

```
{{{F "problem.tsr" problem
...   LIB
}}}
```

Tho fold **LIB** ohows the following two folds:

```
{{{   LIB
...   Library version date of validation
...   SC application PROCs, inputter, element and outputter }}}
```

The contents of the **SC fold** are:

```
{{{   SC application PROCs, inputter, element and outputter
...F "appl.tsr" application PROCs, inputter, element and outputter
...F "appl.dcd" code
...F "appl.dds" descriptor
...F "appl.ddb" debug
}}}
```

The first of these folds contain the PROCs:

```
{{{F "appl.tsr" application PROCs, inputter, element and outputter
#USE header

...   PROC inputter (CHAN OF string input, CHAN OF letters output)
...   PROC element  (CHAN OF letters input, output)
...   PROC outputter(CHAN OF letters input, CHAN OF string output)
}}}
```

The operation of inputter is to input a **string** and then supply it as a sequence of **letters** to a pipeline of **element** processes. The outputter procedure reads the resultant stream of letters and packs the letters back into a string for communication onwards. The **string** communication is far more efficient for link communication as the link can communicate all the data before attempting to gain more processing time.

The design of these three procedures is such that they should all be instanced as parallel processes, communicating with one another using occam channels.

5.10.3   The 'monitor.tsr' library fold

The final library involved in this example is one that contains the interface with the TDS. This is a procedure called **monitor** that supplies the values of keystrokes on the host keyboard to the application while in parallel conveying data and result outputs to the host screen.

This procedure is also contained in a library fold:

```
{{{   LIB
...   Library version date of validation
...   SC monitor.tsr
}}}
```

The SC fold contains the filed fold **monsrce.tsr**, which contains the screen and keyboard handler.

```
{{{F "monsrce.tsr" monsrce
#USE header
#USE strmhdr
#USE streamio

PROC monitor (CHAN OF KS keyboard,
              CHAN OF SS screen,
              CHAN OF string  app.in, app.out,
              VAL BOOL  using.subsystem)

  ...  PROC keyboard.handler
  ...  PROC screen.handler

  CHAN OF INT echo:
  PAR
    keyboard.handler (keyboard, echo, app.in)
    screen.handler   (app.out, echo, screen)
  :
}}}
```

The procedure **monitor** converts the keyboard and screen I/O from the TDS into simple strings of bytes for the application. By using this monitor one can edit text strings in advance of sending them to the application. This means that the application program itself need not concern itself with erroneous strings, multiple carriage returns or case sensitivity. All these functions can be filtered out by the keyboard handler. The keystrokes made at the keyboard are sent down the channel **echo**. The **screen.handler** can distinguish between keys typed by the user and strings supplied by the application.

The use of the screen handler process enables it to be the only parallel process that needs to communicate using the TDS protocol. This makes the application more portable. If it becomes necessary to mount the application in a different system environment then it is only the **monitor** that needs to be changed.

The keyboard handler

```
PROC keyboard.handler (CHAN OF KS in,
                       CHAN OF INT out,
                       CHAN OF string data)
  ...  variables
  SEQ
    going := TRUE
    length := 0
    WHILE going
      SEQ
        in ? char
        CASE char
          stopch
            ...  terminate monitor and application if appropriate
          return
            ...  pass string to application if non zero in length
          ft.del.chl
            ...  user has typed the backspace key
          ELSE
            ...  buffer char, all letters map to lower case
  :
```

This is a good opportunity to note how folds may be used to show the structure of the occam text. As can be seen the keyboard handler procedure is a **CASE** construct repeated many times within a **WHILE**.

Termination of parallel programs is the duty of the programmer. The termination of the monitor process is achieved by the user entering the **stopch** at the keyboard. The keyboard handler must then pass this character to the screen handler so that it will also terminate. This is done because an OCCAM program can only terminate when all of its constituent parallel processes have terminated and in the monitor process the keyboard and screen handlers are running in parallel. This termination request will normally be passed on to the application process as well.

The screen handler

The screen handler is contained in a separate fold:

```
{{{   PROC screen.handler
PROC screen.handler    (CHAN OF string data,
                        CHAN OF INT in,
                        CHAN OF SS out)
   ...   constants, procedures and variables
  SEQ
     ...   initialise
     ...   body
     ...   finish
  :
}}}
```

The main part of the screen handler is contained in the fold **body**:

```
{{{   body
WHILE going.in OR ((NOT using.subsystem) AND going.data)
  SEQ
    clock ? waketime
    waketime := waketime PLUS one.hundredth.of.a.second
    ALT
      going.in & in ? char
        ...   print keyboard character on screen
      going.data & data ? length::string
        ...   print data from application on screen
      monitoring & clock ? AFTER waketime
        ...   if monitoring is TRUE, poll subsystem error pin
    draw.cursor (kb.window)
}}}
```

The screen handler is repeatedly searching for one of three alternatives. Either keyboard characters are echoed, a string of data comes from the application or a timeout happens if neither of the other two have occurred in one hundredth of a second. The timeout is relevant if the **monitor** is monitoring an application running on another transputer. This is discussed in more detail in chapter 7, where the monitor will be used in this way.

This section has shown the structure of the libraries required by the pipeline sorter example. These are used in chapters 6,7 and 8, where the program is run in the three different environments.

5.11   The implementation of occam

This section describes some details of the implementation of OCCAM by the compiler in the TDS. It can be omitted in a first reading of the manual.

It discusses three aspects of the implementation:

- Implementation decisions, such as data representation, for OCCAM on the transputer.

- The layout of code and data in memory.

- Some restrictions of the usage checker.

5.11.1   The transputer implementation of occam

This section defines the implementation of occam for the transputer, supported by the compiler in the TDS. It describes the way certain implementation dependent decisions have been made in the compiler.

Data representation

- The size of an **INT** (word) on an IMS T414, IMS T425, IMS T800, IMS T801 or an IMS T805 is 32 bits.

- The size of an **INT** (word) on an IMS T212, IMS M212, IMS T222 or IMS T225 is 16 bits.

- Scalar variables are always allocated on a word (**INT**) boundary and occupy an integral number of words.

- **BOOL** and **BYTE** variables in arrays occupy 8 bits each. A declared array is aligned on a word boundary, and occupies space rounded up to the next word boundary. Note that an abbreviation of part of such an array might not begin on a word boundary.

- Protocol tags are represented by 8-bit values. The compiler allocates such values from **0 (BYTE)** upwards in order of declaration.

- A **RETYPES** specification is invalid unless the alignment and size of the right-hand side is the same as for the left-hand side. Note especially that an array of **BOOL** or **BYTE** variables specified by an abbreviation (e.g. passed as a parameter) may have any alignment and so can not in general be retyped.

Channel communication

The strict rules of occam protocol require the two processes involved in a communication to agree on the lengths of the blocks communicated and use of strict channel protocols will ensure this. It is also possible to use anarchic protocols (**CHAN OF ANY**) to avoid these compiler checks. The behaviour of communications where the sender and receiver have different expectations of the block length differs on the transputer between hard channel (transputer links) and soft (internal) channels. On hard channels each byte is acknowledged individually and so it is only necessary for the total number of bytes received to match the number sent. On soft channels each block is an acknowledged unit of communication and the size actually communicated will be that required by the second of the two processes to become ready. In occam the lengths of blocks are determined by the types of the expressions sent and variables used as destinations. Users of anarchic protocols need to be aware of these differences if a program is adapted from hard to soft channels or vice-versa. Anarchic protocols should be avoided whenever possible.

An input from the event channel may be of any type. An indeterminate value will be received.

Hardware dependencies and configuration

- The number of priorities supported by the transputer is 2, so a **PRI PAR** may have two component processes. Nested **PRI PAR**s are invalid; the compiler checks this within a procedure, but does not check across procedure boundaries. A runtime check is done to compensate for this; if the program attempts a **PRI PAR** while at high priority, the error flag is set. Future releases of the compiler may check for nested **PRI PAR**s properly.

- The low priority clock increments at a rate of 15 625 ticks per second, or one tick = 64 microseconds.

- The high priority clock increments at a rate of 1 000 000 ticks per second, or one tick = 1 microsecond.

- The numbers used as **PLACE** addresses are word offsets from the bottom of address space (see chapter 11).

- The syntax of the **PROCESSOR** statement is extended so that one of the keywords **T8**, **T4**, **T425**, **T2**, **T414**, **T800** or **T212** must follow the processor number (see chapter 7).

- On the transputer, **TIMER**s cannot be **PLACE**d.

Language extensions accepted by the TDS compiler

- Statements beginning **#USE** are library references (as described in section 5.8) and introduce declarations from a library at the point in the source code where they appear.

- **PLACE** *name* **IN VECSPACE** and **PLACE** *name* **IN WORKSPACE**. These were described earlier in section 5.7.4.

- The koyword **GUY** introduces a section of transputer assembly code (see chapter 11).

- For each transputer target type there is a set of predefined procedures and functions which may be used without declaration.

- Statements beginning **#COMMENT** or **#OPTION** are ignored.

See appendix C for a complete list of reserved keywords which the compiler will not allow programs to redeclare.

The compiler's use of folds

- The compiler reads the text contained within the source fold of a compilation unit being compiled.

- The heading (comment) on a fold line is not passed to the compiler.

- The contents of a comment fold (see section 5.3.2) are not passed to the compiler.

Any further differences between the language defined in the occam 2 Reference Manual and the language accepted by the compiler are described in the Delivery Manual accompanying the TDS release. That manual also describes any other restrictions imposed by the implementation.

5.11.2   Memory allocation by the compiler

Code

The compiler generates code so that any nested procedures are placed at lower addresses (i.e. nearer **MOSTNEG INT** on a transputer) than the code for the enclosing procedure. Nested procedures are placed at increasingly higher addresses in the order in which their definitions are completed. If a unit contains a nested **SC** then the code for this **SC** is loaded at a lower address. If a unit contains more than one nested **SC** then the code for the last textually declared **SC** is loaded at the lowest address. Libraries are linked in at a higher address than the code within the program, except for the compiler's real arithmetic handling library, which (if used) is linked in at the low end of the code.The code for the whole system occupies a contiguous section of memory.

Workspace

Workspace is allocated from higher to lower address (i.e. the workspace for a called procedure is nearer **MOSTNEG INT** than the workspace for the caller). In a **PAR** or **PRI PAR** construct the last textually defined process is allocated the lowest addressed workspace. In a replicated **PAR** construct the instance with the highest replication count is allocated the lowest workspace address.

When the **separate.vector.space** option is enabled, arrays (apart from those explicitly placed in the workspace) are allocated in a separate data space. The allocation is done in a similar way to the allocation of workspace, except that in vector space the data space for a called procedure is at a *higher* address than the data space of its caller.

When a program is run within the TDS or loaded onto a transputer in a network, memory is allocated in the following order, with the workspace nearest **MOSTNEG INT**.

- workspace

- code

- separate vector space

This allows the workspace (and possibly some of the code, starting with the real arithmetic handling library) to be given priority usage of the on-chip RAM.

The variables within a single process (or procedure) are allocated so that the textually first variable is given the highest address in the current workspace.

From within a called procedure the parameters appear immediately above the local variables. When an unsized vector is declared as a formal procedure parameter (e.g. **[]INT**) an extra **VAL INT** parameter is also allocated to store the size of the array passed as the actual parameter. This size is in the units of the array, not in bytes (unless it is a byte array). One extra parameter is supplied for each dimension of the array unsized in the call, in the order in which they appear in the declaration. See section 11.3.1.

If a procedure requires separate vector space, it is supplied by the calling procedure. A pointer to the vector space supplied is given as an additional parameter after all the actual parameters of the call.

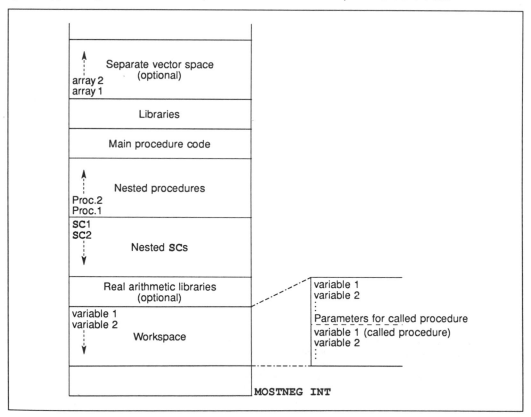

Figure 5.1 Memory allocation

5.11.3 Implementation of usage checking

This section describes some restrictions in the implementation of usage checking.

Usage rules

The usage checker is attempting to check the following rules of occam 2:

- No variable assigned to, or input to, in any component of a parallel may be used in any other component.

- No channel may be used for input in more than one component process of a parallel. No channel may be used for output in more than one component of a parallel.

Checking of non-array elements

Checking of variables and channels which are not elements of arrays is performed correctly.

The usage checker insists that a channel parameter or free channel of a procedure is not used for both input and output.

Checking of arrays of variables and channels

When an array of variables or channels is used in a program the usage checker, where possible, treats each element of the array as a separate variable or channel. This makes it possible, for example, to assign to the first and second elements of an array in parallel.

For the usage checker to operate in this way, it must be possible for the compiler to evaluate all possible subscript values when an array is used. The compiler is capable of evaluating expressions consisting entirely of constant values and operators (but not function calls). Where a replicator index is used in an expression the compiler can evaluate the expression for each possible value of the index provided that the replicator's base and count can be evaluated. However, there are certain problems with parallel replicators which are described later.

Where an array subscript contains variables, a function call or the index of a replicator where the compiler cannot evaluate the base or the count, then the usage checker will assume that all possible subscripts of the array may be used. This may cause the usage checker to give an error message where there is no real problem. For example, consider the following program fragment:

```
x := 1
PAR
  a[0] := 1
  a[x] := 2
```

The usage checker will report the assignment to **a[x]** as a usage error. However, the fragment could be changed to:

```
VAL x IS 1:
PAR
  a[0] := 1
  a[x] := 2
```

Here the checker would accept it because **x** can be evaluated at compile time.

The checker handles segments of arrays similarly to simple subscripts. Where the base and count of a segment can be evaluated by the checker, the checker behaves as if each segment has been used individually. Where the base or count cannot be evaluated by the checker, the checker behaves as if the whole array has been used. For example, the checker will accept

```
PAR
  [a FROM 4 FOR 4] := x
  a[8] := 2
  [a FROM 9 FOR 3] := y
```

without generating an error.

Arrays as procedure parameters

Any variable array which is the parameter of a procedure is treated as a single entity. That is, if any element of the array is referenced, the checker treats the whole array as being referenced. Similarly, if any variable array, or element of a variable array is used free in a procedure then the checker treats it as if every element were used. For example, the usage checker will generate an error on the following program

```
PROC p([]INT a)
  a[1] := 2
:
PAR
  p(a)
  a[0] := 2
```

because it considers every element of **a** to have been used when **p(a)** occurred.

Similarly, where one element of an array of channels has been used for input within a procedure, the checker treats the array as if all elements had been used for input, and, where one element has been used for output within a procedure, the checker treats the array as if all elements had been used for output. For example, the usage checker will generate an error on the following program

```
PROC p()
  c[1] ! 2     -- c free in p
:
PAR
  p()
  c[0] ! 1
```

because it considers an output to have been performed on every element of **c** when **p()** occurred.

Abbreviating variables and channels

The usage checker treats an element which is abbreviated in an element abbreviation as if it had been assigned to, whether or not it is actually updated. If this causes the checker to reject an apparently correct program the program should be altered to use a **VAL** abbreviation. For example, the following program will cause a usage error

```
PAR
  a IS b :
  x := a
  y := b
```

because the first component of the **PAR** is assumed to assign to **b**. This could be changed to:

```
PAR
  VAL a IS b :
  x := a
  y := b
```

Where a channel is used which is an abbreviation of a channel array element, the checker behaves as if the whole of the channel array had been used unless the element is an array element with a single, constant subscript, a constant segment of an array (i.e. with constant base and count) or a constant segment with a single, constant subscript. For example:

```
PAR
  c IS a[1][2] :
  c ! 1
  a[0][1] ! 2
```

is rejected by the usage checker, as it considers the whole of the array **a** to have been used for output when **c ! 1** occurred since **a[1][2]** contains two subscripts.

However,

```
PAR
  c.slice IS a[1] :
  c       IS c.slice[2] :
  c ! 1

  a[0][1] ! 2
```

is accepted, since each abbreviation has just one, constant subscript.

Problems with replicators

The usage checker has the following problems in its handling of replicators:

1 Parallel accesses to an array inside a replicator loop may be incorrectly checked against each other and flagged as errors. For example, in

```
SEQ i = 0 FOR 10
  PAR
    a[i] := 1
    a[i + 1] := 2
```

the checker will flag the seond assignment as an error even though this program does not break the usage rules. (The reason for this is that the array elements which will be assigned to by the first assignment during the execution of the **SEQ** replicator will overlap those assigned to by the second assignment). The only way to avoid this problem in the current compiler is to switch off usage checking.

2 Replicated **PAR** loops are not checked properly.

The checker permits any usage of an array element within a replicated **PAR** provided the replicator index occurs within the subscript expression.

The following two programs are examples of incorrect programs accepted by the checker:

```
PAR i = 0 FOR 10
  a[i - i] := 1

PAR i = 0 FOR 4
  SEQ
    a[i] := 1
    a[i + 1] := 1
```

6 Running programs within the TDS

Chapters 4 and 5 have described how to load code into the TDS, and how to create, compile and run a simple program. This chapter reviews the steps in running a program, and then describes the interfaces available to user programs. These interfaces are presented over channels connecting the running program to the TDS. This chapter introduces these channels, their protocols, and a number of procedures from the input/output libraries supporting communications on these channels.

This chapter concludes by showing the pipeline sorter (discussed in the previous chapter) adapted to run in the TDS, and some simple examples using the input/output procedures.

6.1   Loading and running an executable program

A user program is contained in an **EXE** compilation fold. Once this has been compiled and linked, the compilation fold includes a fold, called a **CODE EXE** fold, containing binary code suitable for loading and running by the system. This fold can be moved from the compilation fold and used directly for loading.

To load a program, the cursor should be placed on a compiled and linked **EXE** fold, or on a **CODE EXE** fold removed from such a compilation fold. The [GET CODE] key should then be pressed, and the code will be loaded ready for execution.

Once it has been loaded, an **EXE** user program may be run by pressing the [RUN EXE] key. The program remains in memory until it is cleared using the [CLEAR EXE] or [CLEAR ALL] key.

A program will either run to completion, fail to terminate (by deadlocking or livelocking), or set the transputer error flag as a result of a runtime error. A program may be interrupted by means of the TDS interrupt key ('control-break' on the IBM PC).

If a program sets the error flag the server will detect this and terminate with an appropriate message.

If the server is interrupted from the keyboard it may also be terminated.

In either of these situations neither the user program nor the TDS are able to proceed.

The user may then restart the TDS, or may have a command file which does this automatically.

Before the TDS is rebooted the user may be given the option of preserving the state of the workspace of the **EXE** in a core dump file . The TDS is then restarted and the Debugger program may be run (see chapter 9) to locate the cause of the error, or to examine the interrupted program.

6.2   The interface for user programs

EXE programs have the form of an occam process. For example:

```
{{{   EXE myprog
{{{F "myprog.tsr" myprog
...   Declarations
SEQ
  ...   Program
}}}
}}}
```

An **EXE** is called by the TDS as if it was an occam procedure with a number of channel parameters connecting it to other components of the TDS. The channel parameters are used for communication between the **EXE** and the processes of the TDS which provide access to the folded data structure, the host terminal and the host filing system. The names of these channels are pre-declared by the compiler, and do not have to be explicitly declared by the programmer.

The parameters implicitly provided by the compiler to an **EXE** are as follows:

Type	Name	Comments
CHAN OF ANY	**keyboard**	Keyboard channel supplying ASCII values and TDS keys.
CHAN OF ANY	**screen**	Screen channel expecting **tt** tags and values.
[max.files]CHAN OF ANY	**from.user.filer**	Array of channels from the user filer.
[max.files]CHAN OF ANY	**to.user.filer**	Array of channels to the user filer.
CHAN OF ANY	**from.fold.manager**	INMOS internal use only.
CHAN OF ANY	**to.fold.manager**	INMOS internal use only.
CHAN OF ANY	**from.filer**	Channel from filer for DOS file access.
CHAN OF ANY	**to.filer**	Channel to filer for DOS file access.
CHAN OF ANY	**from.kernel**	Used to test $\boxed{\text{SET ABORT FLAG}}$.
CHAN OF ANY	**to.kernel**	Used to test $\boxed{\text{SET ABORT FLAG}}$.
CHAN OF SP	**from.isv**	Channel from **iserver**.
CHAN OF SP	**to.isv**	Channel to **iserver**.
[] INT	**freespace**	Remaining free memory within the TDS.

The value of **max.files** is 4.

The **keyboard** and **screen** parameters passed to an executable procedure by the TDS are channels to and from the terminal. By using these channels a program is able to communicate data to and from these devices. The user filer channels provide access to files which are part of the folded data structure visible through the editor. The **.isv** channels provide access to the host filing system and other services by way of **iserver**. The protocol used on **.isv** channels is declared in the library **strmhdr**, for which an implicit **#USE** is supplied when an **EXE** is compiled.

When an **EXE** program runs it may communicate with the TDS on the channels listed above. Communication using these channels must obey a set of protocols set out in chapter 16 (in sections 16.6 and 16.7). Some channels have been declared as **CHAN OF ANY**, rather than with proper occam protocols, to maintain compatibility with previous releases of the TDS and to allow existing programs to continue to run unchanged. Normally, users can make use of procedures from the I/O libraries to handle communications on these channels, and so should not need to use the protocols directly. However, if writing programs to use the channel protocols directly, it is possible to create an occam **PROTOCOL** description matching the protocol on the channel (see section 6.3.2 for an example of this). The channel declared **CHAN OF ANY** may then be passed as an actual parameter to a procedure with a formal parameter declared with an occam **PROTOCOL**.

In order to help programmers to use the channels a number of I/O libraries are provided. By calling procedures from these libraries, it is possible to write programs which perform input and output in a way that is familiar to most programmers using other high level programming languages. Procedures are also provided for conversions between text strings and the numeric types of occam. The I/O libraries are discussed in full in chapter 14.

According to the occam communications model, one end of a channel may not be shared between processes running in parallel. This has implications for I/O in user programs. For example, only one concurrent process in a system may access the **screen** channel. If it is necessary for more than one process to output to the screen, then the programmer must build in explicit multiplexing processes. Some support is provided in the I/O libraries for multiplexing communications on the system interface channels.

Users of INMOS occam toolset products will notice that the last three parameters of an **EXE** match those of a typical stand-alone occam program supported by **iserver**. By restricting channel usage to the server channels it is possible to write **EXE** programs which are trivially portable to the occam toolsets, both at source code level and as bootable programs.

6.3   The channel parameters and their protocols

When a program is run within the TDS, it is run in parallel with, and connected to, certain components of the TDS. These are:

- The 'terminal handler' process, connected via the channels **screen** and **keyboard**

- The 'user filer' process, connected by four pairs of **from.user.filer** and **to.user.filer** channels.

- The 'fold manager' process, connected via the channels **from.fold.manager** and **to.fold.manager**.

- The 'filer' process, connected via the channels **from.filer** and **to.filer**.

- The 'kernel' process, connected via the channels **from.kernel** and **to.kernel**.

These processes within the TDS run in parallel with the **EXE** and communicate both with the **EXE** and with the TDS server running on the host computer. When communicating with the server, the TDS multiplexes these channels, and the explicit **iserver** channels, onto the pair of channels supported by the INMOS link connection.

6.3.1   The explicit iserver channels

The channels **from.isv** and **to.isv** provide an interface between the **EXE** and the **iserver** program running on the host. All communications are buffered within the TDS, and the following commands are intercepted by the TDS: *GetKey, PollKey, CommandLine* and *Exit*, (see chapter 16). In programs intended to be supported by a server independently of the TDS and whose requests for service are channelled through a single process it is desirable to use these channels both for communications with the host terminal's keyboard and screen and with the host filing system. The filing system interface resembles that provided in most conventional sequential operating systems and can support an arbitrary number of simultaneously active files.

The protocol used on these channels is called **SP** and is defined in appendix D. This protocol is common to all implementations of the INMOS server **iserver**.

6.3.2   The keyboard and screen

The **keyboard** channel produces a sequence of integer values corresponding to keys pressed at the terminal. Values are normally either ASCII values for simple keys, or special values for TDS function keys. These values are discussed in more detail in section 16.6. User programs will normally use the input procedures from the library **streamio**, introduced in section 6.5.

The **screen** channel accepts a sequence of screen control commands. Each command consists of a **BYTE** tag identifying the command, followed by the data for the command. These commands are discussed in more detail in section 16.6. User programs will normally use the output procedures from the library **streamio**, introduced in section 6.5. These procedures specify the **screen** channel to have the protocol **SS** whose declaration is in the library **strmhdr**, which is automatically referenced when an **EXE** is compiled.

6.3.3   Communicating with the user filer

The user filer allows a program to read filed folds produced within the TDS, and to write to filed folds so that the output of the program may be read from within the TDS. The user filer has a view of the filing system similar to that already introduced for utilities (see section 4.6.4); data may be read and written at the position in the fold structure given by the current line when RUN EXE is pressed.

For example, the cursor may be on a filed fold, in which case the user program may read that file, or write to it (if it is empty). Alternatively, the cursor may be on a fold 'bundle' consisting of a fold with a number of filed folds inside it. As a specific example, the following might be the view on opening up a fold bundle after running a user program that reads two data files, and outputs two data files.

```
{{{   fold bundle
...F "if1.tsr" input file 1
...F "if2.tsr" input file 2
...F "of1.tsr" output file 1
...F "of2.tsr" output file 2
}}}
```

Communication with the user filer by a user program has two main stages. Firstly the program issues one or more user filer commands to identify the filed fold to be read or written. Folds are identified by their number in the fold bundle. Secondly, once a filed fold has been opened, the program enters a stage where it reads or writes a data stream (possibly including folds) by communicating with the user filer.

The user filer commands, and the data stream communications are described in detail in section 16.7. There is also a library **ufiler** to support user filer communications. However, for most purposes, the most convenient way of using the user filer facilities are to input data as if from a keyboard, and output data as if to a screen, and use interface processes from the library **interf** to convert the data into user filer communications. These are introduced later in this chapter, in section 6.5.

6.3.4   The fold manager

The fold manager channels, **from.fold.manager** and **to.fold.manager**, communicate with the component of the TDS which stores the folded document being edited using the TDS. The protocols on these channels are not documented, and the channels should not be used by user programs.

6.3.5   Communicating with the filer

The channels **to.filer** and **from.filer** and the associated **msdos** library are provided for compatibility with earlier versions of the TDS only. They are supported by protocol conversion code within the TDS which maps the obsolete **.tkf** commands on to appropriate communications with **iserver** using **SP** protocol.

New programs requiring access to host files and other services should use the explicit **iserver** channels.

6.3.6   The kernel channels

The channels **to.kernel** and **from.kernel** have only one useful function to user programs running within the TDS; they allow a user program to test whether the [SET ABORT FLAG] key has been pressed (see section 12.1 for a description of this key). The code to do this is as follows:

```
#USE krnlhdr
INT result:
SEQ
  to.kernel ! k.get.abort.state
  from.kernel ? result
  IF
    result = 0
      ...   not pressed
    result <> 0
      ...   pressed, so abort
```

6.4   Memory usage within the TDS

The memory on the host transputer is shared between the TDS itself and any currently loaded **EXE**s and **UTIL**s. The code of the TDS and its own workspace occupy a fixed space at the top end of memory. The remaining memory is divided in a constant proportion between areas known as the fold manager buffer and the user area. The fold manager buffer is used by the TDS to hold filed folds as they are required by the editor and some other tools. The user area holds the code of all the currently loaded **EXE**s and **UTIL**s and is used for the workspace of any one **EXE** or **UTIL** when it is run.

In order to optimise the use of on-chip RAM, which is at the lowest end of transputer address space, the TDS moves the code of an **EXE** or **UTIL**, when it is about to be run, to an address that is as low as possible, allowing room for its workspace below it. If the program has been compiled with separate vector space, then this is allocated above the code. Above that, and below any other currently loaded programs, is an area of memory which the program may address as the array **freespace**. The size of this array is therefore dependent on how many other programs are currently loaded. See figure 6.1

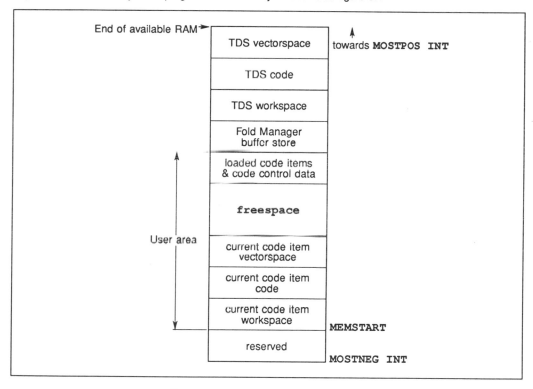

Figure 6.1 Memory usage within the TDS

The size of available RAM and the size of the Fold Manager buffer store are determined by parameters supplied to the TDS when it is loaded. See section 16.2

Programs may themselves subdivide the **freespace** array by abbreviation or retyping (taking appropriate precautions to avoid alignment errors), for example:

```
VAL freesize IS SIZE freespace:
VAL one.fifth.free IS freesize / 5:
VAL one.quarter.free IS freesize / 4:

-- allocate a fifth of freespace for integers
[]INT int.store IS
   [freespace FROM 0 FOR one.fifth.free]:

-- allocate a quarter of freespace for bytes
[]BYTE byte.store RETYPES [freespace FROM
                         one.fifth.free FOR one.quarter.free]:

-- allocate anything else for long reals
VAL rest.start IS one.fifth.free + one.quarter.free:
VAL rest.free IS freesize - rest.start:
VAL double.rest IS rest.free /\ #FFFFFFFE: -- round down to
                                      -- multiple of 2 words
[]REAL64 long.real.store RETYPES [freespace FROM
                         rest.start FOR double.rest]:
```

This proportional allocation technique is used by the occam compilation utilities and so the size and complexity of program unit that may be compiled is dependent both on the size of memory available and the other programs that are loaded at the time it is run.

6.5   The occam input/output procedures

This section describes some I/O procedures which are in libraries provided with the TDS software. These are procedures which are either called in sequence within the user program to carry out a set of communications on a channel, or in parallel with part of a user program to convert a stream of communications in one format to a stream of communications in another format.

When using these procedures the distributed nature of the occam model of communication must be kept in mind. The I/O procedures require access to a channel accepting the appropriate protocol. It is the responsibility of the programmer to ensure that channels of the right protocol are supplied. Any multiplexing of communication streams must be done explicitly within a program. Procedures are available in the libraries to assist in this.

Some of the I/O procedures are based on lower-level procedures for number conversions and similar operations. These lower-level procedures are an essential part of the language implementation and are described in the occam 2 reference manual.

The input/output procedures may be used to facilitate the coding of simple sequential inputs and outputs from and to the external world. The external world is typically a keyboard and a screen and a filing system, but some of these procedures are applicable to arbitrary devices. The procedures hide many of the detailed features of the protocols on the channels to the run-time environment. Programmers whose requirements are less straightforward may use the full facilities of the programming interface described in chapter 16: 'System Interfaces'.

If a program using the I/O procedures is to be run on a transputer network, it is necessary to supply the program with a set of channels accepting the appropriate protocols, and routing messages within the network as required. Some interface procedures are provided in the libraries **interf**, **ssinterf** and **spinterf** to aid in the multiplexing and routing of these protocols.

When a program has been loaded onto a network from the TDS, an **EXE** can be run within the TDS to communicate with the program in the network and supply a run-time environment consisting of screen, keyboard and filing system.

6.5.1   The input/output models

Three models of input/output are supported by appropriate sets of procedures.

The first model, the **hostio** model, is based on the conventional use of a filing system on a host computer within which multiple files may be opened and then accessed using a stream number to identify them. In this model the keyboard and screen are considered as special cases of files.

The second model, the **streamio** model, is a simple model of input and output which is applicable both to an interactive terminal and to sequential text files, is based on a sequence of lines of text separated by carriage return characters. The input from the terminal is called a 'key stream'. Output to the terminal is called a 'screen stream'. This model is also appropriate for communication between the processes of an occam program, if the information being sent is essentially a sequential text stream.

The third model is the 'folded stream' model, which allows files in the hierarchical fold structure within the TDS to be read or written.

The three models are not mutually exclusive, and in some programs an appropriate mixture of library procedures from the three groups may be used. The TDS itself provides an **EXE** with channels for all the models. Programs designed to run eventually without run-time support from the TDS should use **hostio** calls for their external I/O requirements as these can be supported by any version of the host server, **iserver**.

6.5.2   The hostio model

This model is supported by procedures in the **hostio** library group. It assumes the existence of a server process, usually on a separate host processor, which provides the terminal access and filing system of a typical operating system, in a way familiar to users of sequential languages such as C or Pascal. All communications with this server are initiated by the program and use a pair of channels. Each transaction consists of a message to the server, followed by a reply from the server. These messages use a simple counted array protocol **SP**, the details of which are given in section 18.3. Each message to the server includes an identifying tag defining the action required, and supplementary data depending on the tag.

Each possible command across this interface is supported by a library procedure, and there are also higher level procedures performing useful sequences of low level transactions.

An important set of procedures is that which provides access to the host filing system. This includes the concept of a stream identifier returned by the server when a file is opened, and passed back to the server to identify the file in each subsequent transaction involving that file. Although the protocol does not limit the number of files that may be accessed simultaneously, the particular implementation of **iserver** to be used may do so.

The library includes procedures for buffering and multiplexing **SP** protocol channels. By use of these it is possible to use the protocol in many processes in parallel, optionally on many processors. Protocol conversion interface procedures are also provided so that occam processes using the **streamio** model can cooperate with those using the **hostio** model.

6.5.3   The streamio model

This model is supported by procedures in the libraries **streamio**, **ssinterf**, **userio** and **interf**.

For historical reasons the procedures in these libraries are coded in two ways. The old way, used in libraries **userio** and **interf** represents keystream protocol as **CHAN OF INT** and screenstream protocol as **CHAN OF ANY**. This is because variant protocols had not been introduced to the occam2 language when these libraries were first included in the TDS. The new way, used in libraries **streamio** and **ssinterf** uses protocols **KS** for keystream, and **SS** for screenstream.

Channels declared as **CHAN OF ANY** may be passed as parameters to procedures whose formals have explicit protocols, and so should be used if for any reason the old style library procedures are used. The preferred set of procedures is those using the strict protocols.

Communications using the **streamio** model use a single channel. When characters are being handled one at a time the keystream protocol is appropriate which handles each character as an integer. The TDS provides a mechanism whereby the multiple codes generated by keyboards are mapped onto a set of special integer values known as 'cooked keys'. The details of this mapping are determined, in the TDS itself, by the contents of the ITERM file (see section 16.3). In the absence of the TDS the user is reposible for performing any such mapping. When characters are being handled in bigger groups, such as lines of text, screenstream protocol may be used. This also provides abstractions of commonly available screen control operations such as cursor control. It therefore allows the body of programs to be written without knowledge of the sequences required by a particular terminal. In the TDS a mapping defined in the ITERM file may be used, or an explicit conversion into the sequences required by particular terminals may be performed by using appropriate interface procedures.

6.5.4   The folded file store model

Procedures for accessing the TDS fold structure are included in the libraries **userio**, **interf**, **ssinterf** and **ufiler**.

This model uses a pair of channels, called user filer channels, coded as **CHAN OF ANY**. Such a pair is provided to an **EXE** by the TDS.

The user filer allows an **EXE** to perform operations on a fold on which the cursor is positioned when it is run. Access may be made to any fold or immediately embedded sub-fold which may be filed. New sub-folds may be created and written into.

6.5.5   Interface procedures

Procedures designed to be called as processes in parallel with other processes, for purposes of buffering, multiplexing and protocol conversion, are collectively known as interface procedures.

The use of these, which may be unfamiliar to users of purely sequential languages, is demonstrated in several examples supplied with the TDS. See for example section 6.7.2.

6.6   The pipeline sorter example

This section continues with the pipeline sorter example, introduced in section 5.10, and prepares it to run within the TDS. More complete instructions are provided with the version of this example included in the software. The three libraries are referenced from inside an **EXE** fold. The parameters **keyboard** and **screen** of the monitor process are connected to the channels which communicate with the appropriate components of the TDS.

```
{{{  EXE harness
{{{F "harness.tsr" harness
#USE header          --  program constants
#USE monitor         --  EXE interface to TDS
#USE problem         --  PROCs used in application

CHAN OF string app.in, app.out:
PAR
  monitor (keyboard, screen, app.in, app.out, FALSE)
  ...  application
:
}}}
}}}
```

The application contained in the fold is:

```
{{{  application
[string.length+1]CHAN OF letters pipe:
PAR
  inputter (app.in, pipe[0])
  PAR i = 0 FOR string.length
    element (pipe[i], pipe[i+1])
  outputter (pipe[string.length], app.out)
}}}
```

The program runs the monitor in parallel with the application. The application itself is made up of inputter, outputter and a replicated instance of the **element** procedure. See figure 6.2.

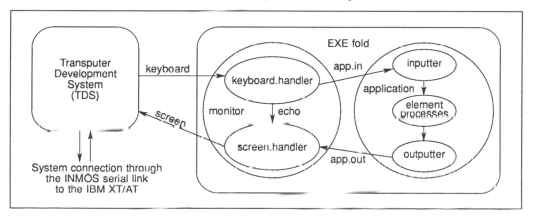

Figure 6.2 Pipeline sorter running in the Transputer Development System

After compiling the program it can be loaded using GET CODE and run using RUN EXE. The screen will clear and the user should enter a string of alphabetic characters followed by RETURN. The string of characters is sorted into alphabetical order and displayed on the next line. The program is terminated by entering the character '%'.

In chapter 7 it is shown how to distribute this application onto a network of transputers.

6.7   Example programs using the I/O libraries

This section presents two example programs using the procedures in the I/O libraries. These example programs are also included with the software, in the directory **\TDS3\EXAMPLES**.

The examples directory contains a number of examples showing how to use the I/O libraries; of these, examples 2 and 4 are listed here. Example 2 shows an example of using the library **streamio** to read from the keyboard and write to the screen. Example 4 shows the same program adapted to take its input from a filed fold, using the interface procedure **ks.keystream.from.fold**. Other examples in this directory show further use of the I/O libraries, such as writing to a file, and reading and writing a folded file.

6.7.1   Keyboard and screen example

This example shows the building up of a table of real numbers using echoed input, followed by a simple output tabulation.

```
{{{   EXE ex2 read a list of real numbers and display it
{{{F "ex2.tsr" ex2
#USE uservals
#USE streamio
SEQ
  -- This example uses keyboard and screen,
  -- with echoed input of real numbers.
  ss.write.nl(screen)
  ss.write.text.line(screen,
   "Type a sequence of real numbers (optionally in hex) *
   *terminated by 0.0")
  ss.write.nl(screen)

  REAL32 x:
  INT kchar:
  [1000]REAL32 ax:
  INT j:
  SEQ
    x := 1.0(REAL32)
    j := 0
    WHILE (NOTFINITE(x) OR (x <> 0.0(REAL32)))
      SEQ
        ss.write.char(screen, '>')
        ks.read.echo.char (keyboard, screen, kchar)
        IF
          kchar = (INT'#')
            INT hexx RETYPES x:
            ks.read.echo.hex.int (keyboard, screen, hexx, kchar)
          TRUE
            ks.read.echo.real32 (keyboard, screen, x, kchar)
        IF
          kchar = ft.number.error
            ss.beep (screen)
          TRUE
            SKIP
        ax[j] := x
        j := j + 1
    ss.write.nl (screen)
    ss.write.text.line (screen, "These are the numbers you typed")
    ss.write.nl (screen)
    SEQ i = 0 FOR j
      SEQ
        ss.write.real32 (screen, ax[i], 10, 10)
        ss.write.nl (screen)

  ss.write.string(screen, "Type ANY to return to TDS")
  INT any:
  ks.read.char(keyboard, any)
  ss.write.nl (screen)
}}}
}}}
```

Note the use of the property of number input procedures which allows the first character to be read before calling an appropriate input procedure.

Note also the need to perform some action (here ringing the terminal bell) if an invalid number is encountered.

6.7.2   Example showing input from file

This example, which is an adaptation of the previous example, shows how a program originally written to use the echoed input procedures may be adapted to take its input from a file in the fold structure and to throw away the echo.

```
{{{  EXE ex4 real numbers from a file
{{{F "ex4.tsr" ex4
#USE uservals
#USE streamio
#USE ssinterf
SEQ
  -- This example is derived from example 2
  -- It takes its input from a file and throws away the echo

  [1000]REAL32 ax:
  INT j:
  INT input.error:
  SEQ
    ss.write.text.line(screen,"Takes from a file a sequence*
                      * of real numbers terminated by 0.0")
    ss.write.nl (screen)

    CHAN OF KS filekeys.
    CHAN OF KS keyboard IS filekeys:
                          -- channel from simulated keyboard
    CHAN OF SS echo:
    CHAN OF SS screen IS echo:
                    -- echo channel with scope local to this PAR only
    PAR
      -----------------------------------------------------------
      SEQ
        ks.keystream.from.fold (from.user.filer[2],
                              to.user.filer[2],
                              keyboard, 1, input.error)
        -- check input.error when real screen accessible again
      -----------------------------------------------------------
      ss.scrstream.sink (screen)    -- consume everything echoed
      -----------------------------------------------------------
      REAL32 x:
      INT kchar:
      SEQ
        j := 0
        ... read a sequence of real numbers
        ss.write.nl (screen)
        ... consume rest of file if any
        ss.write.endstream (screen) -- terminate scrstream.sink
      -----------------------------------------------------------
    ... test input.error, if OK tabulate
  ss.write.string(screen, "Type ANY to return to TDS")
  INT any:
  ks.read.char (keyboard, any)
  ss.write.nl (screen)
}}}
}}}
```

The contents of the three folds in the program are as follows:

The fold headed **read a sequence of real numbers**:

```
{{{   read a sequence of real numbers
kchar := 0
x := 1.0(REAL32)
WHILE (NOTFINITE(x) OR (x <> 0.0(REAL32))) AND
      (kchar <> ft.terminated)
  SEQ
    ss.write.char(screen, '>')
    ks.read.echo.char (keyboard, screen, kchar)
    IF
      kchar < 0
        SKIP
      kchar = (INT'#')
        INT hexx RETYPES x:
        ks.read.echo.hex.int (keyboard, screen, hexx, kchar)
      TRUE
        ks.read.echo.real32 (keyboard, screen, x, kchar)
    IF
      kchar = ft.terminated
        SKIP
      TRUE
        SEQ
          IF
            kchar = ft.number.error
              ss.beep (screen)
            TRUE
              SKIP

    ax[j] := x
    j := j + 1
}}}
```

The fold headed **consume rest of file if any**:

```
{{{   consume rest of file if any
IF
  (kchar >= 0) OR (kchar = ft.number.error)
    ks.keystream.sink (keyboard)
                    -- consume the rest of the keyboard file
  TRUE
    SKIP  -- keyboard file has terminated or failed
}}}
```

The fold headed **test input.error, if OK tabulate**:

```
{{{  test input.error, if OK tabulate
IF
  {{{  input error
  input.error <> 0
    SEQ
      ss.write.full.string (screen, "File reading error: ")
      ss.write.int (screen, input.error, 0)
      ss.write.nl (screen)
  }}}
  TRUE
    SEQ
      ss.write.text.line (screen,
                        "These are the numbers you typed")
      {{{  write the table of j real numbers
      ss.write.nl (screen)
      SEQ i = 0 FOR j
        SEQ
          ss.write.real32 (screen, ax[i], 10, 10)
          ss.write.nl (screen)
      }}}
}}}
```

Note that as a file can only be read to its end (using these simple procedures), the interface procedure **keystream.sink** is called after the application code to ensure that the procedure **keystream.from.file** will terminate.

This example avoids the need to systematically change the names of the parameter channels **keyboard** and **screen** by means of channel abbreviations renaming locally declared channels with these same names.

7 Configuring programs and loading transputer networks

7.1   Introduction

To make effective use of transputer networks, an application must be expressed as a number of parallel processes. Once this has been done, performance requirements can be achieved by adapting the application to run on a number of transputers. To do this the programmer adds information describing the link topology and describes the association of code to individual transputers. This is called 'configuration'. This chapter describes how to configure a program and how to load it onto a transputer network.

7.2   The transputer configuration and loading utilities

This section describes the utilities which enable the user to configure an occam program for a network of transputers, and then load the code into the network for execution.

A section of occam to be allocated onto a processor must be contained within one or more SC folds. The initial step in creating a configuration is separate compilation of each procedure which is to be loaded as the code for a transputer. The resulting SCs and code calling them must then be collected together into a filed fold, to which the MAKE FOLDSET utility is applied with the parameter set to PROGRAM. This makes a PROGRAM foldset.

This PROGRAM then requires the necessary configuration statements to be added to describe the interconnections and to call the required procedures on the desired processors. The configuration language is described in the next section.

These steps must be followed, even if the network contains only a single processor. In the case of a single processor, the procedure loaded may have no formal parameters; in all other cases, the procedure loaded on any processor must have at least one channel parameter which corresponds to a transputer link to enable code to reach that processor. Alternative methods of building single processor applications are given in chapter 8.

The utilities used for configuring and loading transputer networks are as follows:

COMPILE checks that an occam PROGRAM is a valid configuration and produces the necessary code to call the individual procedures to be loaded on each processor.

COMPILATION INFO, when applied to a configured PROGRAM fold, creates a fold containing a list of the inter-processor link connections of the target transputer network, the boot order of the processors in the network and the memory map on each processor.

EXTRACT extracts and links all the code in an occam PROGRAM or SC into a single fold.

LOAD NETWORK loads a transputer network with a previously configured program.

7.3   The configuration description

The allocation of code to processors in a transputer network is achieved using two occam language extensions:

>   PLACED PAR
>
>   PROCESSOR *number transputer.type*

These configuration constructs, and the mapping of inter-process channels onto transputer links, enable the configuration utility, the configurer, to identify the code destined for a specific processor and to check that the network described can be loaded.

The code for any processor consists of one or more procedures, each contained within an **SC** fold, and the code which calls those procedures. Code outside an **SC** should be kept to a minimum and cannot include references to libraries containing code. Such code becomes a process running in parallel with other similar processes on other processors in the network. The inter-processor channels are mapped onto transputer links. One of the processors in the network is connected to the TDS, to allow the system to be loaded. This is known as the 'root processor'. There must be a route, via transputer links, from the root processor to all other processors, to allow the network to be loaded.

The processor *number* is the logical identifier of that processor and may be any value in the integer range. These numbers just identify the processor in messages from the TDS software; they serve no purpose in the allocation.

The root processor of any network must always be the first processor declared in the configuration.

The *transputer.type* part of the **PROCESSOR** statement specifies which type of transputer is placed at this node in the network. This information is used by the configurer to check that the process allocated to processor has been compiled for the correct target transputer. Valid transputer types are **T8** or **T800** (IMS T800, T801, T805), **T4** or **T414** (IMS T414), **T425** (IMS T425) and **T2** or **T212** (IMS T212, T222, T225 or IMS M212). Transputer classes are not permitted.

An **SC** procedure may be allocated to any number of processors in the network. A procedure is exported from the host to the network once, each recipient processor taking a copy of the code. Only those procedures in the **PROGRAM** which are actually allocated to a processor are exported to the network.

The **PLACE** statement is used to tie occam channels to processor links. A channel which is placed at a link twice must be placed at an input link address on one processor and at an output link address on a different processor. A channel placed only once is a 'dangling' link to the environment outside the configuration being described. The configuration utility produces a warning message if a dangling link is detected. For example, the link connecting the TDS to the network program may be specified as a dangling link to allow the program to communicate with an **EXE** running within the TDS. Link addresses are held in the system library **linkaddr**.

If there is a requirement to connect a processor to itself via formal channel parameters of the process allocated to it, a 'soft' channel must be used. A soft channel is a declared channel, which is not placed at a link address, it may only be used by a single processor. Soft channels are useful for providing loop back termination of a pipeline or for filling unused link parameters.

A configuration has the form:

> *Configuration-level declarations*
> *Placed PAR*

A *Placed PAR* has the form described in the occam 2 reference manual, with the extension that the **PROCESSOR** part has the form:

> **PROCESSOR** *number transputer.type*
> *Processor-level declarations*
> *instance*

where: *transputer type* =   **T2 | T4 | T5 | T8 | T212 | T414 | T425 | T800**

Configuration-level declarations may include:

- **SC** folds containing one or more procedures.

- Constant definitions using **VAL**.

- **PROTOCOL** definitions.

- **#USE** lines referring to libraries containing only constant and protocol definitions. Any logical name referring to such a library must be valid for a **T4** target.

- Channel declarations for placement as links between processors.

Processor-level declarations may include:

- Placement of configuration-level channels at link addresses.

- Constant definitions.

- Variable declarations.

- Placement of variables.

- Abbreviations and retypes of variables.

- Channel declarations for use as 'soft channels' on this processor.

- Any other occam code that does not, explicitly or implicitly, reference library code.

Note that procedures to be used at configuration level may not be taken from a library; libraries used at configuration level may only contain constant and protocol definitions.

Configuration examples

The structure required for loading a single processor system is:

```
{{{   PROGRAM single processor
{{{F "source.tsr" source
...   SC  example.sc

PROCESSOR 0 T800
  example.sc ()
}}}
}}}
```

The structure required for loading a system consisting of eight processors in a pipeline, seven of which contain the same program is:

```
{{{   PROGRAM pipeline
{{{F "source.tsr" source
...   SC  element  (CHAN OF INT32 in, out, VAL INT board.no)
...   SC  pipe.end (CHAN OF INT32 in, out, VAL INT board.no)
VAL last        IS 7 :
VAL input.links  IS [5, 7, 6, 7, 5, 7, 6, 7] :
VAL output.links IS [0, 2, 1, 1, 0, 2, 1, 1] :
[last + 1]CHAN OF INT32 links :

PLACED PAR
  PROCESSOR 0 T800
    PLACE links[last]   AT   input.links[0] :
    PLACE links[0]      AT   output.links[0] :

    pipe.end  (links[last], links[0], 0)

  PLACED PAR i = 1 FOR last
    VAL   in  IS i - 1 :
    VAL   out IS i :
    PROCESSOR i T800
      PLACE links[in]   AT   input.links[i] :
      PLACE links[out]  AT   output.links[i] :

      element (links[in], links[out], i)
}}}
}}}
```

7.4   Configuring a program

A program is configured using the COMPILE or RECOMPILE utility applied to a **PROGRAM** fold describing the configuration. The utility will compile any nested compilation units which need to be compiled, and link the **SC** for each processor. It will then check the configuration statements to ensure that they are consistent, and will generate the loading and running information for each processor.

The utility COMPILATION INFO can be used to see the results of configuration. Applied to a configured **PROGRAM** fold, it creates another fold in the foldset which can be opened and viewed, listing the processors and their connections, and giving a memory map for each processor.

After configuration has been completed, the network can be loaded. There is an additional utility EXTRACT, which will collect together all the code within a program into a single filed fold, called a **CODE PROGRAM** fold. It is not necessary to use this utility before loading; it is provided so that the user can make a self-contained code file and separate it from the program source (when, for example, developing the source code further while keeping a backup copy of the last loadable code file produced). It is also used for creating a 'standalone program' (see chapter 8).

For configurations containing different processor types, the COMPILE utility should be applied to each processor **SC**, supplying the appropriate processor type as a parameter. Then RECOMPILE should then be applied to the **PROGRAM** fold to configure the network.

7.5   Connecting a network to the TDS

Before an application can be loaded onto a transputer network from the TDS, the network must be connected to the TDS. This section outlines how to do this; for a detailed description of the connections from the board running the TDS, see the appropriate board manual.

The transputer network is connected together by transputer links; the topology of the network must match that described in the configuration description, otherwise the loading will fail. The network is loaded via a link out of the host transputer (the transputer running the TDS) to one of the transputers in the network: the 'root transputer'. The TDS need only be connected to this one transputer; it will boot this transputer over the link, and send loading information to it. The root transputer will boot the transputers connected to it, and route loading information to them; these will in turn boot and load other transputers in the network, until the whole network has been booted and loaded.

Any of the links out of the host transputer may be used to load the network, apart from the link connecting the host transputer to the host computer. The use of such links to provide run-time support for a network program is discussed in section 7.8.2.

As well as the link connection, INMOS boards also provide system control functions to monitor and control the state of the transputer network. The system control connections on boards are chained together to allow the whole of the network to be controlled from the host. The control connection consists of three signals:

Reset This is a signal from the host transputer to the network, which will reset all the transputers in the network, ready for loading.

Analyse This is a signal from the host transputer to the network, which will bring all of the transputers in the network to a controlled halt, so that their state can be examined.

Error This is a signal from the network to the host transputer, indicating that one of the transputers in the network has set its error flag.

For a more detailed description of system control connections, see the appropriate board manual. For a detailed description of the effect of the **Reset**, **Analyse** and **Error** signals on the transputer, and a description of how a transputer boots, see the Transputer Reference Manual.

7.6   Loading a network

A network is loaded using the LOAD NETWORK utility. The utility may be used on a **PROGRAM** fold, or on a **CODE PROGRAM** fold which has been extracted. Among the parameters for the LOAD NETWORK utility are the link out of the host transputer that should be used for loading, and what type of board the TDS is running on (to tell the utility where the subsystem is). As the network is loaded, messages are displayed to indicate the loading stage.

A detailed description of the loading mechanism is given in INMOS technical note 34 'Loading Transputer Networks'. An outline of the mechanism is included here, for information.

A communication protocol exists between the host transputer and a target transputer network to direct the loading of code to the desired place in each transputer. The communication consists of bootstrap packets, routing information, address information, load information, code packets and execute items.

The bootstrap code for each transputer in the network is sent first. The bootstrap code is loaded at the lowest available address (nearest to **MOSTNEG INT**). The bootstrap loads the distributing loader at the first available addresses above itself. After all the transputers in the network are booted, the code of each of the procedures allocated to processors in the configuration description is exported to the network preceded by the necessary routing and loading information. Following this, the code which calls the procedures (the main body) generated by the configurer is sent to each processor in turn and then each processor is told to start executing the loaded program.

7.7   Using the transputer network tester

When configuring an application, and loading it onto a transputer network, it is important that the network is connected in the configuration expected by the loader, otherwise the loading will fail. It is equally important to be sure that the hardware in the network is all working properly, and that there are no communication problems due to (for example) poor connections, electrical noise, or links set to the wrong speed.

Even with the messages produced while the network is being loaded, it may still be hard to track down the cause of the error.

A program called **nettest** is provided to aid in investigating problems of this kind. It is described in detail in section 15.2. Some of the facilities provided by this program are as follows:

- Explore a network of transputers and establish its topology, displaying the type of each transputer in the network.

- Check the actual connected topology of the network against the topology specified by the configuration description in a **PROGRAM** fold, and report any differences between the two.

- Test the memory of each transputer in the network.

- Reset or analyse all the transputers in the network.

The transputer network tester can be used to establish that the transputer network is functioning correctly, and that it matches the configuration expected by the programmer. This allows the programmer to reduce or eliminate the possibility of hardware faults when investigating problems in loading and running an application on a network.

The transputer network tester uses a program called a 'worm' which distributes itself through all transputers in the network. For an introduction to how worms work, see INMOS technical note 24 'Exploring Multiple Transputer Arrays'.

7.8   Running the pipeline sorter on a target transputer

Returning to the pipeline sorter example described in the previous two chapters, this section describes how to run the example on a second transputer, loaded from the host transputer running the TDS. The host transputer will be used to monitor the behaviour of the target system.

The occam code for the application must be separated from the code used for monitoring. This has already been planned for by defining the code modules in separate procedure declarations.

This example is contained in the directory \TDS3\TUTOR, in EXAMPLES.TOP, so while reading this section it will be useful to start up the TDS in this directory and follow the instructions given.

A later section of this chapter shows how to configure the application to run on multiple transputers.

7.8.1   Creating a PROGRAM fold

A PROGRAM fold describes the configuration of a system and the placement of occam procedures onto distinct processors.

For a single target transputer the PROGRAM fold contains the filed fold prog2.tsr which in turn contains one SC, and configuration information about the target hardware.

```
{{{   PROGRAM prog2
{{{F "prog2.tsr" prog2
#USE header
...   SC app.tsr
...   configuration
}}}
}}}
```

The SC contains the application code described in the previous discussion of this example. Note that the application code had to be executed in parallel with other processes in order to be able to move the code to another processor.

```
{{{   PROGRAM prog2
{{{F "prog2.tsr" prog2
#USE header
{{{   SC app.tsr
{{{F "app.tsr" app
#USE header
#USE problem

PROC application (CHAN OF string in, out)

  [string.length+1]CHAN OF letters pipe:
  PAR
    inputter (in, pipe[0])
    PAR i = 0 FOR string.length
      element (pipe[i], pipe[i+1])
    outputter (pipe[string.length], out)
:
}}}
}}}
...   configuration
}}}
}}}
```

The application code is the same as the TDS version (see chapter 6) although there is now a procedure declaration around it. The procedure is needed to provide a name, implementation detail and parameters for placing this section of code on a processor.

Notice that all the information needed for the application code must be contained inside the **SC** fold, including the library references. The library **header** is used at the start of the **PROGRAM** fold so that the compiler can understand the protocol **string** used in the **SC** procedure's parameter list.

The configuration fold looks like:

```
{{{   PROGRAM prog2
{{{F "prog2.tsr" prog2
...   SC app.tsr
{{{   configuration
...   link constants
CHAN OF ANY app.in, app.out:

PROCESSOR 0 T4
  PLACE app.in  AT link0in:
  PLACE app.out AT link0out:
  application (app.in, app.out)
}}}
}}}
}}}
```

The configuration places the **SC** procedure **application** onto a transputer which has been given the logical number 0. The transputer type is **T4**, denoting an IMS T414. The type of the transputer is needed for the system to know how to initialise it at boot time. The system also checks to make sure that **application** was compiled with the compiler parameter **target.processor** set to **T4**.

The instance of **application** has two actual channel parameters, **app.in** and **app.out**. These correspond to the formal channel parameters **in** and **out**. The **PLACE** statement is used to map these occam channels onto the transputer's serial link hardware. The addresses **link0in** and **link0out** are contained in the fold **link constants**. The communication on **app.in** and **app.out** has been directed onto transputer link zero (the link supports two occam channels, one input channel and one output channel).

In this configuration, link 0 is a 'dangling link'. Once the **PROGRAM** has been loaded into the target transputer, it will run until the first communication made on **app.in** or **app.out**. It is up to the programmer to connect a system to this link which will communicate with **application** in order for it to continue; otherwise it will wait forever. In this example the **monitor** process will be run within the TDS to communicate with the target system.

7.8.2   Monitoring the target with an EXE

To monitor the target system a monitor program must be run as an **EXE**. This may be as follows:

```
{{{   EXE interface
{{{F "interfac.tsr" interface
#USE header
#USE monitor
#USE linkaddr
CHAN OF string app.in, app.out:
PLACE app.in   AT link2.in:
PLACE app.out  AT link2.out:

monitor (keyboard, screen, app.in, app.out, TRUE)
}}}
}}}
```

The **EXE** consists of an instance of the library procedure **monitor** with its **keyboard** and **screen** parameters connected to the TDS keyboard and screen channels, and the other channels connected to the application, over link 2 of the host transputer.

The **monitor** procedure has its parameter **using.subsystem** set to **TRUE**. This enables **monitor** to give the programmer an error message should, for any reason, the target transputer set its error flag.

To show how this is done, it is necessary to look in more detail at the body of the screen handling process, in the **monitor** procedure. The monitor procedure was described earlier, in Chapter 5, but there the details of what happens when **using.subsystem** is **TRUE** were not discussed.

The main part of the screen handler looks like this:

```
{{{  body
WHILE going.in OR ((NOT using.subsystem) AND going.data)
  SEQ
    clock ? waketime
    waketime := waketime PLUS one.hundredth.of.a.second
    ALT
      going.in & in ? char
        ...  print keyboard character on screen
      going.data & data ? length::string
        ...  print data from application on screen
      monitoring & clock ? AFTER waketime
        ...  if monitoring is TRUE, poll subsystem error pin
    draw.cursor (kb.window)
}}}
```

The screen handler is repeatedly waiting for one of three alternatives. Either keyboard characters are echoed, a string of data comes from the application or a timeout happens should neither of the other two have occurred in one hundredth of a second. If the timeout occurs then the program tests the subsystem error pin. If this indicates an error then a message is sent to the user, after which the user can terminate the monitor and use the TDS for subsequent analysis (e.g. running the debugger).

If the TDS is executing on an IMS B004 or IMS B008 board then the subsystem logic is decoded through a PAL that can be accessed by software. The subsystem reset and error are at machine address zero (in the middle of the transputer's address space). occam addresses start from zero and are word aligned so a program can access the subsystem by placing a port at **#20000000**.

This can be done by the following declarations:

```
VAL subsys.error.locn IS #20000000:
PORT OF BYTE subsys.error:
PLACE subsys.error AT subsys.error.locn:
```

Reading from this port, and finding bit zero set, detects the assertion of the subsystem error pin. This can be done with the following occam code:

```
BYTE error:
SEQ
  subsys.error ? error
  IF
    (error /\ 1) = 1 (BYTE)
      ...  Error flag set!
    TRUE
      SKIP
```

7.8.3   Configuring and running the example

The following steps are now required, in the following order, to run the application as described on a two transputer network.

1 Run COMPILE on ...   **EXE interface**.

2 Configure the **PROGRAM** by running COMPILE on ...   **PROGRAM prog2**. This will also compile and link the **SC application**.

3 Connect link two on the host transputer to link zero on the target transputer.

4 Connect the 'Up' port from the target transputer board to the subsystem connection on the host transputer board.

5 Load the **PROGRAM**. To do this invoke the $\boxed{\text{LOAD NETWORK}}$ utility on the **PROGRAM** fold. This will extract the code from the **PROGRAM** fold and transmit it to the network. It will prompt for a parameter indicating which host transputer link to use for the loading. The required value is link two, which is also used by the **monitor** program to monitor the target from the host.

6 Get the **EXE monitor**, using $\boxed{\text{GET CODE}}$ and run it, using $\boxed{\text{RUN EXE}}$. This establishes communication between the two transputers, so that the user can now supply data to the running application. Note that the synchronisation on link communication holds up the **PROGRAM** until the **EXE** outputs some data.

The next section shows how to distribute the application over multiple transputers.

7.9 Running the pipeline sorter on a four transputer network

This section shows how the code for the pipeline sorter example can be distributed over four transputers in a network. The assumption made here is that the four transputer target network is an IMS B003 transputer evaluation board. In the IMS B003, the system control lines are preconnected so that the host board can automatically reset all the transputers simultaneously. Every transputer on the IMS B003 has two links available on the edge connector (links 0 and 1) while the other two are preconnected in a square array (links 2 and 3).

7.9.1 A PROGRAM for four transputers

The **PROGRAM** fold appears as follows:

```
{{{  PROGRAM prog3
{{{F "prog3.tsr" prog3
#USE header
...  SC PROC interface
...  SC PROC worker
...  link constants
-- number of transputers must match value used inside SCs
VAL number.of.transputers   IS 4:
...  configuration
}}}
}}}
```

This example has two separately compiled procedures: **interface** and **worker**.

The procedure **interface** connects to the monitor as well as doing **string** to **letter** protocol conversions and some **element** processes.

The procedure **worker** is a number of **element** processes running in a pipeline.

The number of **element** processes on each transputer depends on the number of transputers available, hence the constant **number.of.transputers**. This constant is needed at configuration level, as will be seen later, and in both **SC** folds. The constant could have been put into a header library. The element processes are divided into four equal sets, and one set is run on each processor. Any processes remaining (in the case where the number of elements is not divisible by four) are run on the root processsor.

See figure 7.1 for a picture of how the pipeline sorter can be split up over four transputers.

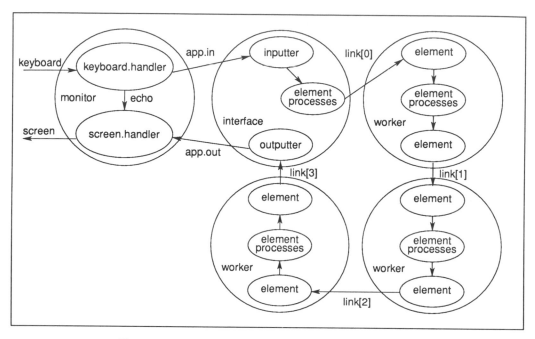

Figure 7.1 Pipeline sorter running on four transputer networker

7.9.2   The root transputer

The procedure **interface** runs on the root transputer in the network. This is as follows:

```
{{{F "interface.tsr" interface
#USE header
#USE problem
{{{  extra constants for configuring for 4 transputers
VAL number.of.transputers   IS 4:
VAL number.of.elements      IS string.length:
VAL elements.per.transputer IS number.of.elements/
                                 number.of.transputers:
VAL remaining.elements IS number.of.elements -
            (elements.per.transputer * number.of.transputers):
}}}
PROC interface (CHAN OF string from.host, to.host,
                CHAN OF letters to.pipe, from.pipe)
  VAL elements IS elements.per.transputer + remaining.elements:
  [elements]CHAN OF letters pipe:
  PRI PAR
    PAR                      -- prioritise processes using links
      inputter (from.host, pipe[0])
      element (pipe[elements - 1], to.pipe)
      outputter (from.pipe, to.host)
    PAR i = 0 FOR elements - 1
      element (pipe[i], pipe[i+1])
  :
}}}
```

The procedure **interface** has three processes at high priority and a number at low priority. The high priority processes are those which communicate with the transputer links whereas the others only use internal channels. This prioritisation of link communication can enhance the throughput of distributed systems. All the above processes, regardless of priority, are running in parallel with each other.

The number of **element** processes in **interface** depends on **number.of.transputers** and how many **element** processes all the other transputers have. The total number of **element** processes in the target system must add up to **number.of.elements**. If the value of **string.length** is divisible by 4 then **interface** will include a quarter of the required **element** processes.

7.9.3   The three other transputers

The three other transputers in the network run copies of the procedure **worker**. This procedure is as follows:

```
{{{F "worker.tsr" worker
#USE header
#USE problem
...    extra constants for configuring for 4 transputers
PROC worker (CHAN OF letters in, out)
  VAL elements IS elements.per.transputer:
  [elements]CHAN OF letters pipe:
  PRI PAR
    PAR                -- prioritise getting the links started
      element (in, pipe[0])
      element (pipe[elements-2], out)
    PAR i = 0 FOR elements - 2
      element (pipe[i], pipe[i+1])
  :
}}}
```

The separately compiled procedure **worker** contains a quarter of the required **element** processes in a pipeline. The two element processes that have channels mapped onto links run at high priority.

7.9.4   Configuration for four transputers

The configuration for the IMS B003 maps **interface** onto the root transputer (it is the first mentioned in the program) and maps **worker** onto all three remaining transputers.

Figure 7.2 shows how the processes are mapped onto the IMS B003.

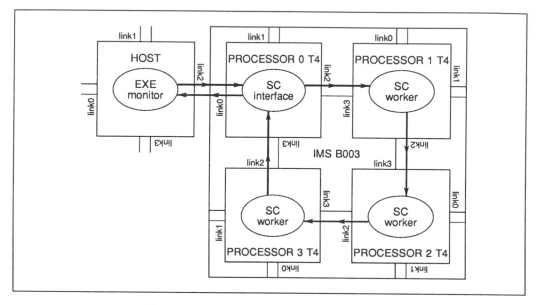

Figure 7.2 Pipeline sorter running on one IMS B003

The configuration is as follows:

```
{{{   configuration
CHAN OF string app.in, app.out:
[number.of.transputers]CHAN OF letters link:

PLACED PAR
  PROCESSOR 0 T4
    PLACE app.in  AT link0.in:
    PLACE app.out AT link0.out:
    PLACE link[0] AT link2.out:
    PLACE link[number.of.transputers - 1] AT link3in:
    interface(app.in,app.out,link[0],
              link[number.of.transputers -1])

  PLACED PAR i = 1 FOR (number.of.transputers - 1)
    PROCESSOR i T4
      PLACE link[i - 1] AT link3.in:
      PLACE link[i]     AT link2.out:
      worker (link[i - 1], link[i])
}}}
```

For the three **worker** processes a replicator has been used with index **i** having the values 1, 2 and 3. All the transputers are of type **T4**.

The interface is connected to the host through **app.in** and **app.out** on link 0 whilst the **worker**s are connected to each other and to interface through the link 2 to link 3 connections provided with the IMS B003 board.

The steps to configure and run the example are the same as in the previous example, where the program was running on one transputer.

8 Standalone transputer programs

8.1   Introduction

The last two chapters have discussed running programs within the TDS, and running programs on a network loaded from the TDS. However, most applications will, once they have been developed, run separately from the TDS. This chapter describes how to export a program from the TDS so that it can be run in a standalone manner.

Programs running on a transputer network separate from the TDS need to be booted onto the network. This can be done in two main ways: either the network is booted from a ROM, contained in one of the transputers in the network, or the network is booted from a host computer connected to the network. Booting from a ROM is discussed in chapter 10; in this chapter we will concentrate on programs booted from a host computer, via a link.

Where a host computer (such as the IBM PC) is used to boot a network, it may also be convenient for the host computer to provide some facilities (such as terminal I/O and filing system support) to the program running in the network. A program which boots a network and provides host support is called a *server*.

The two servers provided with previous versions of the Transputer Development System have been replaced with a single server **iserver**, which is also used by INMOS toolset products.

iserver can boot a program from a host file into an arbitary transputer network, and then provide support using a special protocol called **SP**. It can also detect the setting of the error flag if the necessary connections exist. In principle **iserver** can be re implemented on any host computer for which add-on transputer boards have been designed and can be optimised to take full advantage of all hardware features of the interface between the host and the transputer.

A stand-alone program which communicates at run-time with a version of **iserver** on a host may be coded in one of two ways. It may be coded as a **PROGRAM** within which the link(s) connecting the program to the server are coded as channels **PLACE**d at suitable hardware addresses. Alternatively it may be coded as a procedure in an **SC** compilation unit with a particular conventional parameter list including server channels and free space vector, and called by a bootstrap procedure from code added to the compiled **SC** using the tool **addboot**. Such a procedure is called a standard hosted procedure.

Either of these methods leads to a pure binary file which may be ported to any host on which there is an implementation of **iserver** for a transputer board. In this way programs developed on a PC can be run from a SUN3, MicroVax or other suitable host.

The advantages of the second method, making a standard hosted PROC, include the ability to use whatever memory space is available, and to communicate with the host via whatever link the program was loaded down. It is for these reasons that INMOS stand-alone tools are coded in this way. Such tools may also be made by using INMOS occam toolset products.

The C sources of **iserver** are provided with the TDS. For details of where to find them, and how they can be recompiled, see the 'Delivery Manual'.

8.2   Using the iserver

Once **iserver** has booted a file into a network, it supports **SP** protocol over the link to the root transputer (the first transputer in the network). This protocol is described in detail in section 16.5, of the 'System interfaces' chapter.

Libraries are provided which support the use of **SP** protocol. These are **sphdr**, **splib**, **solib**, **sklib** and **spinterf**.

These libraries are described in chapter 14.

8.3   Creating a parameterless standalone program

The steps in creating a parameterless standalone program are as follows:

Firstly write the program as a **PROGRAM** configured for the required transputer network. Even if the program is to be run on a single processor it must be described as a configuration. Use COMPILE or RECOMPILE to compile and configure the program for the network.

Secondly, use EXTRACT to extract all the code and loading information for the network into a single file. It is important that EXTRACT is used with the parameter **output.fold** set to **BOOTABLE** (not **DIAGNOSTIC**) and the parameter **first.processor.is.boot.from.link** set to **TRUE**. If either of these are wrong, the host file server will fail to boot the program into the network.

Thirdly, the extracted file needs to be exported from the TDS into a standard host operating system file. The EXTRACT utility will leave a **CODE PROGRAM** fold as the last item in the **PROGRAM** foldset. Use WRITE HOST from the file handling utilities to write this out to a DOS file. The resulting file may be used by the host file server. The format of such a code file is defined in appendix G.

8.4   Creating a standard hosted PROC

The steps in creating a standard hosted procedure are as follows:

1 Write the program as a procedure with the heading:

```
#USE strmhdr
PROC program (CHAN OF SP from.iserver,
              to.iserver, []INT free.space)
  ...
:
```

Put this procedure in an **SC** foldset using MAKE FOLDSET. Use COMPILE to compile the foldset.

2 Use EXTRACT to extract the code into a **CODE SC** fold.

3 Use GET CODE to get the **addboot** tool from the toolkit fold. Apply this tool to the **SC**. A new host file will be created with a name supplied by the user.

4 This file may then be loaded and run using **iserver**:

```
iserver /se /sb  filename
```

8.5   The pipeline sorter

This section describes how to configure the sorting application described in chapter 5 so that it can be run as a standalone program from DOS.

The source of this program is contained in the tutorial fold structure, in the fold marked:

Running the example as a standalone program

The contents of this fold are as follows:

```
{{{   PROGRAM prog5
{{{F "prog5.tsr" prog5
#USE header
...   SC app.tsr
...   configuration
}}}
}}}
```

The source of the SC application is as follows:

```
#USE strmhdr   -- SP, SS and KS protocols
#USE sphdr     -- SP constants (sps.success)
#USE userio    -- user io procedures (write.endstream)
#USE interf    -- user io interface procedures (keystream.sink)
#USE splib     -- hostio procedures (so.multiplexor, so.exit)
#USE spinterf  -- SP interface procedures
#USE header    -- application constants
#USE problem      application procedures
#USE monitor   -- application environment

PROC application (CHAN OF ANY from.host, to.host)

  [string.length+1] CHAN OF letters pipe:
  CHAN OF string app.in, app.out:
  CHAN OF INT keyboard:
  CHAN OF ANY screen:
  [2]CHAN OF SP from.isv, to.isv:
  CHAN OF BOOL stopper:
  CHAN OF BOOL mstopper:
  VAL dont.use.subsystem IS FALSE:
  VAL one.hundredth.of.a.second IS 156:
  PAR
  --===============================================
    SEQ
      PAR   -- these processes should terminate in the order written
      ---------------------------------------------------
        inputter (app.in, pipe[0])
      ---------------------------------------------------
        PAR i = 0 FOR string.length
          element (pipe[i], pipe[i+1])
      ---------------------------------------------------
        outputter (pipe[string.length], app.out)
      ---------------------------------------------------
        SEQ
          so.scrstream.to.ANSI(from.isv[0], to.isv[0], screen)
          stopper ! TRUE
      ---------------------------------------------------
        so.keystream.from.kbd(from.isv[1], to.isv[1], keyboard,
                    stopper, one.hundredth.of.a.second)
      ---------------------------------------------------
        SEQ
          monitor(keyboard, screen, app.in, app.out,
                dont.use.subsystem)
          write.endstream (screen)
          ks.keystream.sink (keyboard)
      ---------------------------------------------------
      mstopper ! TRUE
  --===============================================
    SEQ
      so.multiplexor(from.host, to.host,
                    to.isv, from.isv,
                    mstopper)
      so.exit(from.host, to.host, sps.success)
  --===============================================
:
```

This program makes use of the standard library **spinterf**, which provides a number of processes which may be run in parallel with an application to convert its input and output into communications with **iserver**.

As before (in chapter 6), where this example was run as an **EXE** within the TDS, the monitor process is connected to the application, and communicates over channels conforming to the TDS screen and keyboard protocols. Since the program is not going to run within the TDS, but with a server, the keyboard and screen channels need to be connected to the terminal facilities provided by the server. This is done by the process **so.multiplexor** (available in the library **splib**) and the processes **so.scrstream.to.ANSI** and **so.keystream.from.kbd** (available in the library **spinterf**). The process **so.scrstream.to.ANSI** converts the TDS screen protocol into a **iserver** commands which will drive the terminal of the host computer. The process **so.multiplexor** communicates with the server over a pair of channels, supplying keys from the keyboard and sending the stream of characters to the screen.

The code for the configuration is as follows:

```
#USE linkaddr
CHAN OF ANY from.host, to.host:

PROCESSOR 0 T800
  PLACE from.host AT link0.in:
  PLACE to.host   AT link0.out:
  application (from.host, to.host)
```

The steps in creating the standalone program are as follows:

1 Both the compiler utility set and the file handling utilities will be needed. If necessary, get them now by pressing the AUTOLOAD key. If using the standard toolkit fold, this should finish with the compiler utilities as the current utility set.

2 Move to the **PROGRAM** fold contained in the fold marked
   Running the application as a standalone program in the tutorial fold structure. Look at the contents of this fold to check that it corresponds to the program text given earlier.

3 Check that the processor type in the configuration matches the processor you are using. Close the **PROGRAM** fold and press the COMPILE key. The standard compiler parameters are needed, so if it prompts for the parameters, just press EXIT FOLD.

4 When the compiler has finished, press EXTRACT to extract all of the code for the configured program into a single file. EXTRACT requires two parameters, in parameter folds **Transputer extractor parameters** and **First processor in network parameters**, within the **Toolkit** fold. Make sure that these are the values given:

   VAL output.fold IS BOOTABLE:

   VAL first.processor.is.boot.from.link IS TRUE

   If these are not the supplied values use SELECT PARAMETER to change them, then press EXIT FOLD after selecting each parameter. The extractor will then run.

5 Now open the PROGRAM fold; it should look something like this:

```
{{{   PROGRAM prog5
...F "prog5.tsr" prog5
...F "prog5.dcd" code
...F "prog5.dds" desc
...F "prog5.cen" CODE PROGRAM prog
}}}
```

   The last line is the CODE PROGRAM fold containing all of the code. Move the cursor down onto this line.

6 Now the file handling utilities are needed; switch to these using NEXT UTIL. It may be appropriate to use the CODE INFO key to check that the right utility set is current.

7 The last utility in the set, WRITE HOST, will be used to write a TDS fold out into a standard DOS file. With the cursor on the **CODE PROGRAM** fold, press WRITE HOST. It will prompt for one parameter with the line:

   VAL HostFileName IS "":

Set the string in this parameter line to be the file name required, for example:

   VAL HostFileName IS "sorter.b4":

Now press EXIT FOLD to let the utility continue. When it has finished it will have written the file **sorter.b4**, into the current directory.

8 Now exit the TDS (Press EXIT FOLD until reaching the top level and then press FINISH).

9 To run the program, invoke the server as follows, ensuring that the following line, including spaces, is typed exactly as shown:

   iserver /sb sorter.b4

The **/sb** flag instructs the host file server to use the file **sorter.b4** as the file with which to boot the transputer.

The sorter application should now run. As before, type strings of letters followed by RETURN to run the sorter, type **%** to terminate the program.

Another version of the pipeline sorter coded as a standard hosted procedure is also supplied. The main procedure **application** is unchanged from example **prog5** with the exception of the dummy third parameter **free.space**. This example may be compiled and run according to the steps enumerated in section 8.4.

9 Debugging

This chapter describes the source-level debugger provided with the TDS. The TDS debugger provides an interactive environment for the post-mortem debugging of occam programs running on transputer networks. It allows a user to inspect the processes which were running on each transputer, both at the occam source level, and at the transputer instruction level. It can also display the contents of variables, channels, and other data items, for any process running on any transputer. The mechanisms which the debugger uses are also described. See section 15.1 for a full description of the debugger.

9.1   Using the debugger

The debugger is provided in the standard **Autoload** fold in the toolkit fold, so it may be loaded using the AUTOLOAD key. It is an **EXE**, so can be run using the RUN EXE key. Before running the debugger, the cursor should be placed on the foldset containing the source of the program to be debugged.

A program to be debugged should be compiled with the compiler parameter **create.debugging.info** set to **TRUE**. It should also be compiled with **error.checking** set to **HALT**. This ensures that if any errors occur while the program is executing, the transputer will halt immediately. The other error modes (**STOP** and **REDUCED**) will not have this effect, and so in these modes the debugger can only be used after a running program has been externally halted; the program will not halt itself when an error occurs.

A running occam program may halt for a number of reasons. Examples of these are:

- A **STOP** process, or a process which behaves like **STOP** (such as an **IF** with no true guards) has been executed.

- An array access is outside the range of the array.

- An arithmetic error, such as overflow or divide by zero has occurred.

- An array element is being aliased at runtime.

See section 15.1.9 for a full list of possible causes of run time errors.

When one of these errors occurs, the debugger can be used to pinpoint the line of occam causing the error, and investigate the state of that process and other processes in the system.

The debugger is not guaranteed to find all current processes; it may not be possible to find processes which have deadlocked waiting for communication. This is discussed in more detail later in the chapter.

9.2   Debugger facilities

The debugger's facilities divide roughly into two sets. The first set is concerned with the occam source code, and allows the user to view the transputer network from the occam high level language level. This requires that the occam program has been compiled with the **create.debugging.info** compiler option set to **TRUE**. The second set of facilities views the transputer network from the assembly code level, and does not require the debugging information produced by the compiler. Either set of facilities may be used on any transputer in the network.

9.2.1   Symbolic facilities

Given any transputer instruction address, the debugger can 'locate' to the corresponding occam source line (i.e. it can find the line in the source fold and display it). In particular, this means that it can display the occam source line corresponding to any of the following:

- The last transputer instruction executed.

- Any process running in parallel.

- A process waiting for a timer.

- A process waiting for communication on a transputer link.

Processes waiting for communication on internal channels may be found by inspecting the contents of that channel, as explained later.

The ability to locate to any occam source line requires the source to be available. When the location is in a library the source code may not be available. However, if the library was compiled with the debugging option enabled, the debugger can discover the line containing the call to the library routine, and will display that line instead.

After 'locating' the source line, the TDS editing environment is available within the debugger, so that the occam source of the program can be browsed, and if required, modified ready to recompile. The extra debugging features are accessed by pressing special function keys, such as BACKTRACE or INSPECT within this environment. The values of constants, variables, parameters, abbreviations, array elements, and channels, which are in scope at the located line, may be inspected. Non-local variables and channels may also be accessed. Values are displayed in hexadecimal, and in any other normal representation for their type.

From any occam location the user can 'backtrace', or discover where its enclosing procedure or function was called from. This works even if the source of a library is not present because the library has been compacted. This can be repeated for each nested procedure or function call, to form a complete stack trace. The values of variables, etc., may be examined at any stage.

The user can also discover the type of any symbol currently in scope, and the address and workspace requirements of any procedure or function.

By inspecting a channel, the debugger can discover the instruction pointer and workspace pointer of any process waiting for communication on that channel. It can also use these values to 'jump' directly to the process which is waiting (i.e. locate the currently active position in that process), and then continue debugging that process.

9.2.2   Lower level facilities

The debugger can display the transputer's state after being analysed: the instruction pointer (program counter), workspace descriptor, process queue pointers, error, and halt-on-error flags. It can read the process and timer queues, to display a list of the instruction and workspace pointers of the processes on the queues. It can also display any processes waiting for communication on the transputer links, or for a signal on the **Event** pin.

Memory can be displayed in ASCII, hexadecimal, or as any other occam type. It can also be displayed as a simple disassembly of transputer instructions. This disassembly simply translates memory contents directly into transputer instructions; it does not insert labels, nor provide symbolic operands. The debugger can also provide a 'memory map' of each transputer in the network, showing the positions of code and workspace. By displaying memory as **CHAN** type, channels waiting for communication may be located.

9.3   Debugging a program running on a network of transputers

When a program has been loaded onto a network of transputers and run, an error may occur in one of the transputers in the network. This may be indicated to the TDS by the **Error** signal on the transputer subsystem. The example program in chapter 7 shows how a monitor process can be run as an **EXE** within the TDS to monitor the state of the network. After an error has been detected, the monitor program can finish and the debugger program may be run to analyse and examine the state of the network.

The monitor process running within the TDS could also be used to assert **Analyse** on the subsystem, to bring the network to a halt even if no error has occurred. In this case the debugger may be used to examine the network, but it should be told not to assert **Analyse** when it starts up.

The debugger is an executable procedure, or **EXE**, which should be run while the cursor is positioned on the compilation fold of an occam **PROGRAM** which has halted, either because an error has occurred, or because of user intervention. It is not possible to restart the occam program once it has been stopped.

The debugger will start by locating to the source line on which the error occurred, or (if no error has occurred) by showing the state of the first processor in the network. The session using the debugger can then proceed.

A network program which does not terminate may be interrupted by asserting **Reset** or **Analyse** on its up port. This may be done by running the debugger or by rebooting the TDS.

9.4   Debugging a program running within the TDS

If an error occurs while running an **EXE** program within the TDS, then the error will be detected by the TDS server, which will display the message:

        Error - iserver - transputer error flag set

This condition can also be forced by interrupting the TDS; this is done using 'control-break' on the standard IBM PC keyboard. The procedure for restarting the TDS is described in the Delivery Manual.

In order to debug the program which has crashed, the data of the program must be saved before the TDS is restarted. When rebooting the TDS, the TDS will offer the user the option of doing a 'core dump'. This saves the memory contents and state of the host transputer as a file on the host filing system.

Once the TDS has been restarted, the debugger can be loaded. If the debugger is then executed while positioned on the compilation fold of the **EXE** that crashed, it can read the core dump file to determine the state of the program when it crashed. The full range of debugging features are then available to debug the **EXE**, as if the program were running on a single transputer in isolation.

9.5   Debugging a standalone program

The debugger can also be used to debug a program which has been developed under the TDS, but is being run as a standalone program with its own server (such as the host file server). Here it is likely that the host transputer, which is going to run the TDS, is also being used as the root transputer in the network, and communicating with the server on the host. So, in order to be able to examine the state of the whole network, the data space of the root transputer must be saved before the TDS is restarted. The rest of the network can be examined over the link to the host transputer in the normal way.

If the standalone program crashes, the TDS should be restarted with the analyse signal asserted (see the Delivery Manual). The TDS will give the option of producing a core dump before it starts. It is necessary to tell it how much memory to dump, as the TDS does not know how much memory was used by the standalone program. The coredump is only needed if the program includes the host transputer.

The debugger can then be run and used in the 'network including host' mode, which reads the core dump file to determine the state of the root processor, and analyses the rest of the network in the normal way.

9.6   A worked example

This section describes an example debugging session. The source of a program to be debugged is supplied as part of the TDS release, in the directory **\TDS3\TUTOR**. Change to that directory and start the TDS before starting this session.

The program should be compiled as a TDS **EXE**, with **error.checking** set to **HALT**, for a **T4** (assuming that you are not running the TDS on an IMS T800), and executed in the normal way.

The program is a (very inefficient) program to calculate the sum of the squares of the first n factorials. It has been structured this way for clarity, and to demonstrate some debugging methods.

```
#USE userio
VAL stop.real    IS -1.0(REAL64) :
VAL stop.integer IS -1 :

REAL64 FUNCTION factorial (VAL INT n)
  REAL64 result :
  VALOF
    IF
      n < 0
        STOP
      TRUE
        SEQ
          result := 1.0(REAL64)
          SEQ i = 0 FOR n
            result := result * (REAL64 ROUND i)
    RESULT result
:

PROC feed (CHAN OF INT in, out)
  INT n :
  SEQ
    in ? n

    SEQ i = 0 FOR n
      out ! i

    out ! stop.integer
:

PROC facs (CHAN OF INT in, CHAN OF REAL64 out)
  INT    x :
  REAL64 fac :
  SEQ
    in ? x
    WHILE x <> stop.integer
      SEQ
        fac := factorial (x)
        out ! fac
        in  ? x
    out ! stop.real
:
```

```
PROC square (CHAN OF REAL64 in, out)
  REAL64 x, sq :
  SEQ
    in ? x
    WHILE x <> stop.real
      SEQ
        sq := x * x
        out ! sq
        in  ? x
    out ! stop.real
:

PROC sum (CHAN OF REAL64 in, out)
  REAL64 total, x :
  SEQ
    total := 0.0(REAL64)
    in ? x
    WHILE x <> stop.real
      SEQ
        total := total + x
        in ? x
    out ! total
:

PROC control (CHAN OF INT keyboard, CHAN OF ANY screen,
              CHAN OF REAL64 result.in, CHAN OF INT n.out)
  REAL64 result :
  INT    n, key, char :
  SEQ
    write.full.string (screen,
                  "Sum of the first n squares of factorials")
    newline            (screen)
    write.full.string (screen, "Please type n . ")
    char := INT '*s'
    read.echo.int      (keyboard, screen, n, char)
    newline            (screen)
    write.full.string (screen, "Calculating factorials ... ")
    n.out         ! n
    result.in ? result
    newline            (screen)
    write.full.string (screen, "The result was : ")
    write.real64       (screen, result, 0, 0)   -- free format
    newline            (screen)
    write.full.string (screen, "Press any key to exit : ")
    keyboard ? key
:

CHAN OF REAL64 facs.to.square, square.to.sum, sum.to.control :
CHAN OF INT    feed.to.facs,   control.to.feed :
PAR
  feed     (control.to.feed,  feed.to.facs)
  facs     (feed.to.facs,     facs.to.square)
  square   (facs.to.square,   square.to.sum)
  sum      (square.to.sum,    sum.to.control)
  control (keyboard, screen, sum.to.control, control.to.feed)
```

9.6.1   Running the example program

When you run this program, it will ask for a value for **n**. If you supply any number less than 100, it will execute successfully.

Type 101; the TDS will fail with the message:

> **Error - iserver - transputer error flag set**

The next action will depend on how the TDS is called from the host operating system. Please consult the Delivery Manual.

9.6.2   Creating a core dump

Reboot the TDS in diagnostic mode. After a short delay, you will see a welcome message followed by:

> **Options :**
>   **c : normal core dump**
>   **f : normal core dump + freespace**
>   **a : standalone core dump - all of memory**
>   **s : standalone core dump - part of memory**
>   **<RETURN> to skip**

Press 'C', to request a core dump. Option '**F**' should only be used if you had used the '**freespace**' buffer in the program. The TDS will then ask :

> **Core dump filename ("core.dmp") ?**

>   Press ENTER to accept the default filename
> or   enter another filename (any filename extension will be replaced by '**.dmp**')

You will then be told:

> **Writing core dump to file** "*filename*.**dmp**" **...**

Finally, the TDS will be restarted.

9.6.3   Using the debugger

Use AUTOLOAD to load the debugger.

Now you will be able to start debugging. Move the cursor to the source of the **EXE**. When positioned on the **EXE** fold line, you should press RUN EXE to start the debugger.

The screen will show:

> **TDS occam 2 Debugger** - *version identifier*

> **Debugging an EXE**

> **Read Core dump file, Ignore core dump, or Quit (C,I,Q) ?**

You should type 'C' here, to indicate that you wish to read a core dump file. (If you type '**I**', you can perform a single locate to the error position, but because the debugger does not know the memory contents, it cannot find the values of variables, etc., nor backtrace down the procedure stack). You will then be asked for the filename:

> **Core dump filename ("core.dmp", or "QUIT") ?**

Press ENTER to accept the default filename

or   enter another filename (any filename extension will be replaced by '.dmp')

or   type 'QUIT' (uppercase) to abort the debugger.

The debugger will then read the file to find out where the error occurred, displaying the following messages one at time:

```
Reading logical name table ...
Reading Core dump file "filename.dmp" ...
Locating ...
Backtracing ...
Location was in LIB dreals, SC 1, offset 1433-Error explicitly set
```

It will display the program source, and leave the editor on the line causing the error. The error was actually caused within a library for **REAL64** arithmetic, but the debugger will locate to the line which the library was called from. In this case it is inside procedure '**square**', on the line:

```
sq := x * x
```

9.6.4   Inspecting variables

You may move the cursor around the screen, and inspect any variable. If, for example, you move the cursor over the '**x**', and press INSPECT, you will be informed:

```
REAL64 'x' has value ...
9.3326215443944096E+155 (#605166C698CF1838) (at #80000360)
```

The debugger can display the type of any occam symbol, and its contents. Here, '**x**' is displayed first in its decimal form, then hexadecimal. Finally its address in memory is displayed.

If you forget which tool key is INSPECT, you may press CODE INFORMATION, which will display a list of keys along the top of the screen.

You will be able to inspect the values of '**sq**', '**square**', '**stop.integer**', '**stop.real**', etc. Any value which is in scope at the error location will be accessible. You can 'inspect' the values of procedures and functions, to find out their address and workspace requirements. You will also be able to enter other folds, and browse through the source code, to determine the context of the error. If you forget where the error actually was, press RELOCATE to return there. (Press CODE INFORMATION again to tell you which tool key it is!)

Instead of moving the cursor to each symbol in turn, you may also inspect a symbol by typing in its name. If you move the cursor to a position where it is not over any symbol, and press INSPECT, the debugger will ask you to type in the name which you are interested in.

9.6.5   Jumping down channels

As well as finding the error location, the debugger can be used to find out which other processes were executing at the same time. If you point at the channel '**out**', for example, then press INSPECT it will display:

```
CHAN 'out' has Iptr:#80000611 and Wdesc:#80000285 (Lo) (at #800004B0)
```

This indicates that there is a process waiting for communication on that channel (the '**Iptr**' and '**Wdesc**' identify it), and the debugger also informs you that it is a low priority process, and gives the address of the channel word in memory.

To find out which occam process is waiting, press CHANNEL (again to find which tool key, press CODE INFORMATION). The debugger will move the cursor to the line where the other process is waiting.

This will be inside the '**sum**' procedure, on the line

    `in ?  x`

As before, you may now point at any symbol and inspect it. You will find that channel '**out**' also has a process waiting. Use the CHANNEL key to 'jump' to that process. This will be in the '**control**' procedure, which is waiting for the final result. Again you may inspect any symbol. You can also discover that channel '**screen**' has a process waiting, but note that there is a star ('*****') on the message line. This indicates that the process which is waiting is not part of your OCCam program — in this case it is the TDS itself, which is listening for output to the screen. Therefore if you try to jump to that process, you will be told

    Cannot jump - Channel points to an invalid location

9.6.6   Retrace and Backtrace

So far the debugger has helped to find three of the five processes which were running in parallel. What about the other two? You can use the RETRACE key to retrace your steps (see CODE INFORMATION). This will take you back to the '**sum**' procedure, then back to the '**square**' procedure. Now you can look in channel '**in**', which you know is connected to the '**facs**' procedure. Unfortunately it is empty, which means that the other process is not waiting to communicate.

The next function to try is BACKTRACE. This key makes the debugger backtrace down the procedure calling stack for one procedure or function call; i.e. it moves the cursor to the line from which the current procedure or function was called. If you press BACKTRACE now, the cursor will move to the line where '**square**' was called. Again, you can inspect any symbol which is in scope at this line. For example, you can look in the channels '**feed.to.facs**' and '**facs.to.square**', but both will be empty. You have already looked in the other channels, but you can do so again.

This means that the other two processes were actually executing in parallel at the time of the error, rather than waiting to communicate. To find them, you need to look at the transputer's active process queues.

9.6.7   Process Queues

The lower level transputer information is accessed by using the MONITOR key. This displays a screenfull of information about that processor, and a list of available commands. The command which displays the processor's active process queues is '**R**' (for 'running' processes). Again you can use CODE INFORMATION to display a summary of what each command does, or type '**?**'.

'**R**' will display a list of the processes on the queue. There will be two processes, identified by two lines containing an **Iptr** and **Wdesc**.

Other useful commands are '**T**' (Timer queues), which displays the processes waiting on the transputer's timers; and '**L**' (Links), which displays the processes waiting for communication on the transputer's links.

9.6.8   Display OCCam

Type '**O**' for OCCam, so that we can display the OCCam for these processes. You will be asked:

    Iptr (#80000766) ?

Here you should type the **Iptr** value shown on the first line on the right hand side of the display produced by command '**R**'. You can either type it in full, or use a special short-hand version where '**%**' is used to replace '**#800..**'. E.g. you could type either '**#8000055A**', or '**%55A**'. Hexadecimal letters do not need to be in uppercase.

You will then be asked for a **Wdesc**, but the debugger will give the associated **Wdesc** as the default, so you can simply type ENTER here. The debugger will then display the OCCam line where the process was running.

You will be left with the cursor in procedure '**feed**', on the line

    out ! i

Because this process is on the process queue, not waiting for communication, it has performed that communication, and is just about to resume executing. You can examine variables, etc., as before.

To find the last process, press $\boxed{\text{MONITOR}}$ again, and use the 'O' command to locate to the second process listed on the queue. The debugger may discover that this process is actually executing code inside one of the **REAL64** arithmetic libraries. As INMOS does not supply the source code for these libraries, the debugger cannot display the relevant line. Instead it will backtrace to the line where the arithmetic is being performed; in this case the line inside the factorial function:

    result := result * (REAL64 ROUND i)

Depending upon exactly how the program was executing when it failed, it may locate to the replicated **SEQ** instead.

Again, you may inspect variables. By inspecting '**i**', you find out how many times that loop has been executed. You can $\boxed{\text{BACKTRACE}}$ to find out where the function was called from.

9.6.9   Finish

To exit the debugger from symbolic mode use $\boxed{\text{EXIT FOLD}}$, then $\boxed{\text{FINISH}}$. Use command 'Q' to exit from the Monitor page.

9.6.10   Other functions

While in the debugger, there are a few more tools available. The $\boxed{\text{TOP}}$ key will return you to the error location, or to the last location selected by an 'O' command from the monitor page.

The $\boxed{\text{LINKS}}$ key displays a summary of which other processors this transputer's links are connected to. This is not useful when debugging an **EXE**, but is useful when debugging a **PROGRAM**.

The $\boxed{\text{INFO}}$ key displays the **Iptr**, **Wdesc**, and priority, of the last position located to, together with the processor type and number.

9.6.11   More information

This worked example should have given you an idea of how to use this post-mortem debugger. Chapter 15 contains a full description of how to use all of these debugging tools, including extensions not listed here, such as inspecting arrays.

9.7   How the debugger works

The following documents describe the way the transputer implements occam for those who are more interested.

- Technical note 21: 'The transputer implementation of occam'

  This note details how such features as **PAR**, **ALT**, **TIMER**, and channels, are implemented on the transputer.

- 'The transputer instruction set — a compiler writer's guide'

  This book describes the transputer's instruction set, but at a lower level, and is not particularly relevant to occam programmers.

9.7.1   How the debugger accesses the network

The technical details of how the debugger analyses the network and examines its state are described in INMOS technical note 33 'Analysing transputer networks'. The method used is outlined briefly here. It can successfully analyse networks consisting of hundreds or thousands of transputers of mixed type.

First the debugger reads the program's configuration details, and uses these to build a picture of the transputer network. It then reloads the network with a program known as an 'analyse worm'. This program allows the debugger to access any transputer in the network, by setting up a message routing system. Obviously, this program will itself corrupt each transputer's memory contents, so before it is loaded, the debugger 'peeks' the portion of memory which will be overwritten into a buffer on the host, along with the saved register contents. This works out to be approximately 700 bytes per processor, so, for example, a 10 processor network would require 7 Kilobytes, or 10 000 processors would require 7 Megabytes. Note that if each of these 10 000 processors had 1 Mbyte of local store, this is minute compared with a total memory size of 10 Gigabytes! When the debugger needs to read any memory contents which are not included in this buffer, it sets up a communication path through the network and requests the required data. In this way it is not necessary to buffer the complete memory contents of the network, so it is quite feasible to debug large networks of transputers.

9.7.2   Debugging information generated by the compiler

An important aspect of this debugging system is that the **create.debugging.info** option of the compiler merely forces the creation of the extra debugging information; it does not affect the compiled transputer code which is produced. Thus a program compiled with debugging enabled will behave identically to the same program compiled with debugging disabled. The option to disable debugging only exists to speed up compilation, and to reduce file space requirements.

The debugging information generated by the occam 2 compiler now includes:

- Workspace offsets for all variables, procedure and function parameters, abbreviations, channels, and arrays, together with their types.

- The types and values of all constants which have been declared.

- The names of all protocols and their variant tags, together with ports and timers.

- The workspace requirements and location of each procedure and function; at the transputer instruction level, occam functions and procedures are identical.

Using this information, together with the configuration details of the program, the debugger can build a complete map describing the locations of any variables currently in use on any processor in the network.

9.7.3   How the symbolic facilities work

Any occam process running within the network can be identified by the transputer number within the network, an instruction pointer, and a workspace descriptor. On any one transputer, there may be many different processes executing the same portion of code, but each will have a different workspace, where all local variables and channels are stored. Of course, the same code may also be executing on other processors in the network.

After analysing the network, the debugger can determine the last instruction executed and the workspace descriptor of each processor in the network. It uses this last instruction pointer, and instruction pointers taken from the active process and timer queues, and the processes waiting on the transputer links, to find occam processes to be examined.

9.7.4   Backtracing

Included in the debug information are details of the workspace requirements and code layout of each procedure or function. Therefore, given an instruction pointer, the debugger can discover which procedure is currently being executed, and its workspace requirements. Using this information, together with that process' workspace descriptor, it can read the return address of that procedure, and hence find the procedure call. The workspace is then adjusted to allow for that used by the procedure, and the space used by the procedure call, to give the workspace descriptor for the calling statement. This is then used, together with the return address, to locate to the occam line containing the procedure or function call.

9.7.5   Inspecting variables

The compiler produces a map showing the workspace offset and type of each variable, parameter, or abbreviation used within that procedure. Thus, given an instruction pointer to indicate which procedure is being executed, and a workspace descriptor for that procedure's local data space, it can calculate the location of any data item in the transputer's memory, and read the data to discover the variable's contents.

Non-local variables must be accessed differently. The debug information includes details of the lexical level of each procedure, so that the lexical level of non-local variables can be found. The lexical level is the level of procedure nesting within the occam source. This is then used to follow the chain of procedure calls to the correct procedure's local data space, and hence to find the correct location of the data.

9.7.6   Jumping down channels

Channels which provide communication between two processes executing on the same transputer are implemented by means of a word in memory. This contains the workspace descriptor of a process waiting for communication on that channel, or a special value to indicate that it is idle. The debugger can examine a channel to see whether a process is waiting, and if so, it can read the process' instruction pointer and workspace descriptor to jump directly to that process.

9.7.7   Analysis of deadlock

If a set of occam processes is deadlocked, there may be no available path into the occam program, from which to start debugging. Internal, or 'soft', channels can only be inspected by the debugger if they are in the scope of an active occam process. This means that a deadlock may be difficult to debug. Note that a deadlock waiting for a communication on a transputer link, or 'hard' channel, is easily debugged by inspecting the process waiting on the link.

However, a simple source modification will allow easy detection of any of the deadlocked processes. Suppose you believe that a certain channel, or a few channels, are causing the deadlock. Then all that need be done is to add a small process in parallel, in such a position that this channel or channels are in scope. The added process does not need to do anything, except be active in some way. For example, it could just wait on the timer for a long time, or loop continuously. Note that on a transputer a process waiting on the timer consumes no cpu resource. However, the debugger can then find its way into the source, to inspect those channels, and jump to the process which is waiting. Any variables which are in scope there may then be examined, and debugging can continue as normal.

Consider this short procedure:

```
PROC deadlocks ()
  CHAN OF INT c :
  PAR
    c ! 0
    INT x :
    SEQ
      c ? x
      c ? x            -- this procedure will deadlock here!
  :
```

When executed, this procedure will deadlock on the internal channel 'c', leaving no active process, and thus prevent the debugger from accessing any variables, etc. It can be changed to:

```
PROC deadlocks.but.debuggable ()
  CHAN OF INT c :
  PAR
    TIMER t :
    INT n :
    VAL one.second IS 15625 :      -- T414B, low priority
    VAL one.day    IS one.second * ((60 * 60) * 24) :
    SEQ
      t ? n
      t ? AFTER n PLUS one.day -- this process will be
                               --     waiting here!
    c ! 0
    INT x :
    SEQ
      c ? x
      c ? x            -- The debugger will jump to here
  :
```

This procedure will still appear to deadlock, and will not set the transputer's error flag, but when it is interrupted by analysing the network, there will be a process on the timer queue.

The debugger can read the timer queue to locate to the delayed timer input, and leave the cursor on that line. The user can then move the cursor to the declaration of channel 'c', and press the [CHANNEL] function key. The debugger will then move the cursor to show the deadlocked input statement; any variables which are in scope can then be examined, to determine the cause of the deadlock.

Obviously, in this simple case it is easy to see what has caused the deadlock by inspecting the source code. In more complicated programs this ability to find deadlocks can be very useful.

An alternative approach to the analysis of deadlock is to display all the workspace as **CHAN** variables and to locate to any waiting channels so found. Only locations containing **Wdesc** values pointing to valid workspace stack frames are included in such a display.

9.7.8   occam scope rules

It is necessary to realise that the debugger can only supply the contents of variables which are in lexical scope at the current occam context. This can best be illustrated by an example:

```
PROC p ()
  INT a :
  PROC q (VAL INT b)
    INT c :
    SEQ
      c := b + a
  :
  PROC r (VAL INT d)
    INT e :
    SEQ
      e := 0
      e := d / e          -- The debugger will locate to here
                          -- after the error
  :
  INT x :
  SEQ
    x, a := 99, 57
    INT y :
    SEQ
      y := 42
      q (y)
    r (x)                 -- And backtrace to here
:
```

In this example, the divide in procedure 'r' would cause an error, and the debugger can locate to that line. Here the variables 'e', 'd', and 'a' may be inspected, but not 'x', 'y', 'c', or 'b', since these are not in scope.

After backtracing, when located at the call of 'r', only variables 'a' and 'x' may be inspected, since the others are all no longer in scope.

10 EPROM programming

10.1 Introduction

The INMOS EPROM software is designed so that programs which have been developed and tested using the TDS may be placed in a ROM without change. This has the advantages that an application need not be committed to ROM until it is fully debugged and the actual production of the ROMs can be done relatively late in the development cycle without the fear of introducing new problems.

Figure 10.1 shows how a network of five transputers would be loaded from the TDS.

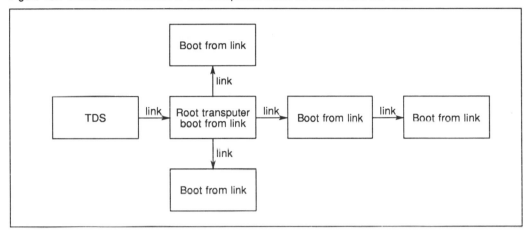

Figure 10.1 Loading a network from TDS

Figure 10.2 shows how the same network of five transputers would be loaded from a ROM accessed by the root transputer. The data being input by the root transputer from the ROM buffer is identical to the data being input by the root transputer in figure 10.1 from the link to the TDS.

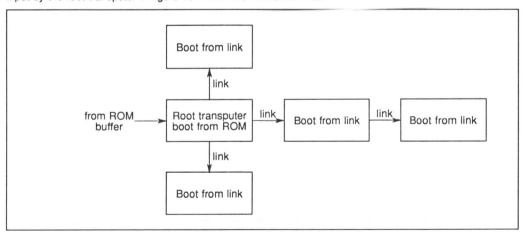

Figure 10.2 Loading a network from ROM

Creating a ROM from a debugged network program is a straightforward sequence of steps using standard TDS utilities and tools. The two components to be put into the ROM are: firstly the debugged application program and secondly an INMOS supplied loader. These two components are placed together in a fold bundle to which the make EPROM tool is applied. The result of applying the tool is a third fold in the bundle which can then be burnt directly into an EPROM.

Details of how to create the fold bundle, how to burn the created output into the ROM and how to create ROMs which have different loading and running requirements from the standard case are described in the following sections.

This chapter introduces three programs which are used in creating ROMs.

EPROM hex program This is the program which is used to convert a working application program into a file suitable for loading into an EPROM.

Hex to programmer program This program takes the output of the EPROM hex program, and sends it to an EPROM programmer. The program interfaces to a GPXP640 EPROM Programmer using Intel Hex format. The sources of the Hex to programmer program are provided so that they may be modified for EPROM programmers expecting a different format.

Write EPROM file program This is similar to the Hex to programmer program but writes to a file.

This chapter also introduces a program which can be used in conjunction with the ROM software.

Memory interface program This is an interactive program which allows the user to explore the effects of changes in the memory interface timing parameters of the IMS T414 and IMS T800 processors. It can produce a memory configuration table which can be included by the EPROM hex program in the file to be burned into the ROM.

Each of these programs is described in detail in chapter 15 of this manual. Section 15.3 describes the Memory interface program in detail. Section 15.4 describes the EPROM hex program. Section 15.5 describes the Hex to programmer program. Section 15.6 describes the Write EPROM file program.

10.2   How to create the fold bundle

An empty fold bundle is created by pressing $\boxed{\text{CREATE FOLD}}$ twice anywhere in the fold structure. Two items need to be placed in this fold bundle; the application program and the loader.

Once the application program has been tested on a target network it should be extracted by running the $\boxed{\text{EXTRACT}}$ utility of the occam compiler utilities on the compiled **PROGRAM** fold set to produce a **CODE PROGRAM** fold. The **CODE PROGRAM** fold produced as a new last item in the **PROGRAM** foldset should be moved into the fold bundle created earlier. When applied to a **PROGRAM** the $\boxed{\text{EXTRACT}}$ utility prompts for two parameters:

```
VAL output.fold IS  BOOTABLE :  --  BOOTABLE | DIAGNOSTIC
```

and

```
VAL first.processor.is.boot.from.link  IS  FALSE :
```

The parameter **output.fold** determines whether the **CODE PROGRAM** fold is to contain load time diagnostic information. **BOOTABLE** is slightly faster and would be used if the processor booting from ROM has no channel to communicate any load failures to. **DIAGNOSTIC** could be used if a channel is available to report failures to and the load is regarded as being unreliable in some way.

The parameter **first.processor.is.boot.from.link** determines how much space in RAM on the first processor should be avoided by code loaded into that processor. The network loader running from ROM has a greater workspace requirement than the equivalent loader run as part of the bootstrap when booting a processor from link. Because the first processor will ultimately be booted from ROM, this parameter should be set to **FALSE** both when developing the application and when extracting the program for burning into EPROM.

The second item to be placed in the fold bundle is the loader. The loader is added to the fold bundle as a **CODE SC** fold. An example loader is provided in the TDS tools source directory **\TDS3\TOOLS\SRC** with the fold comment:

SC multiboard eprom loader (no diagnostics) 17th March 1988

This should be compiled and extracted to create a **CODE SC** fold, which can be moved and placed into the fold bundle without modification.

The fold bundle is now complete and appears as follows:

```
{{{   fold bundle for EPROM
...F CODE PROGRAM application
...F CODE SC multiboard eprom loader (no diagnostics) 17th March
}}}
```

10.3   Creating the ROM file

Having created the fold bundle containing the application and the loader, the next step is to create from this a file suitable for burning into an EPROM. The EPROM hex program **epromhex** (see section 15.4) performs this function. The EPROM hex program must first be loaded from the **Tools** fold in the toolkit fold by using GET CODE and then run on the fold bundle created as a result of the actions described in the previous section. The result is a new last fold in the bundle with the fold comment **EPROM hex** so the fold bundle now appears as follows:

```
{{{   fold bundle for EPROM
...F CODE PROGRAM application
...F CODE SC multiboard eprom loader (no diagnostics) 17th March
...F EPROM hex
}}}
```

The first line in the **EPROM hex** fold holds the start address of the ROM code in the processor's address space and identifies the processor type. The remainder of the fold consists of a sequence of hexadecimal bytes to be placed in ROM from the specified address onwards.

The EPROM hex program will prompt the question:

Insert copy for analyse (y/n)

This is described fully in section 10.6, 'ROMs which load from a host computer'. For the example considered here, the answer should be **n** (no).

10.4   Burning the ROM

The fold created by the previous section is now ready for sending to an EPROM programmer for burning into the ROM. The Hex to programmer program performs this function. The Hex to programmer program must first be loaded from the **Tools** fold in the toolkit fold by using GET CODE.

The Hex to programmer program **hextoprg** (see section 15.5) should be run with the cursor on the **EPROM hex** fold produced by the EPROM hex program described in the previous section. It produces output in a form suitable for controlling an EPROM programmer via **COM1** on the the IBM PC. The file **COM1** is treated by DOS as a communications port. Therefore, to connect an EPROM programmer to an IBM PC requires a serial card installed as **COM1**. It may be necessary to use the DOS **MODE** command to configure the serial card to the correct baud rate, parity, etc., for the EPROM programmer.

The procedure used depends on the width of the memory interface on the board for which the EPROMs are intended. The Hex to programmer program reads the first line of the **EPROM hex** fold to determine the processor type and hence the number of ROMs required. The IMS T414, IMS T425, IMS T800, IMS T801 and IMS T805 have a 4 byte wide memory interface and therefore require 4 byte-wide ROMs. The IMS T212, IMS T222 and IMS T225 have a memory interface which can be configured dynamically to be 1 or 2 bytes wide. If the code is intended for an IMS T212, IMS T222 or IMS T225 the program will ask whether the ROM is being accessed in byte mode (1 ROM required) or word mode (2 ROMs required). If more than one ROM is required they must be programmed separately and the user must identify which is being programmed. The Hex to programmer program will select the appropriate bytes from the **EPROM hex** fold.

The start address of the code within the processor's address space is also read from the first line of the **EPROM hex** fold. This, and the size of the ROM being programmed, are used to calculate the start address of the code within the ROM. The ROM size is entered by the user.

An alternative program **promfile** may be used to send ROM images to host files (see section 15.6).

10.5   Execution from ROM instead of RAM

Earlier sections of this chapter described how to make an EPROM suitable for booting a network of transputers with an application which is run in RAM on all processors in the network. In certain circumstances it may be desirable to execute the application code while it is resident in ROM rather than loaded into RAM. It may be the case that the application running on the processor booted from ROM is the only processor in the network or the processor booted from ROM may boot all the other processors as in the earlier example before continuing with the application code executed from ROM.

Single transputer with application in ROM

In the first case, where the application program is running from ROM as a standalone embedded system on a single transputer, the fold bundle is simplified to contain just a **CODE SC** fold. The **SC** implements the required application.

The application should be developed under the TDS as an **SC** compilation unit containing one procedure and tested as the only processor in a network loaded by the TDS. The EPROM hex program sets up values for a standard set of parameters for the **SC** to be included in the ROM, the parameters are values required by the loader described earlier.

```
PROC EPROM.SC( INT            entry.point,
               [60]BYTE       buffer,
               VAL [600]BYTE  memory.copy,
               VAL []BYTE     program.buffer)
    ...  application
  :
```

The application needs to have this form. The best way to achieve this is to develop it under the TDS with only those parameters necessary for loading from the TDS, and then move the developed **SC** into an **SC** of the above form for final compilation and extraction, as follows:

```
{{{   SC eprom source
{{{F eprom source
...  EPROM.SC              -- standard parameter list
  ...  SC application
  application ()
  :
}}}
}}}
```

When running the EPROM hex program, the question **Insert copy for analyse?** should be answered **n** (no). The EPROM hex program may produce the warning message:

```
WARNING: total RAM space requirement exceeds maximum
         allowed for a loader (limit = 560 bytes)
```

This message is significant only when the ROM is loading code into local RAM and so may be ignored in this case.

Load network then continue in ROM

In the second case, the ROM processor will boot the rest of the network as in the original example, but will then continue executing from ROM.

Two different options are again possible. For the first option, the application to be run on the processor booted from ROM is developed independently from the network; for example as an **EXE** running within the TDS interfacing to the network program. For the second option, the application to be run on the processor booted from ROM is developed as the root processor in the network program.

In the first option, the code running from ROM must emulate the action of the TDS in booting the rest of the network and then carry on with the developed application.

```
...  EPROM.SC              -- standard parameter list
  CHAN OF ANY boot.link:
  PLACE boot.link AT 2:
  SEQ
    boot.link ! program.buffer -- load network
    ...  SC application
    application()
:
```

In the second option, the code running from ROM on the root processor must load the rest of the network as in the original example but ignore all code directed to be loaded into RAM on the root processor. After the load is completed, control should continue within the SC rather than terminate in the manner of the network loader.

```
...  EPROM.SC              -- standard parameter list
  ...  SC modified network loader
  ...  SC application
  SEQ
    load.network (program.buffer)
    application()
:
```

When running the EPROM hex program, the question **Insert copy for analyse?** should be answered **n** (no) in both of the options described above. The EPROM hex program may produce the warning message:

```
WARNING: total RAM space requirement exceeds maximum
         allowed for a loader (limit = 560 bytes)
```

This message is significant only when the ROM is loading code into local RAM and so may be ignored in both of the above options.

10.6   ROMs which load from a host computer

For some applications it may be desirable to create a ROM which is capable of loading a network of transputers from a host computer using a non-link interface (such as RS232). An example of this type of ROM is the monitor program on INMOS evaluation boards which include serial RS232 ports.

This type of system is a variation of the single application running from ROM, in which the application is a loader, and the ROM fold is created in a similar manner. The source of the INMOS monitor program is provided in the TDS tools directory with the fold comment:

```
SC B00x.monitor      (24th February 1988)
```

for modification by users to match their particular hardware. The TDS uses additional handshaking sequences and, if necessary, byte encoding, when loading a network via RS232.

An outline of the INMOS monitor is given below.

```
PROC B00x.monitor (INT              entry.point,
                   [60]BYTE         buffer,
                   VAL[600]BYTE     memory.copy,
                   VAL[]BYTE        program.buffer)
  ... link placements
  ... constants
  ... load
  ... analyse
  SEQ
    ... respond to wake up character
    read.char (line, char)   -- not encoded
    IF
      ... 'B' : straight binary
      ... 'H' :  encoded hex
      ... otherwise bad protocol
    get.char (line, char)    -- encoded
    char := char /\ #7F
    IF
      char = (INT 'A')
        analyse ()
      char = (INT 'L')
        load ()
      TRUE
        ... bad protocol
:
```

The INMOS monitor can also interface to the TDS for analysing networks. Analysing and debugging software accessing a processor needs to examine the contents of workspace as it was when the previous execution of a program was halted (probably through the error flag being set). The low addressed part of RAM, which is likely to be of interest to a debugger, is the area which the ROM employs for workspace. If the response to the prompt by the EPROM hex program **Insert copy for analyse?** is given as **y** (yes), the program in ROM will copy this area to the high addressed part of the available RAM. If it is not necessary for the ROM to be used to interface to host software analysing the network then a **n** (no) response is suitable.

The TDS loading and analysing protocols and the special requirements for using serial lines are described in INMOS technical notes 33 and 34.

Workspace for this type of ROM must be kept small to make certain that the loader is not directed to load code to an area occupied by ROM workspace. The EPROM hex program produces the warning message

```
WARNING: total RAM space requirement exceeds maximum
         allowed for a loader (limit = 560 bytes)
```

if the workspace used by the ROM extends into areas to which the loader may be directed to load code. Note that compiling the code without a separate vector space in general reduces the code's total workspace requirement.

10.7   Adding a memory configuration to the EPROM

IMS T414, IMS T425, IMS T800 and IMS T805 transputers can configure their memory interface from a table of words stored at the most positive addresses in their memory space. These addresses are within the area occupied by an EPROM which can be used to boot a transputer. The EPROM hex program can include such a configuration table into the output file at the correct configuration addresses. To cause this to occur it is simply necessary to include the memory configuration table as an additional filed fold in the fold bundle on which the EPROM hex program is run. The required fold is labelled **...   (configuration)** and must be filed, using FILE/UNFILE The order of the folds is unimportant.

The configuration table expected by the EPROM hex program is in the format output as a result of running the Memory Interface program **memint** (see section 15.3). Such a fold may be generated by hand, if desired, the main requirement being that the fold is complete (i.e. all address-value pairs are present).

11 Low level programming

This chapter describes a number of features of the OCCAM 2 compiler in the TDS which support low-level programming of transputers. These are as follows:

Allocation This allows a channel, a variable, an array or a port to be placed at an absolute location in memory.

Code insertion This allows sections of transputer machine code to be inserted into OCCAM programs.

Dynamic code loading A set of compiler library procedures allow an OCCAM program to read in a section of compiled code (from a file, for example) and execute it.

Extraordinary use of links A set of library procedures allow link communications which have not completed to timeout or be aborted by another part of the program.

11.1 Allocation

allocation = **PLACE** *name* **AT** *expression* **:**

The **PLACE** statement in OCCAM allows a channel, a variable, an array or a port to be placed at an absolute location in memory. This feature may be used for a number of purposes, for example:

- Mapping OCCAM channels onto the 'hard channels' implemented by transputer links, from within an OCCAM program.

- Mapping arrays onto particular hardware, such as video RAM.

- Accessing devices (such as UARTs or latches) mapped into the transputer's address space.

The **PLACE** statement may *not* be used to force critical arrays or variables onto on-chip RAM. The OCCAM compiler allocates memory according to the scheme outlined in chapter 6; it does not take account of data placed at some arbitrary position in the memory it is trying to allocate. So placing data within the data space allocated by the compiler will interfere with other data placed there by the compiler. To make the best use of on-chip RAM, use the 'separate vector space' facility of the compiler described in chapter 6.

The address of a placed object is derived by treating the value of the expression as a subscript into an **INT** array mapped onto memory. Thus **PLACE n AT 1:** would cause **n** to be allocated address #80000004 on a 32-bit transputer. Addresses are calculated in this way so that the transputer links can be accessed using word length independent code (the links are addresses 0, 1 up to 7).

Translation from a machine address to the equivalent OCCAM **[]INT** subscript value can be achieved by the following declaration:

```
VAL occam.addr IS (machine.addr><(MOSTNEG INT)) >> w.length:
```

Where **w.length** is 1 for a 16-bit transputer and 2 for a 32-bit transputer.

Some useful allocations are given below:

```
CHAN OF ANY    in.link0, out.link0 :
CHAN OF ANY    in.link1, out.link1 :
CHAN OF ANY    in.link2, out.link2 :
CHAN OF ANY    in.link3, out.link3 :
CHAN OF ANY    in.event :

PLACE    out.link0 AT 0:
PLACE    in.link0  AT 4:

PLACE    out.link1 AT 1:
PLACE    in.link1  AT 5:

PLACE    out.link2 AT 2:
PLACE    in.link2  AT 6:

PLACE    out.link3 AT 3:
PLACE    in.link3  AT 7:

PLACE    in.event  AT 8:

[4]CHAN OF ANY out.links, in.links :

PLACE    out.links AT 0:
PLACE    in.links  AT 4:
```

All placed objects must be word aligned. If it is necessary to access a **BYTE** object on an arbitrary boundary, or an **INT16** object on an arbitrary 16-bit boundary, the object must be an element of an array which is placed on a word address below the required address. For example, to access a **BYTE** port called **io.register** located at physical address #40000001 on a T4 the following must be used:

```
[4]PORT OF BYTE io.regs.vec :
PLACE io.regs.vec AT #30000000 :
io.register IS io.regs.vec[1] :
```

Placement may be used on transputer boards to access board control functions mapped into the transputer's address space. For example, on the IMS B004, the subsystem control functions (**Error**, **Reset** and **Analyse**) are mapped into the address space, and can be accessed from occam as placed ports. The following code will reset subsystem on the IMS B004, an IMS B008 or compatible board:

```
PROC reset.b004.subsystem()
  VAL subsys.reset   IS (0 >< (MOSTNEG INT)) >> 2:
  VAL subsys.analyse IS (4 >< (MOSTNEG INT)) >> 2:
  VAL subsys.error   IS (0 >< (MOSTNEG INT)) >> 2:
  PORT OF INT reset, analyse, error:
  PLACE reset   AT subsys.reset:
  PLACE analyse AT subsys.analyse:
  PLACE error   AT subsys.error:
  TIMER clock:
  INT time:
  SEQ
    analyse ! 0                    -- set reset and initialise low
    reset ! 0
    reset ! 1                      -- hold reset high
    clock ? time
    clock ? AFTER time PLUS 78:    -- 5 ms is ample
    reset ! 0                      -- reset subsystem
:
```

The error and analyse functions can be controlled from occam in a similar way. The pipeline sorter example described in chapter 7 shows an example of monitoring the subsystem error flag from a program (the **monitor** program) running on the IMS B004.

11.2 Code insertion

This section describes the facilities provided by the occam 2 compiler code insertion mechanism.

The code insertion mechanism enables the user to access the instruction set of the transputer directly within the framework of an occam program. Symbolic access to occam variable names is supported, as is automatic jump sizing. More details on the instruction set may be found in the INMOS document 'The transputer instruction set — a compiler writer's guide'.

Code insertion may be employed to perform tasks not possible from occam, or for particularly time-critical sections of a program. There are several reasons, however, which should encourage the user to refrain from using code insertion as a solution to problems which may, with some thought, be solved using occam. Paramount among these is that the validity of a system consisting entirely of occam can be checked by the compiler. A compiler can check usage of channels, access to variables, communication protocols and range violations. A single code insert prevents the compiler from performing these checks adequately. A second reason for not using code insertions is that the transputer instruction set is suited for use by a high level language, particularly occam, and algorithms which are simple to code and easy to debug in occam become difficult and obscure when coded in the transputer instruction set directly.

11.2.1 Using the code insertion mechanism

An occam 2 code insertion is introduced by the construct GUY. The context of the GUY construct is determined, as with all occam constructs, by its indentation. The transputer instructions which follow the GUY must be indented and there may only be one instruction per line. Lines may be terminated by a comment, which is introduced by the -- symbol as in occam. The transputer instructions are upper case versions of the standard mnemonics listed in INMOS documentation. The code insert is terminated by the matching outdent.

A compiler parameter code.inserts determines which instructions may be used within sections of code insertions, in the unit being compiled. If the value is NONE, no code insertions are allowed and the compiler will flag the first such instruction as invalid. If the value is RESTRICTED, then the instructions allowed are a restricted set of instructions which are sufficient for time-critical sections of sequential code. If the value is ALL, then all transputer instructions are allowed including OPR for the creation of arbitrary instruction codes. Since the inclusion of some instructions may have an unexpected effect on the occam program (for example, instructions which move the workspace pointer), instructions outside of the restricted set must be used with great care. A list of the restricted set of transputer instructions is given in appendix H.

For example, to perform a 1's complement addition we can write the following occam:

```
INT carry, temp:
SEQ
  carry, temp := LONGSUM (a, b, 0)
  c := carry PLUS temp
```

However, if this occurs in a time-critical section of the program we might replace it with:

```
GUY
  LDC 0
  LDL a
  LDL b
  LSUM
  SUM
  STL C
```

which would avoid the storing and reloading of carry and temp.

Values in the range MOSTNEG INT to MOSTPOS INT may be used as operands to all of the direct functions without explicit use of prefix and negative prefix instructions. Access to non-local occam symbols is provided automatically without explicit indirection.

A more complex example, which sets error if a value read from a channel is not in a particular range, takes advantage of both these facilities:

```
INT    a :
...   other stuff
PROC get.and.check.index (CHAN OF INT c)
   SEQ
     c ? a
     GUY
        LDL      a    -- push value of free variable onto stack
        LDC     512   -- push 512 onto stack
        CCNT1         -- if NOT (0 < a <= 512) then set error
   :
```

If there is a requirement for the code insertion to use some work space, then the work space may be declared before the **GUY** construct, in which case, the work space locations are accessed just like any other occam symbols.

```
INT    a :
SEQ
   INT    b, c :
   GUY
      LDL      a    -- push value in a onto stack
      STL      b    -- pop value from stack into b
      ...   more code
```

11.2.2   Labels and jumps

To insert a label into the sequence of instructions, put the name of the label, preceded by a colon, on a line of its own. Then when the label is used in an instruction, precede the name with a full stop. For example:

```
GUY
   ...   some instructions
   :FRED
   ...   some more instructions
   CJ  .FRED
```

A restriction of the compiler is that the same label name may not be defined more than once within an occam procedure.

11.3   Dynamic code loading

Introduction

The transputer development system permits the dynamic loading and execution of code, using the procedures described in this section. The procedures are listed in section 14.2.6.

The procedures described allow the programmer to write an occam program that reads in a compiled occam procedure and then calls it. The called procedure may be compiled and linked separately from the calling program and may even be generated by a compiler outside the TDS. It may be read into an array of the calling program from a transputer link or from a file in a host filing system. Alternatively it may be converted into an occam table by using as source code the **wocctab** tool and included in the calling program. It is possible to pass parameters to the procedure, which must have at least 3 formal parameters.

11.3.1   The call

The occam compiler recognises calls of a procedure **KERNEL.RUN** with the following parameters:

```
PROC KERNEL.RUN(VAL []BYTE code,
                VAL INT entry.offset,
                []INT workspace,
                VAL INT no.of.parameters)
```

The effect of a call of **KERNEL.RUN** is to call the procedure in the **code** buffer, starting execution at the location **code[entry.offset]**. The **workspace** buffer (see figure 11.1) is used by the called procedure for its local scalar data. The required size of this buffer and the code buffer must be derived from information in the code file. The parameters passed to the called procedure should be placed at the top of the **workspace** buffer by the calling procedure before the call of **KERNEL.RUN**. The call to **KERNEL.RUN** returns when the called procedure terminates. If the called procedure requires a separate vector space, then another buffer of the required size must be declared, and its address placed as the last parameter at the top of **workspace**. As calls of **KERNEL.RUN** are handled specially by the compiler it is necessary for **no.of.parameters** to be a constant known at compile time. This imposes some restrictions on the way **KERNEL.RUN** is used.

Note that as, in general, a compiled procedure will include word-aligned constant tables it is important to ensure that the code buffer is word aligned.

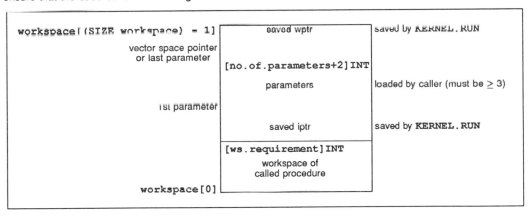

Figure 11.1 **workspace** buffer

The workspace passed to **KERNEL.RUN** must be at least:

```
[ws.requirement + no.of.parameters + 2]INT
```

where **ws.requirement** is the size of workspace required, determined when the called procedure was compiled, and stored in the code file and **no.of.parameters** includes the vector space pointer if it is required. The parameters must be loaded before the call of **KERNEL.RUN**. The parameter corresponding to the first formal parameter of the procedure should be in the word adjacent to the saved iptr word, and the vector space pointer or last parameter should be adjacent to the top of workspace where the wptr will be saved.

11.3.2   Loading parameters

There are a number of library procedures to set up parameters before the call. These are:

```
LOAD.INPUT.CHANNEL            (INT here, CHAN OF ANY in)
LOAD.INPUT.CHANNEL.VECTOR     (INT here, []CHAN OF ANY in.vec)
LOAD.OUTPUT.CHANNEL           (INT here, CHAN OF ANY out)
LOAD.OUTPUT.CHANNEL.VECTOR    (INT here, []CHAN OF ANY out.vec)
LOAD.BYTE.VECTOR              (INT here, []BYTE b.vec)
```

The variable **here** is assigned the address of the second parameter. Note that when passing vector parameters, if the formal parameter of the **PROC** called is unsized then the vector address must be followed by the number of elements in the vector, for example:

```
LOAD.BYTE.VECTOR(param[0], buffer)
param[1] := SIZE buffer
```

Thus an unsized vector parameter requires 2 parameter slots. The size must be in the units of the array (not in bytes, unless it is a byte vector, as above). For multi-dimensional arrays, one parameter is needed for each unsized dimension, in the order the dimensions were declared.

All variables and arrays should be retyped to byte vectors before using **LOAD.BYTE.VECTOR** to obtain their addresses, using a retype of the form: **[]BYTE b.vector RETYPES variable:**. **LOAD.BYTE.VECTOR** may also be used to set up the address of the separate vector space. The size of separate vector space does not have to be supplied, but must be adequate for the called procedure.

11.3.3   Examples

Example 1: load from link and run

This is a simple procedure to load a code packet from a link and run it. The type of the packet is given by the protocol **PROTOCOL CODE.MESSAGE IS INT::[]BYTE; INT; INT**
The code is sent first, as a counted array, followed by the entry offset and workspace size. **KERNEL.RUN** requires the called procedure to have at least 3 parameters, but if, as in this case, none of these are used, then the declaration of the called procedure can have no formal parameters. If the called procedure uses separate vector space then at least 2 other (possibly dummy) parameters must be specified.

```
PROC run.code (CHAN OF CODE.MESSAGE input, []INT run.vector,
               []BYTE code.buffer)
  VAL    no.parameters   IS   3 :   -- smallest allowed
  INT    code.length, entry.offset, work.space.size :
  INT    total.work.space.size :
  SEQ
    input  ?  code.length::code.buffer;
                            entry.offset; work.space.size
    total.work.space.size :=
                      (work.space.size + no.parameters) + 2 :
    []INT  work.space IS
               [run.vector FROM 0 FOR total.work.space.size] :
    KERNEL.RUN (code.buffer, entry.offset,
                work.space, no.parameters)
:
```

Example 2: loading arbitrary compiled procedures

This extended example, which is included in full in the software, shows how a program may be written which can load separately compiled procedures created by the TDS or by INMOS toolset products. A compiled procedure produced by COMPILE and EXTRACT in the TDS, is stored in a **CODE SC** filed fold. The structure of such a fold is described in appendix G. Note that the representation of the fold structure is pairs of bracketting bytes around the records containing the header information and the code. If a compiled procedure in a **CODE**

SC fold is written out to a DOS file by WRITE HOST, the bracketting bytes are removed, but the records within them are unchanged. The resulting host code file has exactly the same structure as one produced by the INMOS toolset products (conventionally in a **.rxx** file).

The procedures **load.tds** and **load.rxx** load the contents of the two kinds of code file respectively. The procedure **call.pgm** may be used to call a procedure loaded from either kind of file. The additional support for programs with stack space in addition to work space and vector space is for programs compiled with the INMOS scientific language compiler toolsets. This example allocates stack space between code space and vector space, it may be more efficient to allocate stack space with workspace. Use of **KERNEL.RUN** gives the user freedom to take such decisions.

It is not in general possible to write code to load and enter procedures whose parameter lists are not known at the time the loader is designed.

```
--{{{  loadpgm
...  #USEs
...  other declarations
...  load.tds
...  load.rxx
--{{{  call.pgm
PROC call.pgm ([]BYTE work.area, VAL INT stack.len, ws.len,
                 params.len, code.len, vs.len, entry)
  INT b.stack.len:
  SEQ
    ...  compute stack length in bytes
    --{{{  abbreviations etc
    VAL b.ws.len    IS ws.len TIMES bpw:
    VAL b.vs.len    IS vs.len TIMES bpw:
    VAL b.params.len IS params.len TIMES bpw:
    VAL b.code.len IS (code.len + (bpw - 1)) /\ (BITNOT(bpw-1)):
    VAL stk.start   IS 0:
    VAL ws.start    IS stk.start + b.stack.len:
    VAL p.start     IS ws.start + b.ws.len:
    VAL c.start     IS p.start + b.params.len:
    VAL v.start     IS c.start + b.code.len :
    VAL f.start     IS v.start + b.vs.len :
    b.workspace     IS [work.area FROM ws.start
                                    FOR b.ws.len + b.params.len] :
    b.codespace     IS [work.area FROM c.start FOR b.code.len] :
    b.stackspace    IS [work.area FROM stk.start FOR b.stack.len] :
    b.vecspace      IS [work.area FROM v.start FOR b.vs.len] :
    VAL f.len       IS (SIZE work.area) - f.start:
    b.freespace     IS [work.area FROM f.start FOR f.len] :
    --}}}
    SEQ
      --{{{  load parameters and run the program
      VAL p.len IS params.len TIMES bpw:
      SEQ
        []INT pspace RETYPES [b.workspace FROM p.start FOR p.len] :
        SEQ
          LOAD.INPUT.CHANNEL (pspace [1], from.isv)
          LOAD.OUTPUT.CHANNEL (pspace [2], to.isv)
          LOAD.BYTE.VECTOR (pspace [3], b.freespace)
          []INT freespace RETYPES b.freespace :
          pspace [4] := SIZE freespace
```

```
                    CASE params.len
                      6  -- no stack, no vector space
                        SKIP
                      7  -- vector space but no stack
                        LOAD.BYTE.VECTOR (pspace [5], b.vecspace)
                      9  -- vector space and stack
                        SEQ
                          LOAD.BYTE.VECTOR (pspace [5], b.stackspace)
                          []INT stackspace RETYPES b.stackspace:
                          pspace [6] := SIZE stackspace
                          LOAD.BYTE.VECTOR (pspace [7], b.vecspace)
                  []INT workspace RETYPES b.workspace :
                  CASE params.len
                    6
                      KERNEL.RUN (b.codespace, entry, workspace, 4)
                    7
                      KERNEL.RUN (b.codespace, entry, workspace, 5)
                    9
                      KERNEL.RUN (b.codespace, entry, workspace, 7)
              --}}}
  :
  --}}}

  --{{{  main
  ...   declarations

  SEQ
      ...   determine code file type
      CASE code.file.type
        --{{{   CODE SC from current fold
        code.file.CODESC.fold
          load.tds (from.user.filer[0], to.user.filer[0],
                code.len, bigspace, prog.loaded,
                ws.len, ps.len, vs.len,
                code.entry.offset, target.type, fresult)
        --}}}
        --{{{  bootable or loadable program file
        ELSE
          load.rxx (from.isv, to.isv, [progfile.name FROM 0
                                                  FOR pfn.len],
                code.file.type = code.file.bootable,
                code.len, bigspace, prog.loaded,
                ws.len, ps.len, vs.len, stack.len,
                code.entry.offset, target.type, fresult)
        --}}}
      IF
        ...   not loaded - message
        TRUE
          call.pgm (bigspace, stack.len, ws.len, ps.len,
                    code.len, vs.len, code.entry.offset)
      ...   conclude
  --}}}
  --}}}
```

Other examples

There are other examples of the use of **KERNEL.RUN** in the loader programs whose source is included with the software.

11.4   Extraordinary use of links

The transputer link architecture provides ease of use and compatibility across the range of transputer products. It provides synchronised communication at the message level which matches the occam model of communication.

In certain circumstances, such as communication between a development system and a target system, it is desirable to use a transputer link even though the synchronised message passing of occam is not exactly what is required. Such extraordinary use of transputer links is possible but requires careful programming and the use of some special occam procedures.

The use of these procedures is described in this chapter. To use them in a compilation unit, the directive **#USE reinit** should be inserted at the top of the source for that unit. See section 14.2.8 for a list of the procedures.

11.4.1   Clarification of requirements

As an example, consider a development system connected via a link to a target system. The development system compiles and loads programs onto the target and also provides the program executing in the target with access to facilities such as a file store. Suppose the target halts (due to a bug) whilst it is engaged in communication with the development system. The development system then has to analyse the target system.

A problem will arise if the development system is written in 'pure' occam. It is possible that when the target system halts, the development system is in the middle of communicating on a link. As a result, the input or output process will not terminate and the development system will be unable to continue. This problem can occur even where an input occurs in an alternative construct together with a timeout (as illustrated below). When the first byte of a message is received the process performing the alternative commits to inputting; the timer guard cannot subsequently be selected. Hence, if insufficient data is transmitted the input will not terminate.

```
ALT
    TIME ? AFTER timeout
      ...
    from.other.system ? message
      ...
```

It is important to note that the problem arises from the need to *recover* from the communication failure. It is perfectly straightforward to *detect* the failure within 'pure' occam and this is quite sufficient for implementing resilient systems with multiple redundancy.

11.4.2   Programming concerns

The first concern of a designer is to understand how to recognise the occurrence of a failure. This will depend on the system; for example, in some cases a timeout may be appropriate.

The second concern is to ensure that even if a communication fails, all input processes and output processes will terminate. As this cannot be achieved directly in occam, there are a number of library procedures which perform the required function. These are described below.

The final concern is to be able to recover from the failure and to re-establish communication on the link. This involves reinitialising the link hardware; again there is a suitable library procedure to allow this to be performed.

11.4.3   Input and output procedures

There are four library procedures which implement input and output processes which can be made to terminate even when there is a communication failure. They will terminate either as the result of the communication completing, or as the result of the failure of the communication being recognised. Two procedures provide input and output where communication failure can be detected by a simple timeout, the other two procedures provide input and output where the failure of the communication is signalled to the procedure via a channel. The procedures have a boolean variable as a parameter which is set **TRUE** if the procedure terminated as a result of communication failure being detected, and is set **FALSE** otherwise. If the procedure does terminate as a result of communication failure having been detected then the link channel can be reset.

All four library procedures take as parameters a link channel **c** (on which the communication is to take place), a byte vector **mess** (which is the object of the communication) and the boolean variable **aborted**. The choice of a byte vector as the parameter to these procedures allows an object of any type to be passed along the channel provided it is retyped first.

The two procedures for communication where failure is detected by a timeout take a timer parameter **TIME**, and an absolute time **t**. The procedures treat the communication as having failed when the time as measured by the timer **TIME** is **AFTER** the specified time **t**. The names and the parameters of the procedures are:

```
InputOrFail.t (CHAN OF ANY c, []BYTE mess,
               TIMER TIME,
               VAL INT t, BOOL aborted)
```

and

```
OutputOrFail.t (CHAN OF ANY c, VAL []BYTE mess,
                TIMER TIME,
                VAL INT t, BOOL aborted)
```

The other two procedures provide communication where failure cannot be detected by a simple timeout. In this case failure must be signalled to the inputting or outputting procedure via a message on the channel **kill**. The message is of type **INT**. The names and parameters to the procedures are:

```
InputOrFail.c (CHAN OF ANY c, []BYTE mess,
               CHAN OF INT kill, BOOL aborted)
```

and

```
OutputOrFail.c (CHAN OF ANY c, VAL []BYTE mess,
                CHAN OF INT kill, BOOL aborted)
```

11.4.4   Recovery from failure

To reuse a link after a communication failure has occurred it is necessary to reinitialise the link hardware. This involves reinitialising both ends of both channels implemented by the link. Furthermore, the reinitialisation must be done after all processes have stopped trying to communicate on the link. So, although the **InputOrFail** and **OutputOrFail** procedures do, themselves, reset the link channel when they abort a transfer, it is necessary to use the fifth library procedure **Reinitialise (CHAN OF ANY c)**, after it is known that all activity on the link has ceased.

The **Reinitialise** procedure must only be used to reinitialise a link channel after communication has finished. If the procedure is applied to a link channel which is being used for communication the transputer's error flag will be set and subsequent behaviour is undefined.

11.4.5   Example: a development system

For our example consider the development system described in section 11.4.1.

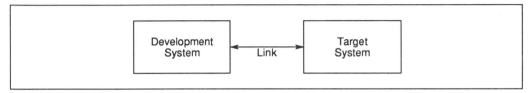

Figure 11.2 Development system

The first step in the solution is to recognise that the development system knows when a failure might occur, and hence the development system knows when it might be necessary to abort a communication.

The process which interfaces to the target system can be sent a message when the development system decides to reset the target causing the interface process to abort any transfers in progress. The development system can then reset the target system (which resets the target end of the link) and reinitialise the link.

The example program below could be that part of the development system which runs once the target system starts executing, until such time as the target is reset and the link is reinitialised.

```
SEQ
  CHAN OF ANY terminate.input, terminate.output :
  PAR
    ...  interface process
    ...  monitor process
    ...  reset target system
  Reinitialise(link.in)
  Reinitialise(link.out)
```

The monitor process will output on both **terminate.input** and **terminate.output** when it detects an error in the target system.

The interface process consists of two processes running in parallel, one which outputs to the link, the other which inputs from the link. As the structures of the two processes are similar only the process which outputs to the link need be shown; the input process is very similar.

If there were no need to consider the possibility of communication failure the process might be

```
WHILE active
  SEQ
    ...
    ALT
      terminate.output ? any
        active := FALSE
      from.dev.system ? message
        link.out ! message
    ...
```

This process will loop, forwarding input from **from.dev.system** to **link.out**, until it receives a message on **terminate.output**. However, if after this process has attempted to forward a message, the target system halts without inputting, the interface process will fail to terminate.

The following program overcomes this problem:

```
WHILE active
  BOOL aborted :
  SEQ
    ...
    ALT
      terminate.output ? any
        active := FALSE
      from.dev.system ? message
        SEQ
          OutputOrFail.c(link.out, message,
                             terminate.output, aborted)
          active := NOT aborted
```

This program is always prepared to input from **terminate.output**, and is always terminated by an input from **terminate.output**. There are two cases which can occur. The first is that the message is received by the input which then sets **active** to **FALSE**. The second is that the output gets aborted. In this case the whole process is terminated because the variable **aborted** would then be true.

11.5   Setting the error flag

The transputer error flag can be set using the predefined procedure **CAUSEERROR()**. This procedure is recognised automatically by the compiler and does not need to be referenced by the **#USE** directive.

CAUSEERROR always halts the program, whatever the mode of the compilation. This is distinct from the occam primitive process **STOP**, which only halts the program if the compilation is in HALT mode.

The reference manual

12 The development environment

12.1 Keys

AUTOLOAD

May be pressed at any position in the program development environment. The toolkit fold is searched for a fold marked:

> **...** **Autoload** *(perhaps with other text here)*

If this fold is found, all runnable code folds which are found in this fold are loaded as though GET CODE had been applied to each fold in turn. If the cursor was on a closed top level fold it is entered.

BOTTOM OF FOLD

Places the cursor on the line displaying the bottom crease symbol of the current enclosing fold.

BROWSE

Used to set the editor into *browse mode,* in which no changes may be made to the document. BROWSE is also used to end browse mode. It switches the set of allowable key functions in the program development environment between the full set and a reduced set which does not allow any form of data input. This function is not available in the toolkit fold or in the code information fold, and it is not possible to enter the toolkit or code information fold when in browse mode.

CALL MACRO

Invokes sequence of keys defined using the DEFINE MACRO key or recovered from a fold by using GET MACRO. If no macro sequence has been defined, the key has no effect.

CLEAR ALL

If the cursor is on a closed top level fold it is entered. Clears all loaded code items, both utilities and user programs.

CLEAR EXE

If the cursor is on a closed top level fold it is entered. Removes the current EXE from the set of current EXEs and selects the next.

CLEAR UTIL

If the cursor is on a closed top level fold it is entered. Removes the current utility set from the set of current utility sets and selects the next.

CLOSE FOLD

Closes the current enclosing fold, and all open folds contained within it. The closed fold line is placed on the line of the screen where the top crease was, unless the top crease was off the top of the screen, in which case the closed fold line appears at the top of the screen. The cursor is positioned on the closed fold line, at the same column position as it was before CLOSE FOLD was pressed. CLOSE FOLD has no effect if the current enclosing fold was opened with an ENTER FOLD operation, but a message is given to remind the user that EXIT FOLD should be used to get out of the current fold.

CODE INFORMATION

May be pressed while in the normal editing environment (not when within the toolkit fold). It creates a display (which appears as a fold structure) showing the following:

1 A 'help' display for the current utility set, which is a sequence of lines listing the utilities in the set and giving a brief explanation of each.

2 A list of the currently loaded code items, both **UTILs** and **EXEs**. Each code item is identified by the text on the fold line when the code was loaded. The current utility set and user program are indicated by a **>** at the start of the line.

3 For each of the loaded code items, there is a fold line which may be opened and viewed. This contains the code size and data requirement for the code, and the 'help' information for a utility.

4 The size of the fold manager buffer and the proportion used.

5 The amount of data space available for running a utility or user program.

While viewing the code information fold the following message is displayed:

Press [EXIT FOLD] to resume editing

Pressing EXIT FOLD to exit the fold returns the editor to the position it was at when CODE INFO was pressed.

Any part of this display may be folded up and saved for later use by means of the CREATE FOLD and MOVE keys.

COPY LINE

Copies the current line and inserts the copy below the current line. If the line is a closed fold then all the text lines and nested folds in the fold are copied. COPY LINE has no effect if the current line is a top or bottom crease. The cursor is placed on the copy.

If the current line is a filed fold, or contains a filed fold, the user is prompted for confirmation, as the operation may take some time. It can be confirmed by pressing COPY LINE again.

File names for files in the copied fold structure are derived from the names of files in the original fold structure, adjusted to avoid clashing with any existing file names in the directory.

COPY PICK

Used to copy a line, which may be a fold line, so that it may be moved to another place in the document. It makes a copy of the current line and appends it to the end of the pick buffer. If the line is a filed fold, or is a fold containing a filed fold, COPY PICK must be pressed again for confirmation, as the copying may take some time.

As COPY PICK has no effect on the document, it may be used to copy portions of a program without forcing the program to require recompilation. It may be used in browse mode.

CREATE FOLD

The first use of CREATE FOLD inserts a new top crease above the current line, at the current column. The second use of CREATE FOLD creates a fold containing the lines between the new top crease and the current line. The fold is closed and the cursor is placed at the end of the fold line marker, where fold header text may be inserted.

Between the two presses of CREATE FOLD all editor functions except cursor movement and scrolling are disallowed.

The indentation of the new fold is determined by the current column on the first use of CREATE FOLD. The lines to be enclosed within the new fold should all be sufficiently indented to fit into a fold at this indentation (i.e. they must not extend to the left of this column).

CURSOR DOWN

Moves the cursor down one line. On the bottom line of the screen it scrolls the screen one line down the current view, if there are lines in the current view below the screen, and the cursor remains in the same position on the screen.

CURSOR LEFT

Moves the cursor left one column, except in the leftmost column on the screen where it may cause the view to pan.

CURSOR RIGHT

Moves the cursor right one column, except in the rightmost column on the screen where it may cause the view to pan.

CURSOR UP

Moves the cursor up one line. On the top line of the screen it scrolls the screen one line up the current view, if there are lines in the current view above the screen, and the cursor remains in the same position on the screen.

DEFINE MACRO

Used to define a sequence of keys (which are commonly going to be used together) and assign the sequence to a single keystroke. Two presses of DEFINE MACRO are needed to define a key sequence; the required keys (which may not include DEFINE MACRO or CALL MACRO) should be pressed between the two presses of DEFINE MACRO. N.B. the keys are obeyed when defining the macro. The sequence may contain up to 64 keys. Any previously defined macro is forgotten. The defined macro sequence may be invoked using the CALL MACRO key, or may be saved in a fold by using SAVE MACRO.

DELETE

Deletes the character to the left of the cursor. The cursor, the character underneath the cursor and all subsequent characters on the line are moved left by one place.

If the cursor is in the leftmost column of the current enclosing fold DELETE concatenates the current line with the line above. The cursor is placed after the end of the first line.

DELETE in the leftmost column has no effect if the current line is a fold line, top crease or bottom crease, or is a line following the top crease of a view, a fold line or bottom crease.

Spaces may be deleted before a closed fold marker symbol to change the indentation of the fold.

DELETE LINE

Removes the current line from the document, and places it in the delete buffer. Anything already in the delete buffer is deleted. All the lines below the current line in the view are moved up by one line. DELETE LINE has no effect if the current line is a top crease or bottom crease.

RESTORE LINE may be used to restore the delete buffer into the current view.

DELETE RIGHT

Deletes the character under the cursor. All the characters to the right of the cursor are moved left by one place. The cursor remains in the same position.

Character deletion has no effect when the character to be deleted is in the top line of the view, is part of a marker symbol, or is to the left of the leftmost column of an open fold.

Spaces may be deleted before a closed fold marker symbol to change the indentation of the fold.

DELETE TO END OF LINE

Deletes all text from the character under the cursor, to the last significant character on the line, inclusive. The cursor remains in the same position.

DELETE WORD LEFT

Deletes the word to the left of the cursor. The deletion is governed by the following rules:

- A symbol is a non-space non-alphanumeric character, or a sequence of alphanumeric characters. A line contains a sequence of symbols, separated by zero or more spaces. A symbol starting position is the position of the first character in a symbol.

- If the cursor is on or to the left of the first significant (non-space) character on the line, the characters from the cursor position to the current indentation are deleted. The cursor will move to the current indentation.

- If the cursor is to the right of the character following (immediately to the right of) the last significant character on the line, the cursor will move to the character following the last significant character on the line.

- In all other cases the cursor will move to the first symbol starting position to the left of the current cursor position, deleting all intervening characters.

DELETE WORD RIGHT

Deletes the word to the right of the cursor. The deletion is governed by the following rules:

- A symbol is a non-space non-alphanumeric character, or a sequence of alphanumeric characters. A line contains a sequence of symbols, separated by zero or more spaces. A symbol starting position is the position of the first character in a symbol.

- If the cursor is to the left of the first significant (non-space) character on the line, all characters between the cursor the first significant character on the line will be deleted.

- If the cursor is on or between the last symbol starting position on the line, and the last significant character on the line, all characters upto and including the last significant character on the line will be deleted.

- If the cursor is to the right of the last significant character on the line, the cursor will not move.

- In all other cases all characters between the cursor and the first symbol starting position to the right of the current cursor position, will be deleted.

END OF LINE

Places the cursor immediately to the right of the last significant character on the current line (i.e. the last non-blank character). If the line is too long for the width of the screen the view will pan if necessary.

ENTER FOLD

When used on a fold line clears the screen and displays the contents of the fold between the top and bottom creases. The display is adjusted to the left so that the top and bottom marker symbols start in the leftmost column. The cursor is positioned in the leftmost column of the second line on the screen. This then becomes the current view and it is not possible to move outside the confines of the fold until a corresponding EXIT FOLD has been done. When a fold has been entered its top crease may no longer be edited.

ENTER TOOLKIT

Clears the screen, and displays the contents of the toolkit fold.

The top crease of the fold will include the name of the file from which this fold has been read. This may be a shared toolkit file located using the search path **TDSSEARCH** or a local one in the current directory.

The editing functions available while the toolkit fold is being edited are the cursor move operations, screen scrolling operations, fold browsing operations, line deletion/undeletion, line moving and copying, code getting, character insertion and deletion.

While editing the toolkit fold the following message is displayed:

Press [EXIT FOLD] to resume editing

Pressing EXIT FOLD to exit the toolkit fold returns the editor to the position it was at when ENTER TOOLKIT was pressed.

EXIT FOLD

Reverses the effect of the most recent ENTER FOLD, closing the fold, and any open folds contained within it. The closed fold line is positioned on the same line of the screen as it was when the ENTER FOLD was done. The cursor is positioned on the first significant (i.e. the first non-blank) character of the closed fold line. This key is also used to return from the toolkit fold, code information fold, the help display, or parameter and output folds displayed by utilities.

FILE/UNFILE FOLD

When pressed on a fold, converts it to a filed fold, writing the contents out to a file. The file name is taken from the fold header; it is the first contiguous alphanumeric sequence of characters in the header. A comment describing the contents of the file may appear after the file name, separated from it by one or more blanks. If a file of that name already exists in the directory, the file name is adjusted to avoid a name clash (If the fold header is blank, a random name is made up). The message **Filed as** *filename* indicates the file name used. The filename is added to the crease line between quotes.

The name may include directory information. The use of directory specifications in file names should be avoided as much as possible, as it makes it difficult to move groups of files between directories. Relative directory specifications using **..** should be avoided.

The system stores the filename along with other information about the fold. All alphabetic characters are forced to lower case.

FILE/UNFILE FOLD acts as a toggle; when pressed on a filed fold, it reads in the contents of the file and makes it into a (non-filed) fold. The file is deleted.

FINISH

May only be used at the very outermost level. It finishes the session and returns to operating system level.

FOLD INFO

Applied to a closed fold line or to a crease line, this function displays a message on the message line giving the attributes of the fold.

These are:

Type: General format of fold.
Contents: What kind of information is in the fold.

See appendix F for a list of the fold attributes.

In addition, when pressed on a closed filed fold, FOLD INFO displays the name of the file corresponding to the fold.

There is one error message associated with FOLD INFO:

Data item has no attributes

The fold information function has no result when applied to a text line.

GET CODE

Applied to a **UTIL** fold, i.e. a closed fold line whose fold contents attribute is **occam2.util** or **occam1.util**, GET CODE reads the code in the fold, and makes it the current utility package. Used on a **EXE** fold, i.e. a fold whose fold contents attribute is **occam2.exe** or **occam1.exe**, it loads the code for the user program into memory, and makes it the current **EXE**.

Note that GET CODE can be applied either to a fold set which contains a **CODE** fold (**UTIL** or **EXE**) produced as a result of compilation, or it can be applied to the **CODE** fold itself.

While code is being loaded the following message is displayed:

Getting *text on fold comment*

For a utility fold, once the code has been successfully loaded, the utility comment line is displayed.

If the **UTIL** or **EXE** has been compiled with the **tds2.style.exe** parameter FALSE a check will be made that its target processor type or class is compatible with the processor on which the TDS is running. If it is not compatible the following message is displayed:

Code not compatible with host processor

GET MACRO

Applied to a fold whose comment starts **Key macro**, copies the keystroke sequence constructed from a sequence of up to 64 integer values in the fold as the current key macro. The sequence may then be invoked using the CALL MACRO key.

HELP

Displays a map of the system function keys. It also displays a system version identity message. See section 16.2 for more information on the file whose contents are displayed.

LINE DOWN

Moves the screen one line down the current view, if there are lines in the current view below the screen.

This function does not affect the position of the cursor on the screen.

LINE UP

Moves the screen one line up the current view, if there are lines in the current view above the screen.

This function does not affect the position of the cursor on the screen.

MAKE COMMENT

Used to 'comment out' a fold so that it will be ignored by the occam checker and complier. Place the cursor on a fold containing some occam source text which is to be commented out.

MAKE COMMENT produces a fold which encloses the source fold. This new fold is given the fold content attribute of **fs.comment.text**. The fold header is prefixed with the letters **COMMENT**, followed by text copied from the original fold header.

The action of MAKE COMMENT may be reversed using the editor function REMOVE FOLD on the fold produced.

MOVE LINE

Used to move a line, which may be a fold line, to another place in the document. A buffer is associated with MOVE LINE. If the buffer is empty, MOVE LINE removes the current line from the document and puts it in the buffer. If there is a line in the buffer, MOVE LINE removes the line from the buffer, puts it into the document on the line above the current line and places the cursor on it.

The move line buffer is shared between all folded displays including the toolkit fold, the debugger editor and compilation info displays and is preserved between sessions.

NEXT EXE

If the cursor is on a closed top level fold it is entered.
Changes the current EXE to the next member from the set of current EXEs.

NEXT UTIL

If the cursor is on a closed top level fold it is entered.
Changes the current utility set to the next member from the set of current utility sets.

OPEN FOLD

On a fold line opens the fold and inserts the contents of the fold into the current view, surrounded by top and bottom creases. The top crease appears on the line of the screen where the closed fold line was before OPEN FOLD was pressed. The position of the cursor on the screen is unaffected.

PAGE DOWN

Moves the screen one page down the current view, or to the bottom of the current view, whichever is the nearest.

This function does not affect the position of the cursor on the screen.

PAGE UP

Moves the screen one page up the current view, or to the top of the current view, whichever is the nearest.

This function does not affect the position of the cursor on the screen.

PICK LINE

Used to pick up a line, which may be a fold line, so that it may be moved to another place in the document. It removes the current line from the document and appends it to the end of the pick buffer.

The pick buffer is shared between all folded displays including the toolkit fold, the debugger editor and compilation info displays and is preserved between sessions.

PUT

Puts down the contents of the pick buffer at the current position in the document. It inserts a fold line at the current line, containing the sequence of lines placed in the pick buffer using PICK LINE and COPY PICK. The pick buffer is cleared. If there are no lines in the pick buffer PUT has no effect on the document, and the terminal beeps.

REFRESH

Repaints the entire screen from the stored representation of the current view.

REMOVE FOLD

On a fold line opens the fold and removes the top and bottom creases, inserting the contents of the fold into the current view at an appropriate indentation. If it is a filed fold, the file associated with the fold is deleted from the directory.

RESTORE LINE

Restores the last line placed in the delete buffer by DELETE LINE, inserting it at the current position in the document. The delete buffer is left empty.

RETURN

Splits a text line in two at the cursor position and creates a new line on which are placed the cursor, the character underneath the cursor and any subsequent characters on the line. The new line is then indented by inserting spaces until the cursor is in the same column as the first significant character of the line above.

RETURN may be used within the text of a top crease line.

RETURN will insert a blank line above the current line when the cursor is before or on the first significant character of a line.

RETURN will insert a blank line below the current line when the cursor is after the last significant character of a line.

RETURN has no effect on a fold line, top crease, or bottom crease if used between the first and last significant characters of the line.

RUN EXE

Runs the current user program.

If no user program has been loaded the following message appears:

No current EXE

The following message may appear when an attempt is made to run the current user program:

Unable to run code – data requirement too large

This indicates that the memory available is not sufficient for the data space required by the program. It is necessary to remove one or more of the loaded code items using the CLEAR UTIL or CLEAR EXE keys. Further steps that may be taken to increase space are described insection 16.2.

SAVE MACRO

The sequence of keys defined using DEFINE MACRO is saved in a comment fold inserted above the current line. The fold comment **Key macro** is written in this fold and the cursor is positioned so that a name may be added to identify this particular saved macro sequence.

SELECT PARAMETER

Enables a user to toggle utility parameter values quickly. It has an effect only on lines of the form:

```
VAL parameter IS value1 : -- value1 | value2 | value3
VAL parameter IS TRUE :
VAL parameter IS FALSE :
VAL parameter IS "string1" : -- "string1" | "string2"
```

If the cursor is not on the first significant character after the **IS** the cursor will be placed on the first significant character after the **IS**. Otherwise, another possible value of the parameter will be substituted for the current value - either the value selected will be chosen in turn from the set given in the comment or **TRUE** will alternate with **FALSE**.

SET ABORT FLAG

Sets a flag in the TDS to indicate that the user wishes the currently running utility or user program to be aborted. Utilities and user programs which have been written to test the value of this flag (this may be done using the **kernel** channels), will be interrupted.

START OF LINE

if the cursor is at or to the left of the indentation of the current enclosing fold, the cursor will move to the extreme left of the line. If the cursor is on or to the left of the first non-space character on the line it will move to the indentation of the current enclosing fold, otherwise places the cursor on the first significant character of the current line. (i.e. the first non-blank character). If necessary the view will pan.

SUSPEND TDS

Can be used anywhere in the normal editing environment to suspend the TDS temporarily and return the user to the host operating system, so that operating system commands can be issued (for example, getting directory listings, or formatting floppy disks). In DOS typing command **exit** returns to the TDS, in the state it was when SUSPEND TDS was pressed. The environment variable **COMSPEC** must be defined.

Operating system commands which reset the transputer board (for example, running a server with another transputer boot file) will cause the state of the suspended session to be lost. When the session is resumed, the current directory must be the same as it was when the session was suspended.

TOP OF FOLD

Places the cursor on the line displaying the top crease symbol of the current enclosing fold.

WORD LEFT

Moves the cursor one symbol left. The move is governed by the following rules:

- A symbol is a non-space non-alphanumeric character, or a sequence of alphanumeric characters. A line contains a sequence of symbols, separated by zero or more spaces. A symbol starting position is the position of the first character in a symbol.

- If the cursor is on or to the left of the first significant (non-space) character on the line, the cursor will move to the current indentation.

- If the cursor is to the right of the character following (immediately to the right of) the last significant character on the line, the cursor will move to the character following the last significant character on the line.

- In all other cases the cursor will move to the first symbol starting position to the left of the current cursor position.

$\boxed{\text{WORD RIGHT}}$

Moves the cursor one symbol right. The move is governed by the following rules:

- A symbol is a non-space non-alphanumeric character, or a sequence of alphanumeric characters. A line contains a sequence of symbols, separated by zero or more spaces. A symbol starting position is the position of the first character in a symbol.

- If the cursor is to the left of the first significant (non-space) character on the line, the cursor will move to the first significant character on the line.

- If the cursor is on or between the last symbol starting position on the line, and the last significant character on the line, the cursor will move to the character following (immediately to the right of) the last significant character on the line.

- If the cursor is to the right of the last significant character on the line, the cursor will not move.

- In all other cases the cursor will move to the first symbol starting position to the right of the current cursor position.

12.2   Messages

12.2.1   Development environment messages

A complete list of the messages which may be produced by the development environment follows. Some error messages may be followed by a result clause, of the form:

(Result = ** *number*)**

This indicates the result produced by the filing system when the error occurred. If the result is **0**, no filing system error has occurred. The explanations of the error numbers are listed in appendix E.

The result number should be quoted if reporting errors associated with the filing system.

All code cleared

This message is displayed after a successful $\boxed{\text{CLEAR ALL}}$.

Autoloading ...

This message will appear briefly during an $\boxed{\text{AUTOLOAD}}$ operation.

Autoload finished

This message will appear briefly at the end of a successful $\boxed{\text{AUTOLOAD}}$ operation.

Cannot copy : *reason*

The editor cannot copy the line for the reason indicated. Any open filed folds not currently in use should be closed; alternatively some new filed folds may have to be made to increase the amount of room available.

Cannot create : cursor must be below first line

This message may appear on the second press of CREATE FOLD. It indicates that the cursor is on or above the top crease inserted at the start of fold creation, and so the fold cannot be made.

Cannot create : folds must not overlap

This message may appear on the second press of CREATE FOLD. It indicates that the current enclosing fold is not the same as it was at the start of fold creation. Since folds cannot overlap, the fold cannot be made.

Cannot create : lines in fold have incorrect indentation

This message may appear on the second press of CREATE FOLD. It indicates that some lines between the top crease and the current line are less indented (more to the left) than the fold indentation given by the top crease. Since all lines in the fold must be at the same or greater indentation than the fold indentation, the fold cannot be made.

Cannot file fold : file cannot be written

This occurs on FILE FOLD. It indicates one of the following:

1 The editor could not open the file with the name given on the fold header for writing. The most common cause of this is an illegal file name.

2 The editor could open the file with the name given on the fold header for writing, but could not complete writing to it. The most common cause of this is running out of available disk space.

Cannot get : not a valid fold

In order to get a utility package, the cursor should be placed on a utility package fold, which has the fold contents attribute of **utility**. In order to get a user program, the cursor should be placed on an appropriate fold which has the fold contents attribute of **executable**.

Cannot get : failed getting code

An error occurred while reading the file containing the code. This message is followed by a number which is the error result produced by the filing system. The possible error values are listed in appendix E.

Cannot get : not a valid fold

GET CODE can only be applied to a **CODE EXE**, a **CODE UTIL** or to a compiled fold set.

Cannot get : not enough room to load code

There is not enough memory to load this code. It may be necessary to use the CLEAR UTIL or CLEAR EXE functions to re-use the memory currently allocated.

Cannot get : there is no valid fold in this bundle

A **CODE EXE** or **CODE UTIL** has not been found while searching the current fold.

Cannot open : file does not exist

The editor could not open the file associated with a filed fold in order to read it in.

Cannot open : file has incorrect format

The file associated with this filed fold is not in the correct format. The system cannot read ordinary text files; they must be converted first, using the READ HOST utility.

Cannot open : fold is not text

It is not possible to open or enter a fold which contains information in a format other than text.

Cannot open : not enough room

The editor could not read in the file associated with a filed fold because of lack of fold manager space. Any open filed folds not currently in use should be closed; alternatively some new filed folds may have to be made to increase the amount of space available.

This message may also appear if an attempt is made to read in a filed fold which exceeds the maximum fold nesting depth (50).

Cannot open : too many open folds

This message will appear if folds have been opened to give a total nesting depth of 50, or if more than 50 folds have been opened above the current line. Some folds should be closed to allow this one to be opened.

Clearing : *EXE or UTIL name*

This message is displayed while a code item is being cleared.

Closing...

This message is displayed when a fold is being closed. Closing a filed fold may take some time as the file may have to be written out.

Copied into pick buffer OK

This message appears after a successful COPY PICK operation.

Copying...

This message is displayed when a fold is being copied.

Creating fold...

This message is displayed between the first and second presses of CREATE FOLD.

Defining macro. Press [DEFINE MACRO] to resume editing.

This message appears between the two presses of DEFINE MACRO, to indicate that the system is currently storing key presses as part of a macro definition.

Entering...

This message is displayed when a filed fold is being entered. This may take some time as the file may have to be read in.

Entering code information fold...

This message is displayed when the code information fold is being entered. This may take some time as the file may have to be read in.

Entering toolkit fold...

> This message is displayed when the toolkit fold is being entered. This may take some time as the file may have to be read in.

Error : cannot open code file

> An error occurred while opening the file containing the code. This message is followed by a number which is the error result produced by the filing system. The possible error values are listed in appendix E.

Error : cannot read file

> Failure to read file during ENTER FOLD or UNFILE.

Error : cannot write file - fold has been unfiled

> In this close operation, one or more filed folds were not closed successfully (i.e. the contents of the fold were not successfully written out to the file). The fold was closed, but converted into an ordinary fold. Refer to the User guide section 4.5.8.

Error : run out of room - no insertions are allowed...

> This message occurs if the room available to store text has run out. No insertions are allowed until some deletions and file operations have been carried out to make some more room. The message will persist, and then disappear when enough room has been made.

Exiting...

> This message is displayed when a fold is being exited. Exiting a filed fold may take some time as the file may have to be written out.

Filed OK as *name*

> This indicates a successful FILE FOLD operation. The *name* may be different to that expected if a file of that *name* already exists.

Filing...

> This message is displayed during a FILE FOLD operation.

Keystroke macro from fold

> The currently defined keystroke macro has been successfully read from a fold.

Keystroke macro saved

> The currently defined keystroke macro has been saved as a sequence of integers in a new fold.

Line restored OK

> This message indicates a successful RESTORE LINE operation.

No Autoload fold present in parameter buffer

> There is no fold in the toolkit fold whose fold comment starts with the word **Autoload**.

No code to run for current utility

>   The current utility does not contain any code.

No current EXE

>   RUN CODE can only be used when there is a current EXE.

No current UTIL

>   This message appears continuously on the top line if there is no currently loaded utility set. If a utility set is current, its fold header comment is displayed instead.

No EXE code to clear

>   CLEAR EXE has been pressed when there is no current EXE.

No UTIL code to clear

>   CLEAR UTIL has been pressed when there is no current UTIL.

Not permitted on top level folds

>   The operation requested is not allowed at the top level (i.e. where only .TOP files can be seen). For example, utilities and programs may not be run at this level.

Only folds can be COMMENTed out

>   MAKE COMMENT can only be used on fold lines, not on text lines.

Opening...

>   This message is displayed when a filed fold is being opened or entered. This may take some time as the file may have to be read in.

Parameter *parameter.name* **is missing**

>   This message, and the following ones, indicate an error in the parameter fold supplied to utility. This indicates that the parameter identified could not be found in the parameter fold.

Parameter *parameter.name* **is specified more than once**

>   This indicates that the same parameter name occurs more than once in the parameter fold.

Parameter *parameter.name* **is not a valid integer**

>   This indicates that the parameter identified is not an integer, where one was expected.

Parameter *parameter.name* **is not a valid string literal**

>   This indicates that the selection parameter identified is not set to one of the expected selection identifiers.

Parameter line *parameter.name* **has a bad string format**

>   This indicates that the line defining a string parameter was not in the expected format, and the string could not be parsed. Note that to make the characters ", ' or * appear in the string they must be preceded by an asterisk (*).

Parameter line *parameter.name* **is not of form VAL <name> IS <value>:**

This indicates that the line defining the parameter was not in the expected format.

Parameter *parameter.name* **is not set to a valid boolean**

This indicates that the boolean parameter identified is not set to either **TRUE** or **FALSE**. Note that these must be in upper case, and the whole word must be typed.

Press [BROWSE] to finish read only

This message appears continuously on the top line after BROWSE has been pressed, to remind the user that any keys which could alter the document are currently disabled. To get out of this mode, press BROWSE again.

Press [ENTER FOLD] to enter outermost fold

This message appears if OPEN FOLD is used on an outermost level filed fold as a reminder that ENTER FOLD must be used to access this fold.

Press [ENTER FOLD] to start session

This message appears on starting up the system as a reminder that ENTER FOLD must bo uocd to access an outermost fold.

Press [EXIT FOLD] to close and exit the enclosing fold

This message appears if CLOSE FOLD is pressed when the current enclosing fold was opened with an ENTER FOLD, as a reminder that EXIT FOLD must be used.

Press [EXIT FOLD] to continue or [SET ABORT FLAG] [EXIT FOLD] to cancel

When in a parameter fold, indicates how to return parameters to the utility or to abort the utility.

Press [EXIT FOLD] to resume editing

Displayed in the toolkit fold, code information display, etc.

Removing filed fold...

This message is displayed when a REMOVE FOLD operation is done on a filed fold. This may take some time as the file may have to be read in.

Suspend failed (Unable to find last process?)

SUSPEND TDS has failed to re-enter the host operating system, perhaps **COMSPEC** is undefined.

Unfiled OK

This message indicates a successful UNFILE FOLD operation.

Unfiling...

This message is displayed when an UNFILE FOLD is being done. The contents of the file are being read in.

Warning : copying filed fold - repeat to copy

The COPY LINE key or COPY PICK key has been pressed on a filed fold, or a fold line containing a filed fold. Since the copy operation will involve file copying, and may take some time, the key press should be repeated to confirm it.

Warning : deleting filed fold - repeat to delete

The DELETE LINE key has been pressed on a filed fold, or a fold line containing a filed fold. Since the operation will cause files to be deleted, the key press should be repeated to confirm it.

Warning : running out of room

The editor is running out of space. Attempts to copy folds or to open filed folds will probably fail through lack of space. Any open filed folds not currently in use should be closed; alternatively some new filed folds may have to be made to increase the amount of space available.

12.2.2   **iserver** termination messages

The messages tabulated in section 16.4.3 may be produced by the server.

13 Utilities

13.1 occam **program development package**

13.1.1 | CHECK |

Used to check the syntax of occam source.

Parameters

The occam checker shares parameters with the compiler. Only the following parameters are relevant to the checker:

```
error.checking
alias.checking
usage.checking
force.pop.up
target.processor
use.standard.libs
code.inserts
tds2.style.exe
ring.bell
```

Description

Place the cursor on a closed and filed occam source fold or on a compilation fold. The occam source inside this fold will be checked.

Occam 2 compiler/configurer *version*   start of run message.

Checking (*target error.mode*) *fold.name*...   checker running message.

Checked (*target error.mode*) *fold.name* **OK**   end of run message.

If an error is found during checking the checker reports the error in the same manner as | COMPILE |.

The checker provides no further checking than that provided with | COMPILE | using the same parameters.

The checker reports an error if it encounters an uncompiled unit within the fold, or if it encounters a unit compiled with a different version of the compiler. The compilation fold is located to and can then be checked and recompiled.

| CHECK | cannot be used on a **LIB** fold to check library text folds. Such folds within a **LIB** may be individually checked if they are self-contained and are filed first..

Error messages

See the section on | COMPILE |.

13.1.2 | COMPILATION INFO |

Provides a readable version of the information kept in a compiled compilation unit. Normally the information is displayed immediately by the utility. When applied to a **PROGRAM** fold, it provides details of the code loading position on each processor, the boot order of the processors and the inter-processor link connections of a configured network. The information is provided as a fold inserted at the end of the **PROGRAM** fold. The created fold is readable and contains folds nested within it.

Parameters

first.processor.is.boot.from.link Only required when the utility is applied to a **PROGRAM** fold. This parameter, if **TRUE**, causes the load address of the code for the first processor to be calculated assuming that it is booted from link. If **FALSE**, the load address is calculated assuming that the processor is booted from ROM, as described in chapter 10.

Description

Place the cursor on a closed compilation fold: an occam 2 **SC** fold, an **EXE** fold, a **PROGRAM** fold, or a **LIB** fold.

The behaviour of this utility differs depending on whether it is applied to a fold which has been compiled (an occam **SC** or an **EXE** fold) or configured (a **PROGRAM** fold), or is a library (**LIB**) fold.

Compilation information

When COMPILATION INFO is pressed on an occam **SC** or **EXE** fold the utility reads information associated with the compilation fold and displays it. A textual version of the descriptor fold is displayed. If the text is longer than the screen it may be scrolled. When EXIT FOLD is pressed the previous context is restored to the screen and the utility finishes.

The descriptor fold contains information used by the compiler and linker giving details of entry points, code sizes, workspace sizes, compiler identity etc.

The first item is the occam 2 title. This is followed by a warning message if the compilation fold has been modified since it was last compiled.

The next three lines give information about the the compiler used to compile the compilation unit. The target processor is the processor the code was compiled for, for example **T4**. The compiler compatibility is used to determine whether compilation units require recompilation because the compiler used to produce them is in some way incompatible. The compiler version indicates the version string of the compiler which compiled this fold.

This is followed by a list of some of the values of the compiler parameters that were used when compiling this unit.

This is followed by a count of the number of nested separate compilation units present and the number of nested non-occam language programs present. These are followed by the code size in bytes of this compilation unit alone and the total code size in bytes of the nested occam compilation units.

This is followed by a list of entry points in this compilation unit, (with parameters for procedures). For each entry point the usage of channel parameters (whether they are used for input or output within this procedure) is shown. Then there is a list of the entry point offsets in this compilation unit, and workspace and vector space requirement in 'slots' (i.e. machine words) for each procedure.

Then there is a list of the libraries used within this compilation unit, along with their version numbers.

Finally, if the compilation unit has been linked, the total linked code size is given.

Any part of this display may be folded up and saved for later use by means of the CREATE FOLD and MOVE keys.

Configuration information

Applied to a valid **PROGRAM** fold the COMPILATION INFO utility produces a special filed fold marked with the comment **CONFIG INFO**. This fold contains three nested folds of information which can be displayed by the editor.

The first fold contains the memory layout for each processor, to be used when the code is loaded into a network. The format is:

```
{{{   Processor Load Map
...   Processor logical.number processor.type
...   Processor logical.number processor.type
{{{   Processor logical.number processor.type
          Memory layout            first byte   last byte

      Work space                   number       number
      Main program                 number       number
      Real Op                      number       number
      SC string                    number       number
      Libraries                    number       number
      Separate vector workspace    number       number
}}}
...   Processor logical.number processor.type
}}}
```

The second fold contains the order in which processors will be booted when the network is loaded. The format is:

```
processor number from processor number link number
```

The third fold contains a list of the link connections between processors as described in the configuration detail of the **PROGRAM** fold:

```
Connect processor number link number to processor number link number
```

Library information

When pressed on a valid **LIB** fold the current library version number, the fold lines of each **SC** compilation unit and a list of entry points and other libraries used for each is displayed. If the text is longer than the screen it may be scrolled. Compacted libraries contain no nested filed folds and so may be too big to read into the fold manager buffer. If this problem is encountered the fold manager buffer may be increased, see section 16.2.

Messages

Creating config info...   displayed when applied to a valid **PROGRAM** fold set.

Fold created OK   end of run message.

13.1.3 COMPILE

Used to compile occam compilation fold sets; it checks the syntax of the program as part of compilation. When applied to a fold describing the configuration of an occam program for a processor network, it generates the information necessary to control the distribution of processes to processors. When applied to a text fold it searches the fold for uncompiled compilation fold sets and compiles them if necessary. This quick method of compiling many units only works if all must be compiled for the same target and error mode.

Parameters

error.checking This selects the type of run time error checking compiled into an occam program. The options are **REDUCED, STOP, HALT** and **UNIVERSAL. HALT** is the most useful for debugging programs. **HALT** causes the entire processor to halt when an error occurs, **STOP** causes the process in which an error occured to stop, and **REDUCED** has undefined behaviour should an error occur. **UNIVERSAL** is the same as **REDUCED** except that units compiled in this mode may be called from units compiled in other modes.

alias.checking When this parameter is **TRUE**, the compiler does full alias checking.

usage.checking When this parameter, and the **alias checking** parameter are **TRUE**, the compiler does full usage checking.

separate.vector.space When this parameter is **TRUE**, the compiler creates separate workspaces for scalars and vectors within the programs being compiled.

create.debugging.info When this parameter is **TRUE** the compiler will create an additional output file. This file is required by the debugger to recognise occam names and the workspace layouts. The parameter should be set to **TRUE** to obtain occam level debugging.

range.checking Setting this to **FALSE** causes the compiler to omit certain run time checking code (for example, array bounds checking). It has no effect when the **error.checking** parameter is set to **REDUCED**, as no checks at all will be inserted in **REDUCED** mode.

compile.all Normally the compiler only recompiles nested compilation units which have changed, or which are in some way incompatible with the current compilation. This parameter, when **TRUE**, forces the compiler to recompile all nested compilation units encountered.

force.pop.up This parameter forces the parameter fold to be displayed whenever the checker or compiler is invoked.

use.standard.libs When this parameter is **TRUE** the compiler will use its standard arithmetic libraries within this compilation. These are **reals**, **dreals**, **ints** and **realpds** (all targets) and **intpds** (not T8). Setting it to **FALSE** will prevent the compiler from compiling any programs with extended arithmetic, and the compiler will also fail to recognise a number of the implicitly defined library procedures.

target.processor This parameter is used to set the target processor type or class when compiling for transputer networks. The following target processors are supported:

> **T8**: IMS T800 and similar transputers.
>
> **T425**: IMS T425 transputer.
>
> **T4**: IMS T414 transputer.
>
> **T2**: IMS T212 and similar transputers.
>
> **TA**: any 32 bit transputer.
>
> **TB**: any 32 bit transputer without hardware floating point.
>
> **TC**: any 32 bit transputer with 2D block move and CRC instructions.

code.inserts This selects whether transputer assembly-level code insertions are allowed. The options are **NONE**, **RESTRICTED** and **ALL**.

When this parameter is **NONE** it prevents any code insertions in compilation units being compiled. **RESTRICTED** allows a restricted set of instructions which are the instructions which may be used in sequential code, without interfering with the occam process model (see appendix H). **ALL** allows the full set of documented instructions to be used.

tds2.style.exe This parameter determines whether on **EXE** or **UTIL** has the extra **iserver** channel parameters **from.isv** and **to.isv**. Setting it **TRUE** will force **EXE**s to have the interface supported in earlier versions of the TDS. Setting it **FALSE** (default) gives access to the **iserver** channels and implicitly #**USE**s the **strmhdr** library. Both styles of **EXE** can be run within the TDS3 environment.

ring.bell this determines whether a bell code is sent to the screen at the end of a compilation, so the programmer may leave the computer during a long compilation. The options are **NEVER**, **ERROR** and **ALWAYS**.

Description

Place the cursor on a compilation fold (a fold which has been created using $\boxed{\text{MAKE FOLDSET}}$). This may be an occam **SC**, an **EXE**, a **LIB** or a **PROGRAM** fold. Besides the source fold there may be data folds left over from a previous compilation; these will be removed or overwritten when $\boxed{\text{COMPILE}}$ runs.

It is also possible to apply the $\boxed{\text{COMPILE}}$ function to any source fold. The utility will search the fold for any compilation units contained within it, and compile each of them in turn, allowing a collection of libraries and separate compilation units to be compiled as a batch.

Compilation

When $\boxed{\text{COMPILE}}$ is pressed on a compilation unit the contents of the source fold are compiled. The compiler adds code and data folds to the contents of the compilation fold that hold the result of the compilation. Any previous code and data folds are deleted.

When applied to a compilation unit the compiler will always recompile that unit. Normally the compiler will not compile units within the fold which have already been compiled, and so are marked with a fold type of **ft.foldset**. If, within the fold being compiled, the compiler encounters any of the following:

- a compilation unit whose fold type is **ft.voidset**,
- a compilation unit whose used libraries have been recompiled and given a new version number,
- a compilation unit compiled for a different processor type,
- a compilation unit compiled in a different error mode (other than **UNIVERSAL**),
- a compilation unit compiled with an incompatible version of the compiler

then the inner compilation unit is automatically recompiled.

If the **compile.all** option is enabled all nested units are recompiled.

Messages

Searching fold...	start of run message, looking for any nested compilation units.
Compiling (*target error.mode*) *fold.name*...	displayed if the compiler finds an **SC** that needs recompiling. The fold is compiled.
Compiled (*target error.mode*) *fold.name* **OK**	end of run message.

If an error is found then an error message is displayed on the screen and the compiler automatically locates to the line on which the error was discovered. It does this by entering the outer level fold and opening any folds necessary to reach the error line. This error line is displayed as close as possible to the middle of the screen.

Only the first error in the program is found and reported.

A successfully compiled compilation fold is given the fold type **ft.foldset**. A compilation fold is 'valid' after compilation as its contents relate to the current version of the source fold. Any changes made to the contents of the fold will make the fold invalid, requiring it to be recompiled.

fold.name **not compiled (searching...)**

> Displayed if the search for nested units finds a unit that needs recompiling. This unit is also searched for constituent units. When this search has finished the nested unit is compiled. If an error is found the error message is displayed and the error line located to.

Compiled (*target error.mode*) *fold.name* (**searching...**)

> Displayed if the compilation is successful. The search for more uncompiled units continues until all necessary units are recompiled. Only then is the enclosing unit compiled.

fold.name **compiled for wrong target (searching...)**

> Displayed if a unit compiled for a different target processor is met during the search.

Similar messages are produced for compilation units which are compiled for a different error mode, which have been compiled with a different version of the compiler, or whose libraries have changed.

If the compiler encounters a library reference referring to a non-existent or invalid library, either in a **#USE** line or in the descriptor of a compiled unit, it will stop compilation and locate to the position where the error was found.

Once compilation is complete, if COMPILE has been invoked on an **EXE**, the compilation unit is automatically linked, into a **CODE EXE** fold which is marked with the time and date of compilation.

Linking code...   start of link run message.

Code linked OK   end of run message.

If COMPILE is compiling a library, once all **SCs** in the library have been compiled, the library is made valid and given a new version number. The time and date of validation are written on the version fold line.

Configuration

When run on a **PROGRAM** compilation fold, COMPILE generates the information necessary to control the distribution of processes to processors. It checks the syntax of the allowed constructs in a **PROGRAM**, and also checks that channels and links are connected correctly and that all the processors are connected in the network. If any **SCs** in the program are not yet compiled, they are compiled and linked before configuration is done. If the **PROGRAM** contains **SCs** for different processor types, RECOMPILE should be used instead of COMPILE.

When configuring a **PROGRAM** fold, COMPILE performs the following tasks:

- Builds a map of the target system in the descriptor fold.
- Generates code which initialises the workspace to the values of actual parameters corresponding to the formal parameters of the **SC** procedures.
- Checks that the link connections are legal by making certain that all channels are connected to no more than one input and one output link.
- Evaluates all constant expressions.
- Allocates work space for declared variables and channels.
- Checks that the system described is loadable from the root processor.

13.1.4   Compiler messages

The compiler produces a large number of messages indicating program errors. Most of these should be self-explanatory, when taken in conjunction with the language manual. Compiler error messages can be of four forms: invalid use of library logical names, program errors, compiler dependent implementation restrictions and catastrophic compiler failures.

There are two places within a compilation when errors attributable to the library logical name system may be reported. They are on analysis of the library logical name fold when inconsistencies and inadequacies in this fold will be reported, and when processing **#USE** lines, or logical names found in compilation descriptors, when a desired translation may be found to be absent from the table.

13.1.5 Library logical name fold errors

Too many libraries
Library name text buffer overflow

> The capacity of an internal table has been exceeded. These capacities depend on the total free workspace available to the utility. If this cannnot be increased then unused libraries should be removed from the fold and/or the names reduced in length.

Library name not a valid name

> An item starting with an occam identifier character contains a character other than a letter, digit or dot.

Braces not matched

> A { is not matched by a } with no intervening space or ".

File name not terminated by "

> A " is not matched by another " with no intervening space or }.

Library line structure error

> The sequence of items in a line is not *[directory.name] file.name keywords logical names.*

No library directory name

> The present line and no previous lines contain no *directory.name.*

No library file name

> There is no *file.name* on the line.

No stopping mode/target

> A logical name has been found before at least one keyword.

Two libraries conflict

> A previous line defines a translation for the same logical name and at least one mode/target as the current line.

Each of the above messages will be followed in the error line by the text of the line on which the error was detected, and the utility will be aborted.

13.1.6 Program errors

The error messages have the general form of:

> **Error** *number*: *qualifier message*

The error message describes which error the compiler has discovered. The error number and qualifier pinpoint exactly which part of the compiler has produced the message.

13.1.7   Implementation limits

These errors produce a message in one of the following forms:

> **Implementation limit:** *description*
> **Implementation restriction:** *description*

These messages refer to limits (usally fixed buffer sizes) in the compiler. In some cases these limits can be bypassed by restructuring the program. In general the following guidelines to writing programs will reduce the possibility of encountering these limits.

- Declare variables and constants for the smallest amount of program that is possible. Variables and constants that are left in scope unnecessarily waste buffer space in the compiler.

- Write many small procedures rather than large pieces of in-line code.

- Nest procedure declarations wherever possible.

- Use **SC** procedures when practicable.

- Split complex expressions by use of abbreviations or temporary variables.

All implementation limits can be avoided by structuring a program so that it can be written as a number of small separately compilable modules.

If the TDS is running on a 1 Mbyte transputer board, and more than one utility set is loaded, the compiler will be unable to compile programs of any reasonable size without hitting one of these limits. On such boards only one utility set should be loaded at any one time.

13.1.8   Compiler errors

These errors produce the following message:

> **COMPILER ERROR - PLEASE REPORT**

Any program that generates an error message of this form should be reported to INMOS.

13.1.9   Configurer error messages

This section describes the error messages which may appear when configuring a **PROGRAM** fold.

Note that in the following messages, when arrays of channels greater than one dimension are declared at configuration level, the configurer has a problem when it tries to report an error involving a particular element of the array. The problem is that the array has been flattened into a one dimensional array by multiplying the appropriate subscripts by the sizes of the dimensions of the array.

e.g. if *c* is declared as **[3] [5] CHAN OF ANY c :** and the user makes an error involving **c [2] [4]** then the configurer will report that channel *c* subscript 14 has been illegally used.

Base value for placed par replicator not of type INT

> The user has written something like **PLACED PAR i = 2.0 (REAL32) FOR 10.**

c is a configuration channel and cannot be placed IN anything

> The user has attempted to **PLACE c IN VECSPACE :** (or **WORKSPACE**)

Cannot place a 2 dimensional array of configuration channels

The user has written something like **PLACE c AT 0 :** where **c** is a 2 dimensional array of configuration level channels. Only single elements of configuration level channels may be placed.

Cannot place an expression

The user has written something like **PLACE x + y AT 10 :**

Cannot place configuration channel at address *n*

Configuration channels must be placed on hardware links (addresses 0 to 7 for T2, T4 and T8)

Channel *c*, **subscript** *n* **has already been placed**

User has attempted to place the same subscript of the same channel array twice on one processor.

Channel *c1* **has already been placed**

User has attempted to place the same channel twice on one processor.

Config channel parameter *n* **is placed for input, not output**

The channel parameter *n* being passed to the procedure is used to pass messages in the opposite direction to that allowed by the hardware.

Config channel parameter *n* **is placed for output, not input**

The channel parameter being passed to the procedure is used to pass messages in the opposite direction to that allowed by the hardware.

Configuration channel must have a constant subscript

The user has written something like **PLACE c [x] AT 0 :** where x is not a constant expression.

Configuration level SCs may not contain functions

A configuration level SC must contain a procedure.

Configuration level SCs may only contain one proc.

A configuration level SC may only have one entry point.

Count value for placed par replicator not of type INT

The user has written something like **PLACED PAR i = 2 FOR 10.0**.

Libraries at configuration level may not contain SCs

Libraries at configuration level may only contain **VAL** definitions and **PROTOCOLs**.

Link pair a/b on processor *n* **is connected to processors** *n1* **and** *n2*

The input and input parts of one hardware link have been connected to different processors. This is not possible.

No load path from root processor to processor *n*

There is no path connecting the root processor to processor *n*.

Only VAL abbreviations / retypes allowed at configuration level

The user has written something like **x IS y :** at configuration level.

Only whole arrays can be placed, not just part of them

The user has attempted to **PLACE [x FROM 2 FOR 3] AT 10** : or something similar.

Parallel inputs on channel c

Usage error of channel c. Two processes are inputting from that channel.

Parallel inputs on channel c, **subscript** n

Usage error of channel array element. Two processes are inputting from that channel.

Parallel outputs on channel c

Usage error of channel c. Two processes are outputting from that channel.

Parallel outputs on channel c, **subscript** n

Usage error of channel array element. Two processes are outputting to that channel.

PLACE address is not constant

The user has written something like **PLACE** c **[2] At x** : where x is not a constant expression.

PLACE address is not of type INT

The user has written something like **PLACE** c **[2] AT 2.0** :

PLACED PAR replicator must have a constant base value

The user has written something like **PLACED PAR i = x FOR 10** where x is not a constant.

PLACED PAR replicator counts must be > 0

The user has written something like **PLACED PAR i = 2 FOR -2**.

PLACED PAR replicator must have a constant value

The user has written something like **PLACED PAR i = 10 FOR x** where x is not a constant.

Processor n1, **link** n2 **is connected to itself**

The output of the link has been wired back on itself.

Processor number cannot be evaluated

The user has written something like PROCESSOR x T4 where x is not a constant.

Processor number must be of type INT

The user has written something like **PROCESSOR 2.0 T4**.

Processor number n **has been used before**

The user has specified the same logical processor number twice in one program.

Processors n1 **and** n2 **may not be connected by a channel with INT in its protocol**

Processors n1 and n2 have different word length and so cannot exchange **INTs** properly on a link between them.

program too big to configure

Implementation limit.

SC PROC *name* **is not compiled for processor type** *processor.type*

The SC is compiled for a different target.

Statement not allowed at configuration level

User has attempted to use **IF, WHILE,** etc. at configuration level.

There is already a channel placed at address *n*

The user has tried to place two different channels at the same address on one processor.

Variables of type *type* **may not be declared at configuration level**

Only channels may be declared at configuration level.

13.1.10 | EXTRACT |

Extracts all the code from a fold set and puts it into a single additional fold placed as a new last item in the fold set. If the fold set is a **PROGRAM** fold, the created fold contains all the necessary routing and bootstrap information for loading the target network of transputers. If the fold set is an **SC** fold, the created fold contains all of the linked code from that fold set as a single contiguous structure.

When applied to a valid **PROGRAM** fold the | EXTRACT | utility is used to produce a **CODE PROGRAM** fold which may be loaded to a transputer network or included as part of an EPROM. In general loading a network from a previously extracted **CODE PROGRAM** fold is quicker than loading from a **PROGRAM** fold which has not been extracted. The time and date of extraction are written on the **CODE PROGRAM** fold line.

When applied to a valid **SC** fold the | EXTRACT | produces a linked **CODE SC** fold. The fold produced can be included as part of an EPROM. Note that extraction is performed automatically on any enclosed **SCs** when a **PROGRAM** is configured.

Parameters — **PROGRAM** fold

output.fold This parameter determines whether the fold produced by the **EXTRACT** utility should contain load time diagnostic information or not. The option **BOOTABLE** will cause the utility to produce a fold which contains only data which is needed to load a transputer network. The option **DIAGNOSTIC** will cause the utility to produce a fold which contains additional information to keep track of the progress of a load. A **DIAGNOSTIC** fold requires the tool which is going to be used to send the contents to the network to understand the format of the fold and keep track of the progress of the load. A **BOOTABLE** fold can be sent directly to a network with no interpretation.

first.processor.is.boot.from.link This parameter, if **TRUE**, causes the load address of the code for the first processor to be calculated assuming that it is booted from link. If **FALSE**, the load address is calculated assuming that the processor is booted from ROM.

Parameters — **SC** fold

None

Description

Place the cursor on a valid compilation fold set.

Applied to a valid **PROGRAM** fold, the | EXTRACT | utility produces a fold containing all of the routing information, the bootstraps and the code necessary to load the network described by the **PROGRAM**. The fold produced may also contain embedded messages, which can be used to keep track of the progress of a future load.

While the fold is being produced various messages are displayed. The fold produced is marked with the fold header string taken from the **PROGRAM** fold, prefixed with the string **CODE.**   If the TDS is hosted by a version of **iserver** which can get the date and time from the host operating system, this will be added to the fold header.

Extracting network...   start of run message.

Extracting *SC.string*   displayed as a particular **SC** is extracted.The string is taken from the fold header of the **SC** being extracted.

Extracting main bodies   displayed after all the **SCs** have been extracted.

Network extracted OK   end of run message.

Applied to a valid **SC** fold, the EXTRACT utility produces a fold containing all of the code from the **SC** and any libraries referenced. The fold produced is marked with the fold header string taken from the **SC** fold, prefixed with the string **CODE.**   the time and date of extraction are added to the end of the header.

The initial message is

> **Extracting** *sc.string*

Where *sc.string* is taken from the fold header of the **SC** being extracted. When extraction is successfully completed the message is changed to:

> *sc.string* **extracted OK**

Error messages — PROGRAM fold

Error messages generated by the EXTRACT utility fall into two groups; errors during extraction of the code from the fold and errors which are generated by filing the extracted **CODE PROGRAM** fold.

Messages in the first group have the form:

> **Extraction error :**   *error.message*

Messages in the second group have the form:

> **Filing error** *error.number : error.message*

The extraction error messages are:

Failure while entering fold

> The extractor has been unable to enter a fold. This may often be overcome by making more of the folds in the current environment into filed folds, and thereby, increasing the space available for entering new folds.

Failure while exiting fold

> This error is unlikely to occur. It normally signifies a problem with writing to the disk.

Incorrect compiler identity string

> The compiler identity of the compiled code does not match the identity of the current utility set. The item referenced should be recompiled with a compatible compiler.

Load path maximum exceeded

> This is an implementation limit. The load path necessary to enable code to reach a specific processor exceeds the buffer size allocated. A processor network structure which branches does not require as long a load path as a linear structure.

Not a compiled PROGRAM fold

> The cursor is not currently placed on a **PROGRAM** fold which has the attribute **fold set**.

Saved code buffer overflow

> This is an implementation limit. It indicates that the portion of procedure code held back for extraction to the network with the main bodies has overflowed the buffer available for it. This message is very unlikely to occur.

Stack overflow

> This is an implementation limit. The stack is used during production of the load path. A processor network structure which branches does not require as large a stack as a linear structure.

Too many PROCESSORs

> This is an implementation limit.

Error messages — SC fold

Error messages generated by the $\boxed{\text{EXTRACT}}$ utility when applied to an **SC** fold are similar to the filing error messages produced for a **PROGRAM** fold.

13.1.11 $\boxed{\text{LIST FOLD}}$

This is the listing utility.

Parameters

DestinationFileName This parameter is the name of the DOS filename to which the listing will be sent. It defaults to '**PRN**' for an on-line printer. This default should be changed if this printer does not exist as attempts to use it may then result in DOS failure.

Topcrease This string parameter determines the representation of top crease lines in the output. This string will be followed by the fold header comment, if any. The default is '**{ { {**   '.

Bottomcrease This string parameter determines the representation of bottom crease lines in the output. The default is '**} } }**'.

Closedfold This string parameter determines the representation of closed folds in the output. This string will be followed by the fold header comment, if any. The default is '**. . .**   '.

Endofcreaseline This string parameter, which defaults to an empty string, determines a string to be added to the end of each fold line or crease printed.

Description

Place the cursor on a fold line or a top crease line. If the fold is closed, all lines and text folds within the fold will be sent to the printer or a listing file. If the fold is open only the lines within the fold which are accessible by cursor move or scroll operations are listed. Any closed fold encountered is not entered but is listed as a closed fold. In this way a listing may be constructed where lower levels of detail are hidden away in folds in exactly the same way as in the current view on the screen.

13.1.12   LOAD NETWORK

Used to send code to a transputer network.

Parameters

link This parameter selects the output link from the host transputer through which the network is to be loaded. All transputers in the network must be set to boot from link.

first.processor.is.boot.from.link This parameter, if **TRUE**, causes the load address of the code for the first processor to be calculated assuming that it is booted from link. If **FALSE**, the load address is calculated assuming that the processor is booted from ROM.

output.or.fail This parameter, if **TRUE**, causes all output by the host to the network to be done using the extraordinary link handling library. This means that the loader is able to recover if it at any point it is unable to output to the network when attempting to load it. The link is reset before use, enabling multiple attempts to load. This parameter, if **FALSE**, causes the loader to use normal channel output instructions. If the loading fails then the loader and the TDS will deadlock and need to be rebooted. The reason for using **FALSE** is that it will not time-out if communication is slowed down significantly. For example, communication might be slowed down by inserting an extra processor into the network to display on a terminal all bytes that pass through it while the network is being loaded for debugging purposes.

host.subsystem This parameter may be set to **B004.8** (for either IMS B004 or TRAM) or **B002** and causes the loader to use the reset subsystem hardware at the appropriate address. This enables the loader to reset the network before loading it.

Description

Place the cursor on a configured **PROGRAM** fold, or on a **CODE PROGRAM** fold produced by EXTRACT.

The LOAD NETWORK utility exports code to the transputer network. It will link and extract that code from the fold set only if it is necessary as described below.

Applied to a PROGRAM fold

Applied to a valid **PROGRAM** fold containing no **CODE PROGRAM**, the LOAD NETWORK utility extracts all the allocated code and exports it along with all the necessary routing information to the transputer network. A map is built of the target network from the configuration information in the fold. Using this map the bootstraps for all the processors in the network are exported so that each processor is ready to receive loading information. After this the utility traverses the source fold structure extracting all the procedure code and exporting it from the selected line along with the necessary routing information to direct it to the correct place on the target processors. Finally all the main program code is exported to the line in the reverse order to the order in which the processors were booted.

Extracting and loading network...	start of run message.
Extracting *sc.string*	displayed when a particular **SC** is being extracted, where *sc.string* is taken from the fold header of the **SC** being extracted.
Extracting main bodies	displayed after all the **SC**s have been extracted.
Network extracted and loaded OK	end of run message.

Applied to a CODE PROGRAM fold

Applied to a **CODE PROGRAM** fold, or a **PROGRAM** fold which contains a **CODE PROGRAM** fold the utility LOAD NETWORK exports the contents to the transputer network. All of the bootstraps and routing information are already contained within the **CODE PROGRAM** fold. If the **CODE PROGRAM** fold has been produced as

BOOTABLE by the EXTRACT utility, the LOAD NETWORK utility simply copies the fold to the selected output line. If the **CODE PROGRAM** fold has been produced as **DIAGNOSTIC** by the EXTRACT utility, the LOAD NETWORK utility interprets the contents of the fold, displaying messages on the screen and sending the actual loading data to the selected output line.

Loading network...   start of run message.
                     If the **CODE PROGRAM** fold has been produced as **DIAGNOSTIC**, this message is replaced by the messages which were displayed at the time the **CODE PROGRAM** fold was produced by the EXTRACT utility. If the **CODE PROGRAM** fold has been produced as **BOOTABLE**, this message will not be replaced.

Network loaded OK   end of run message.

Error messages

Error messages generated by the LOAD NETWORK utility fall into three groups; errors during extraction of the code from the fold, filing errors during extraction of the code from the **CODE PROGRAM** fold and errors with communication to the target transputer network. Messages in the first group have the form:

Extraction error : *error.message*

Messages in the second group have the form:

Filing error *error.number : error.message*

Messages in the third group have the form:

Communication error : *error.message*

The extraction error messages are similar to the extraction error messages described for the EXTRACT utility.

The filing error messages report failures received from the TDS filer. The *error.number* will correspond to the error numbers described in appendix E.

The communication error messages are:

Failed to output boot code for processor *number*
Failed to output boot terminator for processor *number*
Failed to output main body code for processor *number*
Failed to output *sc.string*
Failed to output saved SC code
Failed to output terminating null message

These messages indicate that an output to the loading link has failed. This will occur for bootstraps if the links are not physically connected correctly, the processor is not ready to boot from link, or some other hardware fault has occurred.

As the code for any processor may pass through a large number of intermediate processors, the failing processor can not be identified exactly in all cases. The bootstrap code consists of eight packets and, as there will be a single code packet on each intermediate processor, the identity of the actual failing processor can often be determined.

13.1.13 MAKE FOLDSET

Used to create a compilation fold around the current fold.

Parameters

This utility uses the parameters fold **make foldset parameters** which has the single parameter **make.foldset.type**. This parameter can be set to one of the following:

SC, **EXE**, **UTIL**, **PROGRAM**, or **LIB** which creates the corresponding compilation fold.

Description

Place the cursor on a filed fold containing some occam source text which is to become a compilation unit.

Pressing MAKE FOLDSET produces a compilation fold which encloses the source fold. This new fold is given an attribute to indicate that it contains an uncompiled occam program. The fold header is prefixed with the type of the compilation unit (**SC**, **EXE**, **UTIL**, **LIB** or **PROGRAM**) followed by text copied from the source fold header. This text may be edited.

The action of MAKE FOLDSET may be reversed using the editor function REMOVE FOLD on the fold produced.

Error messages

Cannot make a fold round this item

The cursor must be on a closed fold line.

Error - fold is not empty

To make a library, the utility must be applied to an empty fold.

Error - library cannot be filed

To make a library, the utility must be applied to an empty (non-filed) fold.

13.1.14 RECOMPILE

RECOMPILE is used to recompile a fold structure containing compilation units which have already been compiled. It must be used to compile libraries, **PROGRAM** folds, or text folds containing **SC**s for mixed processor types or error modes.

Parameters

The recompilation function uses **compile.all**, **ring.bell** and **force.pop.up** parameters from the compiler parameter fold (other parameters may be displayed but are not used); for a **PROGRAM** fold it also uses other compiler parameters.

Description

Place the cursor on a compilation fold (a fold which has been created using the MAKE FOLDSET utility). This may be an occam **SC**, an **EXE**,a **LIB** or a **PROGRAM** fold. Besides the source folds within this fold structure there must also be descriptor folds left over from a previous compilation.

It is also possible to apply the RECOMPILE function to any source fold. The utility will search the fold for any compilation units contained within it, and recompile each of them in turn, allowing a collection of libraries and programs to be compiled as a batch.

The RECOMPILE function behaves like COMPILE, but for each unit compiled it takes the compiler parameters from the descriptor left by the previous compilation. Unlike COMPILE, which recompiles any compilation unit it finds which has been compiled for a different target or error mode than the current set of parameters (see 14.5.3), RECOMPILE may be applied to a fold structure containing compilation units compiled for different targets or in different error modes. RECOMPILE will only recompile units with uncompiled foldsets or changed libraries, unless **compile.all** is **TRUE**, in which case it will recompile everything. If any inner compilation fold encountered is found not to have a descriptor fold, the compilation halts, and the compiler locates to the compilation fold in error.

Error messages

The same messages as for COMPILE are generated, plus

No descriptor fold present in foldset

> This message indicates that there was no descriptor fold in the foldset for the compilation unit indicated, so RECOMPILE could not proceed.

13.1.15 | REPLACE |

REPLACE replaces one string with another, once or repeatedly to the end of the current view.

Parameters

REPLACE shares a parameter fold with SEARCH. It contains the following parameters:

search.string This is a string parameter, and is the string to be matched. It may include spaces, as it is delimited by double quote marks ("). occam rules for strings should be followed. In particular, to make the characters quote (') , double quote (") or asterisk (*) appear in the string they should be preceded by an asterisk.

replace.string This string parameter is the string used to replace the search string if a match is found.

case.sensitive This boolean parameter determines whether the string matching is case-sensitive. If it is **TRUE**, an exact match for the string, with all letters in the same case as the search string, must be found for the match to succeed. If it is **FALSE**, a string which differs from the search string only in the case of one or more letters will match. The default is **TRUE**.

forward.replace This boolean parameter determines where the cursor ends up after the matched string has been replaced. If **forward.replace** is **TRUE**, the cursor is moved forward to the character following the new replaced string. If **forward.replace** is **FALSE**, the cursor remains in the same column, on the first character of the new string. The default is **TRUE**.

global.replace This boolean parameter determines whether only the current occurrence of the search string is replaced by the replace string or all subsequent ones in the current view. If it is **FALSE**, the cursor must be on a string matching the search string and this will be replaced. If it is **TRUE**, the cursor may, or may not, be on a matching string and all subsequent matching strings in the current view will be replaced.

Description

Place the cursor on the first character of the string to be replaced or anywhere for a global replace. If SEARCH has been pressed this will have already been done. **N.B.** REPLACE does not do any searching if **global.replace** is **FALSE**.

REPLACE may be used on its own, with an empty search string, to insert a string of text at multiple positions in the document.

REPLACE may be used anywhere it is legal to insert text using the editor, except for inserting spaces to the left of a fold line marker.

When $\boxed{\text{REPLACE}}$ is pressed the utility attempts to match the string at the current position (starting with the character under the cursor) with the search string.

Cannot replace: no match for search string no match found.

Replaced OK   If a match is found the characters of the matching string are replaced with the replace string. If **forward.replace** is **TRUE**, the cursor is moved forward to the character following the new replaced string. This allows searching to continue to the next occurrence without examining the newly inserted string. If **forward. replace** is **FALSE**, the cursor remains in the same column, on the first character of the new string. This is useful if using $\boxed{\text{REPLACE}}$ in conjunction with cursor keys.

Error messages

Cannot replace : on invalid item

The replace utility may not be used to insert text to the left of the leftmost column of a fold, or on a fold or crease marker symbol. Since text may not be inserted on a bottom crease, line replacement may not be done anywhere on a bottom crease line.

13.1.16 $\boxed{\text{SEARCH}}$

$\boxed{\text{SEARCH}}$ is the string searching utility. It can search anywhere in the current view for an exact match of a string provided as a parameter.

Parameters

$\boxed{\text{SEARCH}}$ shares a parameter fold with $\boxed{\text{REPLACE}}$. It contains the following parameters:

search.string This is a string parameter, and is the string to be searched for. It may include spaces, as it is delimited by double quote marks (**"**). occam rules for strings are followed. In particular, to make the characters quote (**'**), double quote (**"**) or asterisk (*****) appear in the string they should be preceded by an asterisk.

replace.string This string parameter may be supplied when $\boxed{\text{SEARCH}}$ is pressed, but it is not used by $\boxed{\text{SEARCH}}$. See the section on $\boxed{\text{REPLACE}}$.

case.sensitive This boolean parameter determines whether the string matching is case-sensitive. If it is **TRUE**, an exact match for the string, with all letters in the same case as the search string, must be found for the search to succeed. If it is **FALSE**, a string which differs from the search string only in the case of one or more letters will match. The default is **TRUE**.

forward.replace This boolean parameter may be supplied when $\boxed{\text{SEARCH}}$ is pressed, but it is not used by $\boxed{\text{SEARCH}}$. See the section on $\boxed{\text{REPLACE}}$.

forward.search This boolean parameter determines the direction of search. If it is **TRUE** the search proceeds forwards (downwards) in the fold structure, entering the fold if the cursor starts on a fold line. If it is **FALSE** the search proceeds backwards (upwards) in the fold structure. In this case the search does not enter the fold if the cursor starts on a fold line.

global.replace This boolean parameter may be supplied when $\boxed{\text{SEARCH}}$ is pressed but is not used by $\boxed{\text{SEARCH}}$. See the entry for $\boxed{\text{REPLACE}}$.

Description

Traverse the fold structure using $\boxed{\text{OPEN FOLD}}$ and $\boxed{\text{ENTER FOLD}}$ until the current view (i.e. the contents of the last entered fold) is the context within which a search is to be done. Place the cursor anywhere on the screen before pressing $\boxed{\text{SEARCH}}$. It will search from the current line forwards (or backwards), including the contents of any folds and their creases.

The searcher looks for a match with the search string, starting with the character following (preceding) the character under the cursor.

Searching *up/down* **for** "*string*"... is displayed.

The searcher will search from the cursor position up/down to and including the first/last line in the current view, or until a match is found. Fold header strings and text lines are examined for a match with the search string. Fold and crease markers, spaces inserted to the left of the leftmost column of a fold, and spaces to the right of the rightmost significant character on a line are not matched.

All folds and filed folds which may be opened by the editor are opened and searched. Data and code folds are not opened.

"*string*" **not found**   is displayed if no match is found.
                                             (The editor position remains as it
                                             was before the searcher was run)

"*string1*" **found ; replace with** "*string2*"   is displayed if a match is found.

The found string is located. If the string is on the screen the cursor is moved to the first character of the string. If a global replacement has been requested the word **all** will appear in this message after **replace**.

If the string is not on the screen, the current position is moved from the position at which the searcher was invoked (old position) to the position at which the string was found (new position). Any folds needed to reach the new position are opened, and the line containing the string is placed as close as possible to the centre of the screen. Any folds which contain the old position but which do not contain the new position are closed. The cursor is placed on the first character of the found string. The view will be panned if necessary.

Error messages

No search string

An empty search string has been supplied and as a result no searching can be done.

The following error message may be generated by ⌑SEARCH⌑, due to failure to read a filed fold:

Failed to open fold

When one of these errors occurs, the current position remains as it was when the searcher was started.

This indicates one of the following conditions:

1 A filed fold was encountered, for which there was not enough room to read in the contents of the file (possibly because other filed folds were still open elsewhere). This is quite a common occurrence, since while a search is being done the filed folds containing the position at which the search was started and the filed folds containing the position which the search has reached are both open.

   If this occurs, it is necessary to move out a few levels to close some surrounding filed folds, and start the search again on a filed fold further down the document.

2 A filed fold was encountered which could not be opened. This indicates that the file could not be opened for reading, and should be treated as a system error, and the most likely cause is that the file does not exist.

3 A filed fold was encountered for which the file does exist, but appears to be in an incorrect format for reading by the system. This should be treated as a system error.

13.2   File handling package

13.2.1   ATTACH

Attaches a file to a fold. It may be used to attach to any type of TDS supported file, such as text, executable code (**UTIL** or **EXE**) and so on but not **.TOP** files (see appendix F.3). The attributes (i.e. type of fold) are determined by the file name extension. For example, a **.tsr** extension will mean that the file contains text, while a **.cex** extension indicates that the content is executable code.

Many folds may be attached to the same file, allowing the contents to be shared with other locations in the fold structure. If a file is shared between different locations it can prevent the separate compilation version control mechanism from functioning correctly. Also, by deleting one attached filed fold the contents of all such attached files will be lost. It is strongly recomended to use DOS file protection facilities to prevent the deletion of an attatched filed fold from inadvertently deleting the underlying file. If it is required to delete an attached fold use DETACH, which removes an attached filed fold, without deleting the file.

DETACH is on the same key; the two utilities are toggled.

Parameters

None.

Description

ATTACH should be invoked on an empty fold. The name of the file to be attached to the fold should be the first word on the fold line. The filename may include a directory specification.

When ATTACH is pressed, the TDS reads the file name from the fold line and checks to see if the file exists and if it does the file is attached to the fold, making its contents accessible.

Attaching file...   start of run message.

Attached file OK   end of run message.

Error messages

Cannot attach file: must be on a fold

> The cursor must be on a (non-filed) fold when ATTACH is invoked.

Cannot attach file: fold is not empty

> A fold cannot be attached to a file unless the fold is empty.

Cannot attach file: file does not exist

> A fold cannot be attached to a non-existent file.

Error attaching file (Result = *n*)

> Where *n* is the filing system error code. The system failed to attach a file to the fold.
> See appendix E for a list of the error numbers.

13.2.2   COMPACT LIBRARIES

Copies the contents of a filed fold, including nested files, to another directory. Any library folds encountered are compacted, that is, all information in the library is written into a single file. A parameter allows source to be removed from the library as it is compacted. The name of a file being written is normally the same as that of the file being read.

Parameters

DestinationFileName gives the full name of the file to be written, including all necessary directory specifications needed to locate it.

DeleteSource is a boolean parameter. If it is set to **TRUE** then source text is removed from occam SC foldsets in the compacted libraries, and certain information is removed from the debug fold. If set to **FALSE** then source text is copied across. The default is **TRUE**. (This parameter has no effect on the original fold to which the utility has been applied).

OverwriteFiles determines what happens when a file name clash occurs in the destination directory. If **OverwriteFiles** is **FALSE** then when a name clash occurs the name of the file being written is modified to make it unique. If the value is **TRUE**, then the old file is overwritten with the new file. The default is **FALSE**.

The parameters are always offered for editing when this utility is run.

Description

Place the cursor on the fold to be copied and compacted.

When ⌴COMPACT LIBRARIES⌴ is pressed the fold under the cursor and all nested files are copied into the destination file. Nested files are copied to new files nested within the destination file. Libraries encountered during the copying are compacted, and the source text is removed (depending on the value of the **DeleteSource** parameter).

Apart from its behaviour with libraries, ⌴COMPACT LIBRARIES⌴ behaves like ⌴COPY OUT⌴

`Copying out and compacting libraries...`   start of run message.

`Copying out:  "string"...`   shows fold header of file being copied.

`Copied from this directory OK`   end of run message.

Error messages

See the section on ⌴COPY OUT⌴.

13.2.3 ⌴COPY ATTACH⌴

Performs a similar action to ⌴ATTACH⌴, but before attaching a file a complete copy of the file and all nested files is made. The copy is attached to the fold under the cursor. File name clashes which occur as a result of the copy are prevented, by making up new names for files derived from the file names in the original.

Parameters

None.

Description

⌴COPY ATTACH⌴ should be invoked on an empty fold of a type that can be opened by the editor. The name of the file to be copied and attached to the fold should be the first word on the fold line.

When ⌴COPY ATTACH⌴ is pressed, the TDS reads the file name from the fold line and checks to see if the file exists and if it does the file and all nested filed folds are copied and the copy is attached to the fold, so that opening the (now) filed fold gives access to the copied file.

Attaching and copying file...   start of run message.

Attached file OK   end of run message.

Error messages

The error messages produced by COPY ATTACH are the same as for ATTACH.

13.2.4   COPY IN

Copies a TDS format file and any nested files from another directory to a fold (in the current directory). All nested folds may be copied or only text folds. It can be used to copy from a floppy disk into the TDS. The name of a file being written is the same as that of the file being read, except where a file name clash would occur. If a name clash occurs the name of the file being written is modified to make it unique. If a file being copied contains a filed fold whose file is located in a directory other than that specified by the COPY IN parameter (i.e. the file name has a directory name prefix) then only the filed fold is copied, not its contents.

Parameters

SourceFileName gives the full name of the file to be copied, including all necessary directory specifications needed to locate it.

TextOnly is a boolean parameter. If set to **TRUE** then only text folds are copied. If set to **FALSE** then text, data and code folds are copied (all folds are copied). The default is **FALSE**.

The parameters are always offered for editing when this utility is run.

Description

The cursor should be pointing at an empty fold, or an empty filed fold.

When COPY IN is pressed the named source file and all nested files are copied into the current fold.

Copying to this directory...   start of run message.

Copying in: "*string*"...   shows fold header of file being copied.

Copied in OK   end of run message.

Error messages

Copy aborted by user

> The copying operation has been aborted by using SET ABORT FLAG.

Copy in failed (Result = *n*) in "*string*"

> The copy operation has failed while it was being carried out.
> *n* is the filing system error result (if any), see appendix E.
> *string* is the fold header of the file being read when the error occurred.

Cannot copy in: *error message* (Result = *n*)

> where *n* is a filing system error code (see appendix E), and *error message* is one of the following messages:

cannot open destination file

> The system cannot open the file that is to be written.

cannot open source file

> The system cannot open the file that is to be read.

filed fold must be empty

> The filed fold that the copy is to be made to must be empty.

must be on a text fold

> The utility was invoked either on a text line or on a fold of a type that cannot be read by the editor.

cannot create filed fold

> The utility was invoked on a non-filed fold and the system could not file it.

file name not given

> No file name has been supplied in the parameters to the utility.

directory name not given

> The file name supplied to the utility does not include a directory path.

13.2.5 COPY OUT

Copies the contents of a fold, including nested files, to another directory. All nested folds may be copied or only text folds. It can be used to copy to a floppy disk from within the TDS. The name of a file being written is the same as that of the file being read, except where a file name clash would occur. If a name clash occurs the name of the file being written is modified to make it unique. If a file being copied contains a filed fold whose file is located in a directory other than that specified by the COPY OUT parameter (i.e. the file name has a directory name prefix) then only the filed fold is copied, not its contents.

Parameters

DestinationFileName gives the full name of the file to be written, including all necessary directory specifications needed to locate it.

TextOnly is a boolean parameter. If set to **TRUE** then only text folds are copied. If set to **FALSE** then text, data and code folds are copied (all folds are copied). The default is **FALSE**.

The parameters are always offered for editing when this utility is run.

Description

The cursor should be pointing at the fold to be copied.

When COPY OUT is pressed the filed fold under the cursor and all nested files are copied into the destination file. Nested files are copied to new files nested within the destination file.

Copying from this directory...   start of run message

Copying out: "*string*"...   shows fold header of file being copied

Copied out to "*filename*" OK   end of run message.

Error messages

Copy aborted by user

>   The copying operation has been aborted by using $\boxed{\text{SET ABORT FLAG}}$.

Copy out failed (Result = _n_) in "_string_**"**

>   The copy operation has failed while it was being carried out.
>   _n_ is the filing system error result (if any), see appendix E.
>   _string_ is the fold header of the file being read when the error occurred.

Cannot copy out: _error message_ **(Result = _n_)**

>   where _n_ is a filing system error code and _error message_ is one of the following messages:

cannot open destination file

>   The system cannot open the file that is to be written.

cannot open source file

>   The system cannot open the file that is to be read.

must be on a text fold

>   The utility was invoked either on a text line or on a fold of a type that cannot be read by the editor.

cannot create filed fold

>   The utility was invoked on a non-filed fold and the system could not file it.

file name not given

>   No file name has been supplied in the parameters to the utility.

directory name not given

>   The file name supplied to the utility does not include a directory path.

13.2.6 $\boxed{\text{DETACH}}$

$\boxed{\text{DETACH}}$ detaches a file from a filed fold, leaving the fold unfiled. All the contents of the fold are removed from the fold structure, but are not deleted at the host (DOS) level.

$\boxed{\text{ATTACH}}$ is on the same key; the two utilities are toggled.

Parameters

None.

Description

$\boxed{\text{DETACH}}$ should be invoked on a filed fold, the contents of which are to be removed.

When $\boxed{\text{DETACH}}$ is pressed the filed fold is unfiled and all its contents are removed from the TDS. The actual file, and nested files (if any), are not deleted. To regain access to a detached file $\boxed{\text{ATTACH}}$ should be used.

`Detaching file...`   start of run message.

`Detached file OK`   end of run message.

Error messages

`Cannot detach file:  must be on a filed fold`

> [DETACH] has been invoked an a line that is not a fold line.

13.2.7   [READ HOST]

[READ HOST] copies the contents of a host (DOS) file into a fold, thus converting it to TDS file format. It is normally used for importing text files to a TDS fold structure.

Parameters

`HostFileName:` This gives the full name of the file to be read.

The parameter is always offered for editing when this utility is run.

Doooription

Place the cursor on an empty fold, or an empty filed fold. When [READ HOST] is pressed the contents of the named file are read by the system and written into the fold.

`Reading host file "filename"...`   start of run message.

`Read aborted by user`   abort message.

`Read host file OK`   end of run message.

Error messages

`Cannot read host file:  must be on a text fold`

> The cursor is either not on a fold line or is on a fold of a type that cannot be opened by the editor.

`Cannot read host file:  fold must be empty`

> A host file cannot be read to a fold that already contains data.

`Cannot read host file:  cannot open filed fold`

> The filed fold pointed to by the cursor cannot be opened.

`Cannot read host file:  cannot create filed fold`

> The utility was invoked on a fold that was not filed and the utility failed to file that fold.

`Cannot read host file:  file name not given`

> The host file name parameter has been supplied without a file name.

The following messages take the form:

>   **Error** *error message* **(Result = *n*)**

where *n* is a filing system error code and *error message* is one of the following messages.

opening file for writing

A file system error occurred opening the filed fold for writing.

writing file

A file system error occurred writing to the filed fold.

reading host file

A filing error occurred reading the DOS file.

13.2.8   RENAME FILE

RENAME FILE allows the name of the file belonging to a filed fold to be changed. It has no effect on the contents of the filed fold, only on the host file name.

Parameters

None.

Description

RENAME FILE should be invoked on a filed fold. The new file name should be the first word, after the old filename in quotes, on the filed fold line following the quoted current filename. When RENAME FILE is pressed, the TDS reads the file name and changes the host file name to match it, provided it is a legal file name and there is no file already existing with the same file name. If a file already exists with the new name the system will alter the name so that it is unique. The file will then be renamed as the system derived name. The revised name will be displayed by the editor in quotes at the beginning of the fold line.

>   **Renaming file...**   start of run message.

>   **Renamed file as** *filename* **OK**   end of run message.

Error messages

Cannot rename file: fold is not filed

RENAME FILE has been invoked on a fold that is not filed.

Cannot rename file: must be on a filed fold

RENAME FILE has been invoked on a line that is not a fold line.

Error renaming file (Result = *n*)

Where *n* is the filing system error code (see appendix E). RENAME FILE failed to rename the file. The most likely cause of this is an illegal file name. If the error code is 0, the most likely cause is using a TDS file extension that is not permitted for the contents of that file. For example, using a **.tsr** file extension on a utility code file (the extension for these files is **.cut**).

13.2.9   ┌─────────────┐
│ WRITE HOST │
└─────────────┘

WRITE HOST copies a fold and any nested folds into a DOS format file, with all fold information removed. This is normally used to convert a TDS fold structure into a DOS text file. It may also be used on data and code folds. When used on a **CODE PROGRAM** fold the DOS file is suitable for loading with **iserver**.

Parameters

HostFileName This gives the full name of the file to be written.

The parameter is always offered for editing when this utility is run.

Description

The cursor should be pointing at the fold to be written. When WRITE HOST is pressed the contents of the fold are copied to the named file. The file is written in DOS format. All nested text folds are expanded in line, including **COMMENT** folds.

Writing host file *"filename"* . . .   start of run message.

Write aborted by user   abort message.

Written host file OK   end of run message.

Error messages

Cannot write host file: must be on a text fold

> The cursor is not on a fold line

Cannot write host file: cannot create filed fold

> The utility was invoked on a fold that was not filed and the utility failed to file that fold.

Cannot write host file: file name not given

> The host file name parameter has been supplied without a file name.

The following messages take the form:

> **Error** *error message* **(Result = ** *n***)**

> where *n* is a filing system error code and *error message* is one of the following messages:

opening file for reading

> A file system error occurred opening the filed fold for reading.

reading file

> A file system error occurred reading from the filed fold.

writing host file

> A filing error occurred writing the DOS file.

14 Libraries

14.1 Introduction to the libraries

Libraries are collections of source text and/or compiled procedures and functions which are suitable for use in a variety of occam programs. Libraries are provided in the TDS for a variety of necessary and commonly used operations. Users may create their own libraries and any such libraries may be handled in the same way as those provided in the TDS.

The libraries provided in the TDS are grouped into four principal groups; This grouping has no fundamental significance but is principally an aid to documentation and packaging for delivery. The lower level grouping of procedures and functions into libraries is governed by a need to keep libraries reasonably small to aid processing at compile time and is also influenced by the requirement that any procedure called from the code of more than one separate compilation unit must itself be in a different library.

Procedures in the compiler libraries are provided compiled for all transputer targets and error modes except universal. Other libraries are supplied compiled in **HALT** mode for **T2** and appropriate 32-bit classes. Source code for these is also provided so that users may recompile for other error modes.

When library code is linked into an occam program, the unit of linking is the individual separate compilation unit (**SC**). Most libraries therefore contain many **SC**s. The grouping of procedures into **SC**s may be determined by applying the COMPILATION INFO utility to the libraries in the TDS.

The library groups are:

1 Compiler and system libraries (complibs) which are libraries containing supporting code for the code generated by the compiler, and code for certain low level transputer operations which may sometimes be required. These libraries include the so-called **standard.libs** whose use may be suppressed by setting a compilation parameter.

2 Mathematical libraries (mathlibs) which are libraries providing a variety of mathematical functions.

3 Host input and output libraries (hostlibs) are designed particularly for use in programs which will be run outside the TDS, but supported by the host file server **iserver**. Code for such programs can be created and tested within the TDS and then with minimal alteration be recompiled for stand alone use.

4 occam channel input and output libraries (iolibs) contain procedures tailored for use in communication with the channel interfaces provided by the TDS to an **EXE** at run time. Many of these procedures may also be used in other contexts where a simple stream oriented model of communication of textual information using a single occam channel is appropriate. There are also a number of general purpose representation conversion procedures which may be used in any program. The procedures in this library are particularly suited for programs built from many concurrent processes and include suitable protocol conversion and multiplexing procedures for use in such programs.

Library group	Library name	Description	Processor
complibs		Multiple length integer arithmetic functions	T2 TA
		Floating point functions	T2 TB
		32 bit IEEE arithmetic functions	T2 TB
		64 bit IEEE arithmetic functions	T2 TB
		2D block move library	T2 T4
		Bit manipulation and CRC library	T2 T4
		Code execution library	T2 TA
		Arithmetic instruction library	T2 TA

Library group	Library name	Description	Processor
complibs	reinit	Extraordinary link handling library	T2 TA
	blockcrc	CRC library	T2 T4 TC
	linkaddr	Hard link addresses	T2 TA
	r64util	Long real arithmetic support	TA
	t2utils	Arithmetic support for **T2**	T2
	reals	32 bit real arithmetic support	T2 TB
	dreals	64 bit real arithmetic support	T2 TB
	ints	Integer arithmetic support	T2 TA
	realpds	Real predefined routines	T2 TB T8
	intpds	Integer predefined routines	T2 TB T8
mathlibs	mathvals	Constants for mathematical algorithms	T2 TA
	snglmath	Single length elementary function library	T2 TB T8
	dblmath	Double length elementary function library	T2 TB T8
	t4math	T414 elementary function library	TB
hostlibs	sphdr	Constants for **hostlibs** procedures	T2 TA
	splib	Low level SP operations	T2 TA
	solib	Higher level SP operations	T2 TA
	sklib	Keyboard input library	T2 TA
	spinterf	SP interface procedures	T2 TA
	afsp	**afserver** to **iserver** convertor	T2 TA
iolibs	userhdr	Constants for TDS terminal interface, etc.	T2 TA
	filerhdr	Constants for TDS user filer interface, etc.	TA
	krnlhdr	Constants for TDS kernel and server interfaces	TA
	strmhdr	Protocols for channels to the TDS	T2 TA
	uservals	Useful subset of **userhdr** and **filerhdr**	T2 TA
＊ afhdr		Constants for the afserver interface	TA
	ioconv	Basic type i/o conversion library	T2 TA
	extrio	Extra type i/o conversion library	T2 TA
	strings	String handling library	T2 TA
	streamio	Keystream and screenstream library	T2 TA
	ssinterf	Keystream and screenstream interface procedures	T2 TA
	userio	General purpose i/o procedure library	T2 TA
	interf	Interface procedure library	T2 TA
＊ slice		Block transfer procedure library	T2 TA
	ufiler	Low level user filer interface support library	TA
＊ msdos		TDS server channel support library	TA
＊ derivio		Byte stream i/o library	T2 TA
＊ afio		Afserver low level protocol library	TA
＊ afiler		Afserver command library	TA
＊ afinterf		Afserver protocol interface and multiplexor	TA
	t4board	Transputer board support library	TA
	t2board	B006 support library	T2

Libraries marked with an '＊' in the above table are provided for compatibility with earlier versions of the TDS only. Procedures from these libraries should not be called in new programs and they are not documented in this manual.

Libraries containing only constants and protocol declarations are tabulated in full in appendix D. The other libraries and the functions and procedures declared in them are described in full in the remaining sections of this chapter. The description of each library includes a table enumerating the procedures (and functions) in the library and the parameters required for each. A description of each of these is also provided.

Libraries contain code compiled for a variety of transputer target types and/or classes and of error checking modes. In the complibs library group where the sources of library procedures are not included in the product, all targets and error checking modes except universal are supported by compiled libraries. In the other library groups source code is included in the product and so users can compile the libraries for modes and targets not supported by the compiled code supplied. Wherever possible library code is compiled for the broadest possible class of transputer target (TA, or TB as appropriate) to avoid unnecessary replication of compiled code.

14.2 Compiler and system libraries (complibs)

This group includes procedures and functions which are predefined by the compiler in the sense that calls to them will be recognised by the compiler and specially handled. Some of these compile into in-line code and so do not correspond to any code in a library file supplied in the TDS. From the point of view of the user it is only necessary to know whether a #USE line is required. This requirement is mentioned after the table enumerating the contents of each library. All procedures and functions in this library group are fully supported for all processor types and error checking modes except universal. The user need only be concerned with which target group a procedure has been compiled for if the user code is being compiled for a transputer class rather than an individual target type. For obvious reasons it is not possible, for example, to compile a program doing real arithmetic for a group such as TA or TC whose members have different ways of achieving this.

The standard libraries for which #USE lines are provided by the system if the compilation parameter **use.standard.libs** is **TRUE** are **reals**, **dreals**, **ints**, **realpds** and **intpds**. It is alternatively possible for these to be explicitly used.

14.2.1 Multiple length integer arithmetic functions

The arithmetic functions provide arithmetic shifts, word rotations and the primitives to construct multiple length arithmetic and multiple length shift operations. Available for all targets.

Result	Function	Parameter specifiers
INT	LONGADD	VAL INT left, right, carry.in
INT	LONGSUB	VAL INT left, right, borrow.in
INT	ASHIFTRIGHT	VAL INT argument, places
INT	ASHIFTLEFT	VAL INT argument, places
INT	ROTATERIGHT	VAL INT argument, places
INT	ROTATELEFT	VAL INT argument, places
INT, INT	LONGSUM	VAL INT left, right, carry.in
INT, INT	LONDIFF	VAL INT left, right, borrow.in
INT, INT	LONGPROD	VAL INT left, right, carry.in
INT, INT	LONGDIV	VAL INT dividend.hi, dividend.lo, divisor
INT, INT	SHIFTLEFT	VAL INT hi.in, lo.in, places
INT, INT	SHIFTRIGHT	VAL INT hi.in, lo.in, places
INT, INT, INT	NORMALISE	VAL INT hi.in, lo.in

This library does not have to be referred to by a #USE statement; the compiler will automatically recognise calls to these routines and will compile them into in-line code.

For further information on the functions provided by this library see the occam 2 Reference Manual.

14.2.2   Floating point functions

The floating point functions include the list of facilities suggested by the ANSI-IEEE standard 754–1985. Available for all targets.

Result	Function	Parameter specifiers
REAL32	ABS	VAL REAL32 X
REAL64	DABS	VAL REAL64 X
REAL32	SCALEB	VAL REAL32 X, VAL INT n
REAL64	DSCALEB	VAL REAL64 X, VAL INT n
REAL32	COPYSIGN	VAL REAL32 X, Y
REAL64	DCOPYSIGN	VAL REAL64 X, Y
REAL32	SQRT	VAL REAL32 X
REAL64	DSQRT	VAL REAL64 X
REAL32	MINUSX	VAL REAL32 X
REAL64	DMINUSX	VAL REAL64 X
REAL32	NEXTAFTER	VAL REAL32 X, Y
REAL64	DNEXTAFTER	VAL REAL64 X, Y
REAL32	MULBY2	VAL REAL32 X
REAL64	DMULBY2	VAL REAL64 X
REAL32	DIVBY2	VAL REAL32 X
REAL64	DDIVBY2	VAL REAL64 X
REAL32	LOGB	VAL REAL32 X
REAL64	DLOGB	VAL REAL64 X
BOOL	ISNAN	VAL REAL32 X
BOOL	DISNAN	VAL REAL64 X
BOOL	NOTFINITE	VAL REAL32 X
BOOL	DNOTFINITE	VAL REAL64 X
BOOL	ORDERED	VAL REAL32 X, Y
BOOL	DORDERED	VAL REAL64 X, Y
INT, REAL32	FLOATING.UNPACK	VAL REAL32 X
INT, REAL64	DFLOATING.UNPACK	VAL REAL64 X
BOOL, INT32, REAL32	ARGUMENT.REDUCE	VAL REAL32 X, Y, Y.err
BOOL, INT32, REAL64	DARGUMENT.REDUCE	VAL REAL64 X, Y, Y.err
REAL32	FPINT	VAL REAL32 X
REAL64	DFPINT	VAL REAL64 X

This library does not have to be referred to by a #USE statement; the compiler will automatically recognise calls to these routines and will compile them into in-line code, or into compiler library calls.

For further information on the functions provided by this library see the occam 2 Reference Manual.

14.2.3   IEEE arithmetic functions

Result	Function	Parameter specifiers
REAL32	REAL32OP	VAL REAL32 X, VAL INT Op, VAL REAL32 Y
REAL64	REAL64OP	VAL REAL64 X, VAL INT Op, VAL REAL64 Y
BOOL, REAL32	IEEE32OP	VAL REAL32 X, VAL INT Rm, Op, VAL REAL32 Y
BOOL, REAL64	IEEE64OP	VAL REAL64 X, VAL INT Rm, Op, VAL REAL64 Y
BOOL, REAL32	IEEE32REM	VAL REAL32 X, Y
BOOL, REAL64	IEEE64REM	VAL REAL64 X, Y
REAL32	REAL32REM	VAL REAL32 X, Y
REAL64	REAL64REM	VAL REAL64 X, Y
BOOL	REAL32EQ	VAL REAL32 X, Y
BOOL	REAL64EQ	VAL REAL64 X, Y
BOOL	REAL32GT	VAL REAL32 X, Y
BOOL	REAL64GT	VAL REAL64 X, Y
INT	IEEECOMPARE	VAL REAL32 X, Y
INT	DIEEECOMPARE	VAL REAL64 X, Y

This library does not have to be referred to by a **#USE** statement; the compiler will automatically recognise calls to these routines for all targets and will compile them into in-line code, or into calls to appropriate functions in the libraries **reals** or **dreals**.

For further information on the functions provided by this library see the **occam 2** Reference Manual.

14.2.4   2D block move library

Procedure	Parameter Specifiers
MOVE2D	VAL [][]BYTE Source, VAL INT sx, sy, [][]BYTE Dest, VAL INT dx, dy, width, length
DRAW2D	VAL [][]BYTE Source, VAL INT sx, sy, [][]BYTE Dest, VAL INT dx, dy, width, length
CLIP2D	VAL [][]BYTE Source, VAL INT sx, sy, [][]BYTE Dest, VAL INT dx, dy, width, length

This library does not have to be referred to by a **#USE** statement; the compiler will automatically recognise calls to these routines. They will be compiled into in-line code for the **T8** or **T425**, or into a call to procedures in the library **intpds** for the **T4** or the **T2**.

MOVE2D

```
PROC MOVE2D(VAL [][]BYTE Source, VAL INT sx, sy, [][]BYTE Dest,
           VAL INT dx, dy, width, length)
```

Moves a block of size **width** by **length** which starts at byte **Source[sy][sx]** to the block starting at byte **Dest [dy][dx]**.

DRAW2D

> PROC DRAW2D (VAL [][]BYTE Source, VAL INT sx, sy, [][]BYTE Dest,
>   VAL INT dx, dy, width, length)

Moves a block of size **width** by **length** which starts at byte **Source[sy][sx]** to the block starting at byte **Dest [dy][dx]**. Only non-zero bytes in the source are transferred to the destination.

CLIP2D

> PROC CLIP2D (VAL [][]BYTE Source, VAL INT sx, sy, [][]BYTE Dest,
>   VAL INT dx, dy, width, length)

Moves a block of size **width** by **length** which starts at byte **Source[sy][sx]** to the block starting at byte **Dest [dy][dx]**. Only zero bytes in the source are transferred to the destination.

14.2.5   Bit manipulation and CRC library

Result	Function	Parameter Specifiers
INT	BITCOUNT	VAL INT Word, CountIn
INT	CRCWORD	VAL INT data, CRCIn, generator
INT	CRCBYTE	VAL INT data, CRCIn, generator
INT	BITREVNBITS	VAL INT x, n
INT	BITREVWORD	VAL INT x

This library does not have to be referred to by a **#USE** statement; the compiler will automatically recognise calls to these routines. They will be compiled into in-line code for the **T8** or **T425**, or into a call to procedures in the library **intpds** for the **T4** or the **T2**.

See INMOS technical note 26 for a discussion of CRC generation.

BITCOUNT

> INT FUNCTION BITCOUNT (VAL INT Word, CountIn)

This function counts the number of bits set in **Word**, and returns this number added to the value **CountIn**.

CRCWORD

> INT FUNCTION CRCWORD (VAL INT data, CRCIn, generator)

This function performs a cyclic redundancy check over 1 word. It is normally used iteratively on a sequence of words to obtain the CRC.

CRCIn	contains initial value or running CRC.
data	contains data on which the CRC is to be performed.
generator	contains CRC polynomial generator.

CRCBYTE

> **INT FUNCTION CRCBYTE (VAL INT data, CRCIn, generator)**

This function performs a cyclic redundancy check over 1 byte. It is normally used iteratively on a sequence of bytes to obtain the CRC. The byte processed is contained in the most significant end of the word **data**).

CRCIn	contains initial value or running CRC.
data	contains data on which the CRC is to be performed.
generator	contains CRC polynomial generator.

BITREVNBITS

> **INT FUNCTION BITREVNBITS (VAL INT x, n)**

This function takes **INT** parameters **x** and **n** and returns an **INT** containing the **n** least significant bits of **x**, in reverse order.

BITREVWORD

> **INT FUNCTION BITREVWORD (VAL INT x)**

This function takes an **INT** **x** and returns an **INT** which is the bit reversal of **x**.

14.2.6   Code execution

Procedure	Parameter Specifiers
KERNEL.RUN	**VAL []BYTE code, VAL INT entry.offset,** **[] INT workspace,** **VAL INT number.of.parameters**
LOAD.INPUT.CHANNEL	**INT here, CHAN OF ANY in**
LOAD.INPUT.CHANNEL.VECTOR	**INT here, [] CHAN OF ANY in.vec**
LOAD.OUTPUT.CHANNEL	**INT here, CHAN OF ANY out**
LOAD.OUTPUT.CHANNEL.VECTOR	**INT here, [] CHAN OF ANY out.vec**
LOAD.BYTE.VECTOR	**INT here, [] BYTE b.vec**

This library does not have to be referred to by a **#USE** statement; the compiler will automatically recognise calls to these routines for all targets and will compile them into in-line code.

The procedures described allow an occam program to read in a compiled occam **PROC** and call it. The called **PROC** may be compiled and linked separately from the calling program and read in from a file. The calling program runs the called **PROC** with a normal sequential **PROC** call mechanism.

The facilities include provision for passing parameters to the called **PROC** before running it.

KERNEL.RUN

```
PROC KERNEL.RUN(VAL []BYTE code,
                VAL INT entry.offset,
                []INT workspace,
                VAL INT number.of.parameters)
```

The effect of this procedure is to call the procedure loaded in the **code** buffer, starting execution at the location **code[entry.offset]**. The **workspace** buffer is used to hold the local data of the called procedure. The parameters passed to the called procedure should be placed at the top of the **workspace** buffer by the calling process before the call of **KERNEL.RUN**. The call to **KERNEL.RUN** returns when the called **PROC** terminates.

See section 11.3 for a full description of how to set up the workspace for **KERNEL.RUN**.

LOAD.INPUT.CHANNEL

```
PROC LOAD.INPUT.CHANNEL (INT here, CHAN OF ANY in)
```

The variable **here** is assigned the address of the second parameter.

LOAD.INPUT.CHANNEL.VECTOR

```
PROC LOAD.INPUT.CHANNEL.VECTOR (INT here, []CHAN OF ANY in.vec)
```

The variable **here** is assigned the address of the second parameter.

LOAD.OUTPUT.CHANNEL

```
PROC LOAD.OUTPUT.CHANNEL (INT here, CHAN OF ANY out)
```

The variable **here** is assigned the address of the second parameter.

LOAD.OUTPUT.CHANNEL.VECTOR

```
PROC LOAD.OUTPUT.CHANNEL.VECTOR (INT here, []CHAN OF ANY out.vec)
```

The variable **here** is assigned the address of the second parameter.

LOAD.BYTE.VECTOR

```
PROC LOAD.BYTE.VECTOR (INT here, []BYTE b.vec)
```

The variable **here** is assigned the address of the second parameter.

14.2.7   Arithmetic instruction library

Result	Function	Parameter Specifiers
INT,INT,INT	UNPACKSN	VAL INT X
INT	ROUNDSN	VAL INT Yexp, Yfrac, Yguard
INT	FRACMUL	VAL INT X,Y

This library does not have to be referred to by a **#USE** statement; the compiler will automatically recognise calls to these routines, when compiling for a **T4**, and will compile them into in-line code. The **FRACMUL** function is available for the **T8**. None of the functions are available for a **T2**.

This library provides access to some of the low-level arithmetic instructions on the transputer.

UNPACKSN and **ROUNDSN** support floating-point arithmetic on the **T4**.

FRACMUL supports fractional arithmetic on the **T4** and the **T8**.

UNPACKSN

> **INT, INT, INT FUNCTION UNPACKSN (VAL INT X)**
>
> **UNPACKSN** unpacks **X**, regarded as an IEEE single-length format binary floating-point quantity, into **Xexp**, the (biased) exponent, and **Xfrac**, the fractional part. It also returns an integer defining the **Type** of **X**. This is:
>
> > 0 if **X** is zero
> >
> > 1 if **X** is a normalised or denormalised number
> >
> > 2 if **X** is Inf
> >
> > 3 if **X** is NaN
>
> The sign of **X** is ignored.
>
> The results are returned in the order: **Xexp**, **Xfrac**, **Type**.

ROUNDSN

> **INT FUNCTION ROUNDSN (VAL INT Yexp, Yfrac, Yguard)**
>
> **ROUNDSN** takes a possibly unnormalised fraction, guard word and exponent and returns the rounded IEEE floating point value it represents. To do this the fraction is normalised, if necessary, then postnormalised and finally rounded to the nearest IEEE value. The exponent should already be biased. If overflow occurs, Inf is returned. Its use is in processes that have operated on unpacked floating point numbers to produce an unpacked result. It takes care of all the normalisation, postnormalisation, rounding and packing of the result. The round mode used is round to nearest.
>
> The function normalises and postnormalises the number represented by **Yexp**, **Yfrac** and **Yguard** into the local variables **Xexp**, **Xfrac** and **Xguard**. It then packs the (biased) exponent **Xexp** and fraction **Xfrac** into the result, rounding using the extra bits in **Xguard**. The sign bit is set to 0. If there is overflow, the result is set to Inf.

FRACMUL

> **INT FUNCTION FRACMUL (VAL INT X,Y)**
>
> **FRACMUL** takes two arguments representing real fractions in the range $[-1,1)$ and returns their product rounded to the nearest available representation. The value of the fractions represented by the arguments and result can be obtained by multiplying their **INT** value by 2^{-31}.

14.2.8 Extraordinary link handling library `reinit`

Procedure	Parameter Specifiers
InputOrFail.t	CHAN OF ANY c, []BYTE mess, TIMER TIME, VAL INT t, BOOL aborted
OutputOrFail.t	CHAN OF ANY c, VAL []BYTE mess, TIMER TIME, VAL INT t, BOOL aborted
InputOrFail.c	CHAN OF ANY c, []BYTE mess, CHAN OF INT kill, BOOL aborted
OutputOrFail.c	CHAN OF ANY c, VAL []BYTE mess, CHAN OF INT kill, BOOL aborted
Reinitialise	CHAN OF ANY c

To use this library a program header must include the line:

> **#USE reinit**

There are four procedures which implement input and output processes which can be made to terminate even when there is a communication failure. They will terminate either as a result of the communication completing, or as a result of the failure of the communication being recognised. Two procedures provide input and output where communication failure can be detected by a simple timeout, the other two procedures provide input and output where the failure of the communication is signalled to the procedure via a channel. The procedures have a boolean variable as a parameter which is set true if the procedure terminated as a result of communication failure being detected, and is set false otherwise. If the procedure does terminate as a result of communication failure having been detected then the link channel can be reset using a fifth procedure.

InputOrFail.t

> **PROC InputOrFail.t (CHAN OF ANY c, []BYTE mess, TIMER TIME,**
> **VAL INT t, BOOL aborted)**

The procedure takes as parameters a link channel **c** (on which the communication is to take place), a byte vector **mess** (which is the object of the communication) and the boolean variable **aborted**. The choice of a byte vector as the parameter to these procedures allows an object of any type to be passed along the channel provided it is retyped first.

The procedure is used for communication where failure is detected by a timeout. It take a timer parameter **TIME**, and an absolute time **t**. The procedure treats the communication as having failed when the time as measured by the timer **TIMER** is **AFTER** the specified time **t**.

OutputOrFail.t

> **PROC OutputOrFail.t (CHAN OF ANY c, VAL []BYTE mess, TIMER TIME,**
> **VAL INT t, BOOL aborted)**

The procedure takes as parameter a link channel **c** (on which the communication is to take place), a byte vector **mess** (which is the object of the communication) and the boolean variable **aborted**. the choice of a byte vector as the parameter to these procedures allows an object of any type to be passed along the channel provided it is retyped first.

This procedure is used for communication where failure is detected by a timeout. It takes a timer parameter **TIME**, and an absolute time **t**. The procedure treats the communication as having failed when the time as measured by the timer **TIME** is **AFTER** the specified time **t**.

InputOrFail.c

> **PROC InputOrFail.c (CHAN OF ANY c, []BYTE mess,**
> **CHAN OF INT kill, BOOL aborted)**

The procedure takes as parameter a link channel **c** (on which the communication is to take place), a byte vector **mess** (which is the object of the communication) and the boolean variable **aborted**. The choice of a byte vector as the parameter to these procedures allows an object of any type to be passed along the channel provided it is retyped first.

This procedure provides communication where failure cannot be detected by a simple timeout. In this case failure must be signalled to the inputting procedure via a message on the channel **kill**. The **kill** message is of type **INT** and can be any value.

OutputOrFail.c

> **PROC OutputOrFail.c (CHAN OF ANY c, VAL []BYTE mess,**
> **CHAN OF INT kill, BOOL aborted)**

The procedure takes as parameters a link channel **c** (on which the communication is to take place), a byte vector **mess** (which is the object of the communication) and the boolean variable **aborted**. The choice of a byte vector as the parameter to these procedures allows an object of any type to be passed along the channel provided it is retyped first.

This procedure provides communication where failure cannot be detected by a simple timeout. In this case failure must be signalled to the inputting or outputting procedure via a message on the channel **kill**. The **kill** message is of type **INT** and can be any value.

Reinitialise

> **PROC Reinitialise (CHAN OF ANY c)**

This procedure may be used to reinitialise the link channel **c** after it is known that all activity on the link has ceased.

Reinitialise must only be used to reinitialise a link channel after communication has finished. If the procedure is applied to a link channel which is being used for communication the transputer's error flag will be set and subsequent behaviour is undefined.

14.2.9   Block CRC library blockcrc

Result	Function	Parameter Specifiers
INT	CRCFROMMSB	VAL []BYTE InputString, VAL INT PolynomialGenerator, INT OldCRC
INT	CRCFROMLSB	VAL []BYTE InputString, VAL INT PolynomialGenerator, INT OldCRC

To use this library a program header must include the line:

> **#USE blockcrc**

CRCFROMMSB

> **FUNCTION CRCFROMMSB (VAL []BYTE InputString,**
> **VAL INT PolynomialGenerator, INT OldCRC)**

The string of bytes is polynomially divided by the generator starting from the most significant bit of the most significant byte in decreasing bit order.

CRCFROMLSB

> **FUNCTION CRCFROMLSB (VAL []BYTE InputString,**
> **VAL INT PolynomialGenerator, INT OldCRC)**

The string of bytes is polynomially divided by the generator starting from the least significant bit of the least significant byte in increasing bit order.

14.3   Mathematical libraries (`mathlibs`)

14.3.1   Single length and double length elementary function library

The elementary functions for any processor are contained in two separate libraries: one for the single length functions, the other for the double length functions. The TB specific version of these functions, which is described in the next section, consists of one library only.

The version of the library described by this section has been written using only floating-point arithmetic and pre-defined functions supported in occam. Thus it can be compiled for any processor with a full implementation of occam, and give identical results.

It will be efficient on processors with fast floating-point arithmetic and good support for the floating-point prede-fined functions such as **MULBY2** and **ARGUMENT.REDUCE**. For 32-bit processors without special hardware for floating-point calculations the alternative version described in section 14.3.2 using fixed-point arithmetic will be faster, but will not give identical results.

A special version has been produced for the IMS T212, which avoids the use of any double-precision arithmetic in the single precision functions. This is distinguished in the notes by the annotation 'T212 special'; notes relating to the version for T8 and TB are denoted by 'standard'.

Result	Function	Parameter specifiers
REAL32	ALOG	VAL REAL32 X
REAL32	ALOG10	VAL REAL32 X
REAL32	EXP	VAL REAL32 X
REAL32	POWER	VAL REAL32 X, VAL REAL32 Y
REAL32	SIN	VAL REAL32 X
REAL32	COS	VAL REAL32 X
REAL32	TAN	VAL REAL32 X
REAL32	ASIN	VAL REAL32 X
REAL32	ACOS	VAL REAL32 X
REAL32	ATAN	VAL REAL32 X
REAL32	ATAN2	VAL REAL32 X, VAL REAL32 Y
REAL32	SINH	VAL REAL32 X
REAL32	COSH	VAL REAL32 X
REAL32	TANH	VAL REAL32 X
REAL32, INT32	RAN	VAL INT32 X

To use the single length library a program header must include the line

```
#USE snglmath
```

Result	Function	Parameter specifiers
REAL64	DALOG	VAL REAL64 X
REAL64	DALOG10	VAL REAL64 X
REAL64	DEXP	VAL REAL64 X
REAL64	DPOWER	VAL REAL64 X, VAL REAL64 Y
REAL64	DSIN	VAL REAL64 X
REAL64	DCOS	VAL REAL64 X
REAL64	DTAN	VAL REAL64 X
REAL64	DASIN	VAL REAL64 X
REAL64	DACOS	VAL REAL64 X
REAL64	DATAN	VAL REAL64 X
REAL64	DATAN2	VAL REAL64 X, VAL REAL64 Y
REAL64	DSINH	VAL REAL64 X
REAL64	DCOSH	VAL REAL64 X
REAL64	DTANH	VAL REAL64 X
REAL64,INT64	DRAN	VAL INT64 X

To use the double length library a program header must include the line

```
#USE dblmath
```

Introduction

This, and the following subsections, contain some notes on the presentation of the elementary function libraries, including the TB version described in section 14.3.2.

These function subroutines have been written to be compatible with the ANSI standard for binary floating-point arithmetic (ANSI-IEEE std 754-1985), as implemented in occam. They are based on the algorithms in:
Cody, W. J., and Waite, W. M. [1980]. *Software Manual for the Elementary Functions.* Prentice-Hall, New Jersey.
The only exceptions are the pseudo-random number generators, which are based on algorithms in:
Knuth, D. E. [1981]. *The Art of Computer Programming, 2nd. edition, Volume 2: Seminumerical Algorithms.* Addison-Wesley, Reading, Mass.

Inputs

All the functions in the library (except **RAN** and **DRAN**) are called with one or two parameters which are binary floating-point numbers in one of the IEEE standard formats, either 'single-length' (32 bits) or 'double-length' (64 bits). The parameter(s) and the function result are of the same type.

NaNs and Infs

The functions will accept any value, as specified by the standard, including special values representing **NaNs** ('Not a Number') and **Infs** ('Infinity'). **NaNs** are copied to the result, whilst **Infs** may or may not be in the domain. The domain is the set of arguments for which the result is a normal (or denormalised) floating-point number.

Outputs

Exceptions

Arguments outside the domain (apart from **NaNs** which are simply copied through) give rise to *exceptional results*, which may be **NaN**, **+Inf**, or −**Inf**. **Infs** mean that the result is mathematically well-defined but too large to be represented in the floating-point format.

Error conditions are reported by means of three distinct **NaNs**:

undefined.NaN

This means that the function is mathematically undefined for this argument, for example the logarithm of a negative number.

unstable.NaN

This means that a small change in the argument would cause a large change in the value of the function, so *any* error in the input will render the output meaningless.

inexact.NaN

This means that although the mathematical function is well-defined, its value is in range, and it is stable with respect to input errors at this argument, the limitations of word-length (and reasonable cost of the algorithm) make it impossible to compute the correct value.

Accuracy

Range Reduction

Since it is impractical to use rational approximations (i.e. quotients of polynomials) which are accurate over large domains, nearly all the subroutines use mathematical identities to relate the function value to one computed from a smaller argument, taken from the 'primary domain', which is small enough for such an approximation to be used. This process is called 'range reduction' and is performed for all arguments except those which already lie in the primary domain.

For most of the functions the quoted error is for arguments in the primary domain, which represents the basic accuracy of the approximation. For some functions the process of range reduction results in a higher accuracy for arguments outside the primary domain, and for others it does the reverse. Refer to the notes on each function for more details.

Generated Error

If the true value of the function is large the difference between it and the computed value (the 'absolute error') is likely to be large also because of the limited accuracy of floating-point numbers. Conversely if the true value is small, even a small absolute error represents a large proportional change. For this reason the error relative to the true value is usually a better measure of the accuracy of a floating-point function, except when the ouput range is strictly bounded.

If f is the mathematical function and F the subroutine approximation, then the relative error at the floating-point number X (provided $f(X)$ is not zero) is:

$$RE(X) = \frac{(F(X) - f(X))}{f(X)}$$

Obviously the relative error may become very large near a zero of $f(X)$. If the zero is at an irrational argument (which cannot be represented as a floating-point value), the absolute error is a better measure of the accuracy of the function near the zero.

As it is impractical to find the relative error for every possible argument, statistical measures of the overall error must be used. If the relative error is sampled at a number of points X_n ($n = 1$ to N), then useful statistics are the *maximum relative error* and the *root-mean-square relative error*.

$$MRE = \max_{1 \leq n \leq N} |RE(X_n)|$$

$$RMSRE = \sqrt{\sum_{n=1}^{N} (RE(X_n))^2}$$

Corresponding statistics can be formed for the absolute error also, and are called MAE and $RMSAE$ respectively.

The MRE generally occurs near a zero of the function, especially if the true zero is irrational, or near a singularity where the result is large, since the 'granularity' of the floating-point numbers then becomes significant.

A useful unit of relative error is the relative magnitude of the least significant bit in the floating-point fraction, which is called one 'unit in the last place' (ulp). This is the relative magnitude of the least significant bit of the floating-point fraction (i.e. the smallest ϵ such that $1 + \epsilon \neq 1$). Its magnitude depends on the floating-point format: for single-length it is $2^{-23} = 1.19 * 10^{-7}$, and for double-length it is $2^{-52} = 2.22 * 10^{-16}$.

Propagated Error

Because of the limited accuracy of floating-point numbers the result of any calculation usually differs from the exact value. In effect, a small error has been added to the exact result, and any subsequent calculations will inevitably involve this error term. Thus it is important to determine how each function responds to errors in its argument. Provided the error is not too large, it is sufficient just to consider the first derivative of the function (written f').

If the relative error in the argument X is d (typically a few ulp), then the absolute error (E) and relative error (e) in $f(X)$ are:

$$E = |Xf'(X)d| \equiv Ad$$

$$e = \left| \frac{Xf'(X)d}{f(X)} \right| \equiv Rd$$

This defines the absolute and relative error magnification factors A and R. When both are large the function is unstable, i.e. even a small error in the argument, such as would be produced by evaluating a floating-point expression, will cause a large error in the value of the function. The functions return an **unstable.NaN** in such cases which are simple to detect.

The functional forms of both A and R are given in the specification of each function.

Test Procedures

For each function, the generated error was checked at a large number of arguments (typically 100 000) drawn at random from the appropriate domain. First the double-length functions were tested against a 'quadruple-length' implementation (constructed for accuracy rather than speed), and then the single-length functions were tested against the double-length versions.

In both cases the higher-precision implementation was used to approximate the mathematical function (called f above) in the computation of the error, which was evaluated in the higher precision to avoid rounding errors. Error statistics were produced according to the formulae above.

Symmetry

The subroutines were designed to reflect the mathematical properties of the functions as much as possible. For all the functions which are even, the sign is removed from the input at the beginning of the computation so that the sign-symmetry of the function is always preserved. For odd functions, either the sign is removed at the start and then the appropriate sign set at the end of the computation, or else the sign is simply propagated through an odd degree polynomial. In many cases other symmetries are used in the range-reduction, with the result that they will be satisfied automatically.

The Function Specifications

Names and Parameters

All single length functions except **RAN** return a single result of type **REAL32**, and all except **RAN**, **POWER** and **ATAN2** have one parameter, a **VAL REAL32** for the argument of the function.

POWER and **ATAN2** have two parameters which are **VAL REAL32**s for the two arguments of each function.

RAN returns two results of type **REAL32, INT32**, and has one parameter which is a **VAL INT32**.

In each case the double-length version of **name** is called **Dname**, returns a **REAL64** (except **DRAN**, which returns **REAL64, INT64**), and has parameters of type **VAL REAL64** (**VAL INT64** for **DRAN**).

Terms used in the Specifications

A and R Multiplying factors relating the absolute and relative errors in the output to the relative error in the argument.

Exceptions Outputs for invalid inputs (i.e. those outside the *domain*), other than **NaN** (**NaN**s are copied direcly to the output and are not listed as exceptions). These are all **Inf**s or **NaN**s.

Generated Error The difference between the true and computed values of the function, when the argument is error-free. This is measured statistically and displayed for one or two ranges of arguments, the first of which is usually the *primary domain* (see below). The second range, if present, is chosen to illustrate the typical behaviour of the function.

Domain The range of valid inputs, i.e. those for which the output is a normal or denormal floating-point number.

MAE and RMSAE The Maximum Absolute Error and Root-Mean-Square absolute error taken over a number of arguments drawn at random from the indicated range.

MRE and RMSRE The Maximum Relative Error and Root-Mean-Square relative error taken over a number of arguments drawn at random from the indicated range.

Range The range of outputs produced by all arguments in the *Domain*. The given endpoints are not exceeded.

Primary Domain The range of arguments for which the result is computed using only a single rational approximation to the function. There is no argument reduction in this range.

Propagated Error The absolute and relative error in the function value, given a small relative error in the argument.

ulp The unit of relative error is the 'unit in the last place' (ulp). This is the relative magnitude of the least significant bit of the floating-point fraction (i.e. the smallest ϵ such that $1 + \epsilon \neq 1$).
N.B. this depends on the floating-point format!
For the standard single-length format it is $2^{-23} = 1.19 * 10^{-7}$.
For the double-length format it is $2^{-52} = 2.22 * 10^{-16}$.
This is also used as a measure of absolute error, since such errors can be considered 'relative' to unity.

Specification of Ranges

Ranges are given as intervals, using the convention that a square bracket '[' or ']'   means that the adjacent endpoint is included in the range, whilst a round bracket '(' or ')'   means that it is excluded. Endpoints are given to a few significant figures only.

Where the range depends on the floating-point format, single-length is indicated with an S and double-length with a D.

For functions with two arguments the complete range of both arguments is given. This means that for each number in one range, there is at least one (though sometimes only one) number in the other range such that the pair of arguments is valid. Both ranges are shown, linked by an 'x'.

Abbreviations

In the specifications, $XMAX$ is the largest representable floating-point number: in single-length it is approximately $3.4 * 10^{38}$, and in double-length it is approximately $1.8 * 10^{308}$.

Pi means the closest floating-point representation of the transcendental number π, ln(2) the closest representation of $\log_e(2)$, and so on.

In describing the algorithms, 'X' is used generically to designate the argument, and 'result' (or RESULT, in the style of occam functions) to designate the output.

ALOG

> **REAL 32 FUNCTION ALOG (VAL REAL32 X)**
>
> **REAL 64 FUNCTION DALOG (VAL REAL64 X)**

These compute: $\log_a(X)$

Domain:	$(0, XMAX]$
Range:	$[MinLog, MaxLog]$ (See Note 2)
Primary Domain:	$[\sqrt{2}/2, \sqrt{2}) = [0.7071, 1.4142)$

Exceptions

All arguments outside the domain generate an **undefined.NaN**.

Propagated Error

$A \equiv 1, \qquad R = 1/\log_e(X)$

Generated Error

Primary Domain Error:	MRE	RMSRE
Single Length(Standard):	1.7 ulp	0.43 ulp
Single Length(T212 special):	1.6 ulp	0.42 ulp
Double Length:	1.4 ulp	0.38 ulp

The Algorithm

> 1 Split X into its exponent N and fraction F.
>
> 2 Find LnF, the natural log of F, with a floating-point rational approximation.
>
> 3 Compute ln(2) $* N$ with extended precision and add it to LnF to get the result.

Notes

1) The term ln(2) $* N$ is much easier to compute (and more accurate) than LnF, and it is larger provided N is not 0 (i.e. for arguments outside the primary domain). Thus the accuracy of the result improves as the modulus of $\log(X)$ increases.

2) The minimum value that can be produced, MinLog, is the logarithm of the smallest denormalised floating-point number. For single length $Minlog$ is -103.28, and for double length it is -745.2. The maximum value $MaxLog$ is the logarithm of $XMAX$. For single-length it is 88.72, and for double-length it is 709.78.

3) Since **Inf** is used to represent *all* values greater than $XMAX$ its logarithm cannot be defined.

4) This function is well-behaved and does not seriously magnify errors in the argument.

ALOG10

> **REAL32 FUNCTION ALOG10 (VAL REAL32 X)**
>
> **REAL64 FUNCTION DALOG10 (VAL REAL64 X)**

These compute: $\log_{10}(X)$

Domain: $(0, XMAX]$
Range: $[MinL10, MaxL10]$ (See Note 2)
Primary Domain: $[\sqrt{2}/2, \sqrt{2}) = [0.7071, 1.4142)$

Exceptions

All arguments outside the domain generate an **undefined.NaN**.

Propagated Error

$$A \equiv \log_{10}(e), \qquad R = \log_{10}(e)/\log_e(X)$$

Generated Error

Primary Domain Error:	MRE	RMSRE
Single Length(Standard):	1.70 ulp	0.45 ulp
Single Length(T212 special):	1.71 ulp	0.46 ulp
Double Length:	1.84 ulp	0.45 ulp

The Algorithm

> 1 Set $temp$:= **ALOG (X)** .
>
> 2 If $temp$ is a **NaN**, copy it to the output, otherwise set result = log(e) $*$ $temp$

Notes

1) See note 1 for **ALOG**.

2) The minimum value that can be produced, $MinL10$, is the base-10 logarithm of the smallest denormalised floating-point number. For single length $MinL10$ is -44.85, and for double length it is -323.6. The maximum value $MaxL10$ is the base-10 logarithm of $XMAX$. For single length $MaxL10$ is 38.53, and for double-length it is 308.26.

3) Since **Inf** is used to represent *all* values greater than $XMAX$ its logarithm cannot be defined.

4) This function is well-behaved and does not seriously magnify errors in the argument.

EXP

> **REAL32 FUNCTION EXP (VAL REAL32 X)**
>
> **REAL64 FUNCTION DEXP (VAL REAL64 X)**

These compute: e^X

Domain: $[-\text{Inf}, MaxLog)$ $= [-\text{Inf}, 88.72)\text{S},$ $[-\text{Inf}, 709.78)\text{D}$
Range: $[0, \text{Inf})$ (See note 4)
Primary Domain: $[-Ln2/2, Ln2/2)$ $= [-0.3466, 0.3466)$

Exceptions

All arguments outside the domain generate an **Inf**.

Propagated error

$$A = Xe^X, \qquad R = X$$

Generated error

Primary Domain Error:	MRE	RMSRE
Single Length(Standard):	0.99 ulp	0.25 ulp
Single Length(T212 special):	1.0 ulp	0.25 ulp
Double Length:	1.0 ulp	0.25 ulp

The Algorithm

1 Set N = integer part of $X/\ln(2)$.

2 Compute the remainder of X by $\ln(2)$, using extended precision arithmetic.

3 Compute the exponential of the remainder with a floating-point rational approximation.

4 Increase the exponent of the result by N. If N is sufficiently negative the result must be denormalised.

Notes

1) $MaxLog$ is $\log_e(XMAX)$.

2) For sufficiently negative arguments (below -87.34 for single-length and below -708.4 for double-length) the output is denormalised, and so the floating-point number contains progressively fewer significant digits, which degrades the accuracy. In such cases the error can theoretically be a factor of two.

3) Although the true exponential function is never zero, for large negative arguments the true result becomes too small to be represented as a floating-point number, and **EXP** underflows to zero. This occurs for arguments below -103.9 for single-length, and below -745.2 for double-length.

4) The propagated error is considerably magnified for large positive arguments, but diminished for large negative arguments.

POWER

REAL32 FUNCTION POWER (VAL REAL32 X, VAL REAL32 Y)

REAL64 FUNCTION DPOWER (VAL REAL64 X, VAL REAL64 Y)

These compute: X^Y

Domain:	[0, Inf] x [−Inf, Inf]
Range:	(−Inf, Inf)
Primary Domain:	See note 3.

Exceptions

If the first argument is outside its domain, **undefined.NaN** is returned. If the true value of X^Y exceeds $XMAX$, **Inf** is returned. In certain other cases other **NaNs** are produced: See note 2.

Propagated Error

$$A = YX^Y(1 \pm \log_e(X)), \qquad R = Y(1 \pm \log_e(X)) \text{ (See note 4)}$$

Generated error

Example Range Error:	MRE	RMSRE	(See note 3)
Single Length(Standard):	1.0 ulp	0.25 ulp	
Single Length(T212 special):	63.1 ulp	13.9 ulp	
Double Length:	21.1 ulp	2.4 ulp	

The Algorithm

Deal with special cases: either argument = 1, 0, +**Inf** or −**Inf** (see note 2). Otherwise:

(a) For the standard single precision:

    1 Compute $L = \log_e(X)$ in double precision, where X is the first argument.

    2 Compute $W = Y \times L$ in double precision, where Y is the second argument.

    3 Compute $RESULT = e^W$ in single precision.

(b) For double precision, and the single precision special version:

    1 Compute $L = \log_2(X)$ in extended precision, where X is the first argument.

    2 Compute $W = Y \times L$ in extended precision, where Y is the second argument.

    3 Compute $RESULT = 2^W$ in extended precision.

Notes

1) This subroutine implements the mathematical function x^y to a much greater accuracy than can be attained using the **ALOG** and **EXP** functions, by performing each step in higher precision. The single-precision version is more efficient than using **DALOG** and **EXP** because redundant tests are omitted.

2) Results for special cases are as follows:

First Input (X)	Second Input (Y)	Result
< 0	ANY	undefined.NaN
0	≤ 0	undefined.NaN
0	$0 < Y \le XMAX$	0
0	Inf	unstable.NaN
$0 < X < 1$	Inf	0
$0 < X < 1$	-Inf	Inf
1	$-XMAX \le Y \le XMAX$	1
1	± Inf	unstable.NaN
$1 < X \le XMAX$	Inf	Inf
$1 < X \le XMAX$	-Inf	0
Inf	$1 \le Y \le$ Inf	Inf
Inf	-Inf$\le Y \le -1$	0
Inf	$-1 < Y < 1$	undefined.NaN
otherwise	0	1
otherwise	1	X

3) Performing all the calculations in extended precision makes the double-precision algorithm very complex in detail, and having two arguments makes a primary domain difficult to specify. As an indication of accuracy, the functions were evaluated at 100 000 points logarithmically distributed over (0.1, 10.0), with the exponent linearly distributed over (−35.0, 35.0) (single-length), and (−300.0, 300.0)

(double-length), producing the errors given above. The errors are much smaller if the exponent range is reduced.

4) The error amplification factors are calculated on the assumption that the relative error in Y is $\pm$ that in X, otherwise there would be separate factors for both X and Y. It can be seen that the propagated error will be greatly amplified whenever $\log_e(X)$ or Y is large.

SIN

REAL32 FUNCTION SIN (VAL REAL32 X)

REAL64 FUNCTION DSIN (VAL REAL64 X)

These compute: sine(X)   (where X is in radians)

Domain:	$[-Smax, Smax]$	$= [-205887.4, 205887.4]$S (Standard),
		$[-4.2 * 10^6, 4.2 * 10^6]$S (T212 special)
		$= [-3.4 * 10^9, 3.4 * 10^9]$D
Range:	$[-1.0, 1.0]$	
Primary Domain:	$[-Pi/2, Pi/2]$	$= [-1.57, 1.57]$

Exceptions

All arguments outside the domain generate an **inexact.NaN**, except $\pm$Inf, which generates an **undefined.NaN**.

Propagated Error

$A = X \cos(X), \quad R = X \cot(X)$

Generated error   (See note 1)

	Primary Domain		[0, 2Pi]	
	MRE	RMSRE	MAE	RMSAE
Single Length(Standard):	0.94 ulp	0.23 ulp	0.96 ulp	0.19 ulp
Single Length(T212 special):	0.92 ulp	0.23 ulp	0.94 ulp	0.19 ulp
Double Length:	0.9 ulp	0.22 ulp	0.91 ulp	0.18 ulp

The Algorithm

1 Set N = integer part of $|X|/Pi$.

2 Compute the remainder of $|X|$ by Pi, using extended precision arithmetic (double precision in the standard version).

3 Compute the sine of the remainder using a floating-point polynomial.

4 Adjust the sign of the result according to the sign of the argument and the evenness of N.

Notes

1) For arguments outside the primary domain the accuracy of the result depends crucially on step 2. The extra precision of step 2 is lost if N becomes too large, and the cut-off $Smax$ is chosen to prevent this. In any case for large arguments the 'granularity' of floating-point numbers becomes a significant factor. For arguments larger than $Smax$ a change in the argument of 1 ulp would change more than half of the significant bits of the result, and so the result is considered to be essentially indeterminate.

2) The propagated error has a complex behaviour. The propagated relative error becomes large near each zero of the function (outside the primary range), but the propagated absolute error only becomes large for large arguments. In effect, the error is seriously amplified only in an interval about

each irrational zero, and the width of this interval increases roughly in proportion to the size of the argument.

3) Since only the remainder of X by Pi is used in step 3, the symmetry $\sin(x + n\pi) = \pm\sin(x)$ is preserved, although there is a complication due to differing precision representations of π.

4) The output range is not exceeded. Thus the output of **SIN** is always a valid argument for **ASIN**.

COS

REAL32 FUNCTION COS (VAL REAL32 X)

REAL64 FUNCTION DCOS (VAL REAL64 X)

These compute: cosine(X)   (where X is in radians)

Domain: $[-Cmax, Cmax]$ $= [-205887.4, 205887.4]$S (Standard),
                                    $[-12868.0, 12868.0]$S (T212 special)
                                $= [-3.4 * 10^9, 3.4 * 10^9]$D

Range: $[-1.0, 1.0]$
Primary Domain: See note 1.

Exceptions

All arguments outside the domain generate an **inexact.NaN**, except $\pm$**Inf**, which generates an **undefined.NaN**.

Propagated Error

$A = -X\sin(X)$, $R = -X\tan(X)$ (See note 4)

Generated error

Range: $[0, Pi/4)$ $[0, 2Pi]$
                       MRE   RMSRE   MAE   RMSAE
Single Length(Standard):   0.93 ulp   0.25 ulp   0.88 ulp   0.18 ulp
Single Length(T212 special): 1.1 ulp   0.3 ulp   0.94 ulp   0.19 ulp
Double Length: 1.0 ulp   0.28 ulp   0.9 ulp   0.19 ulp

The Algorithm

1 Set N = integer part of $(|X| + Pi/2)/Pi$ and compute the remainder of $(|X| + Pi/2)$ by Pi, using extended precision arithmetic (double precision in the standard version).

2 Compute the sine of the remainder using a floating-point polynomial.

3 Adjust the sign of the result according to the evenness of N.

Notes

1) Inspection of the algorithm shows that argument reduction always occurs, thus there is no 'primary domain' for **COS**. So for all arguments the acuracy of the result depends crucially on step 2. The standard single-precision version performs the argument reduction in double-precision, so there is effectively no loss of accuracy at this step. For the T212 special version and the double-precision version there are effectively K extra bits in the representation of $\pi(K = 8$ for the former and 12 for the latter). If the argument agrees with an odd integer multiple of $\pi/2$ to more than k bits there is a loss of significant bits from the computed remainder equal to the number of extra bits of agreement, and this causes a loss of accuracy in the result.

2) The difference between **COS** evaluated at sucessive floating-point numbers is given approximately by the absolute error amplification factor, A. For arguments larger than $Cmax$ this difference may

be more than half the significant bits of the result, and so the result is considered to be essentially indeterminate and an **inexact.NaN** is returned. The extra precision of step 2 in the double-precision and T212 special versions is lost if N becomes too large, and the cut-off at $Cmax$ prevents this also.

3) For small arguments the errors are not evenly distributed. As the argument becomes smaller there is an increasing bias towards negative errors (which is to be expected from the form of the Taylor series). For the single-length version and X in $[-0.1, 0.1]$, 62% of the errors are negative, whilst for X in $[-0.01, 0.01]$, 70% of them are.

4) The propagated error has a complex behaviour. The propagated relative error becomes large near each zero of the function, but the propagated absolute error only becomes large for large arguments. In effect, the error is seriously amplified only in an interval about each irrational zero, and the width of this interval increases roughly in proportion to the size of the argument.

5) Since only the remainder of $(|X|+Pi/2)$ by Pi is used in step 3, the symmetry $\cos(x+n\pi) = \pm \cos(x)$ is preserved. Moreover, since the same rational approximation is used as in **SIN**, the relation $\cos(x) = \sin(x + \pi/2)$ is also preserved. However, in each case there is a complication due to the different precision representations of π.

6) The output range is not exceeded. Thus the output of **COS** is always a valid argument for **ACOS**.

TAN

REAL32 FUNCTION TAN (VAL REAL32 X)

REAL64 FUNCTION DTAN (VAL REAL64 X)

These compute: $\tan(X)$ (where X is in radians)

Domain:	$[-Tmax, Tmax]$	$= [-102943.7, 102943.7]$S(Standard),
		$[-2.1 * 10^6, 2.1 * 10^6]$S(T212 special),
		$= [-1.7 * 10^9, 1.7 * 10^9]$D
Range:	$(-\text{Inf}, \text{Inf})$	
Primary Domain:	$[-Pi/4, Pi/4]$	$= [-0.785, 0.785]$

Exceptions

All arguments outside the domain generate an **inexact.NaN**, except $\pm$Inf, which generate an **undefined.NaN**. Odd integer multiples of $\pi/2$ may produce **unstable.NaN**.

Propagated Error

$A = X(1 + \tan^2(X))$, $R = X(1 + \tan^2(X))/\tan(X)$ (See note 3)

Generated error

Primary Domain Error:	MRE	RMSRE
Single Length(Standard):	1.44 ulp	0.39 ulp
Single Length(T212 special):	1.37 ulp	0.39 ulp
Double Length:	1.27 ulp	0.35 ulp

The Algorithm

1 Set N = integer part of $X/(Pi/2)$, and compute the remainder of X by $Pi/2$, using extended precision arithmetic.

2 Compute two floating-point rational functions of the remainder, $XNum$ and $XDen$.

3 If N is odd, set $RESULT = -XDen/XNum$, otherwise set $RESULT = XNum/XDen$.

Notes

1) R is large whenever X is near to an integer multiple of $\pi/2$, and so tan is very sensitive to small errors near its zeros and singularities. Thus for arguments outside the primary domain the accuracy of the result depends crucially on step 2, so this is performed with very high precision, using double precision $Pi/2$ for the standard single-precision function and two double-precision floating-point numbers for the representation of $\pi/2$ for the double-precision function. The T212 special version uses two single-precision floating-point numbers. The extra precision is lost if N becomes too large, and the cut-off $Tmax$ is chosen to prevent this.

2) The difference between **TAN** evaluated at sucessive floating-point numbers is given approximately by the absolute error amplification factor, A. For arguments larger than $Smax$ this difference could be more than half the significant bits of the result, and so the result is considered to be essentially indeterminate and an **inexact.NaN** is returned.

3) Tan is quite badly behaved with respect to errors in the argument. Near its zeros outside the primary domain the relative error is greatly magnified, though the absolute error is only proportional to the size of the argument. In effect, the error is seriously amplified in an interval about each irrational zero, whose width increases roughly in proportion to the size of the argument. Near its singularities both absolute and relative errors become large, so any large output from this function is liable to be seriously contaminated with error, and the larger the argument, the smaller the maximum output which can be trusted. If step 3 of the algorithm requires division by zero, an **unstable.NaN** is produced instead.

4) Since only the remainder of X by $Pi/2$ is used in step 3, the symmetry $\tan(x + n\pi) = \tan(x)$ is preserved, although there is a complication due to the differing precision representations of π. Moreover, by step 3 the symmetry $\tan(x) = 1/\tan(\pi/2 - x)$ is also preserved.

ASIN

> **REAL32 FUNCTION ASIN (VAL REAL32 X)**
>
> **REAL64 FUNCTION DASIN (VAL REAL64 X)**

These compute: $\text{sine}^{-1}(X)$ (in radians)

Domain: $[-1.0, 1.0]$
Range: $[-Pi/2, Pi/2]$
Primary Domain: $[-0.5, 0.5]$

Exceptions

All arguments outside the domain generate an **undefined.NaN**.

Propagated Error

$A = X/\sqrt{1 - X^2}$, $R = X/(\sin^{-1}(X)\sqrt{1 - X^2})$

Generated Error

	Primary Domain		$[-1.0, 1.0]$	
	MRE	**RMSRE**	**MAE**	**RMSAE**
Single Length:	0.58 ulp	0.21 ulp	1.35 ulp	0.33 ulp
Double Length:	0.59 ulp	0.21 ulp	1.26 ulp	0.27 ulp

The Algorithm

1 If $|X| > 0.5$, set $Xwork := \textbf{SQRT}\,((1 - |X|)/2)$. Compute $Rwork = \arcsin(-2 * Xwork)$ with a floating-point rational approximation, and set the result $= Rwork + Pi/2$.

2 Otherwise compute the result directly using the rational approximation.

3 In either case set the sign of the result according to the sign of the argument.

Notes

1) The error amplification factors are large only near the ends of the domain. Thus there is a small interval at each end of the domain in which the result is liable to be contaminated with error: however since both domain and range are bounded the *absolute* error in the result cannot be large.

2) By step 1, the identity $\sin^{-1}(x) = \pi/2 - 2\sin^{-1}(\sqrt{(1-x)/2})$ is preserved.

ACOS

> **REAL32 FUNCTION ACOS (VAL REAL32 X)**
>
> **REAL64 FUNCTION DACOS (VAL REAL64 X)**

These compute: $\text{cosine}^{-1}(X)$ (in radians)

Domain: $[-1.0, 1.0]$
Range: $[0, Pi]$
Primary Domain: $[-0.5, 0.5]$

Exceptions

All arguments outside the domain generate an **undefined.NaN**.

Propagated Error

$$A = -X/\sqrt{1-X^2}, \quad R = -X/(\sin^{-1}(X)\sqrt{1-X^2})$$

Generated Error

	Primary Domain		$[-1.0, 1.0]$	
	MRE	**RMSRE**	**MAE**	**RMSAE**
Single Length:	1.06 ulp	0.38 ulp	2.37 ulp	0.61 ulp
Double Length:	0.96 ulp	0.32 ulp	2.25 ulp	0.53 ulp

The Algorithm

1 If $|X| > 0.5$, set $Xwork := \textbf{SQRT}\,((1-|X|)/2)$. Compute $Rwork = \text{arcsine}(2 * Xwork)$ with a floating-point rational approximation. If the argument was positive, this is the result, otherwise set the result $= Pi - Rwork$.

2 Otherwise compute $Rwork$ directly using the rational approximation. If the argument was positive, set result $= Pi/2 - Rwork$, otherwise result $= Pi/2 + Rwork$.

Notes

1) The error amplification factors are large only near the ends of the domain. Thus there is a small interval at each end of the domain in which the result is liable to be contaminated with error, although this interval is larger near 1 than near -1, since the function goes to zero with an infinite derivative there. However since both the domain and range are bounded the *absolute* error in the result cannot be large.

2) Since the rational approximation is the same as that in **ASIN**, the relation $\cos^{-1}(x) = \pi/2 - \sin^{-1}(x)$ is preserved.

ATAN

> **REAL32 FUNCTION ATAN (VAL REAL32 X)**
>
> **REAL64 FUNCTION DATAN (VAL REAL64 X)**

These compute: $\tan^{-1}(X)$ (in radians)

Domain: [−Inf, Inf]
Range: $[-Pi/2, Pi/2]$
Primary Domain: $[-z, z]$, $z = 2 - \sqrt{3} = 0.2679$

Exceptions

None.

Propagated Error

$A = X/(1 + X^2)$, $R = X/(\tan^{-1}(X)(1 + X^2))$

Generated Error

Primary Domain Error:	**MRE**	**RMSRE**
Single Length:	0.56 ulp	0.21 ulp
Double Length:	0.52 ulp	0.21 ulp

The Algorithm

1 If $|X| > 1.0$, set $Xwork = 1/|X|$, otherwise $Xwork = |X|$.

2 If $Xwork > 2 - \sqrt{3}$, set $F = (Xwork * \sqrt{3} - 1)/(Xwork + \sqrt{3})$, otherwise $F = Xwork$.

3 Compute $Rwork = \arctan(F)$ with a floating-point rational approximation.

4 If $Xwork$ was reduced in (2), set $R = Pi/6 + Rwork$, otherwise $R = Rwork$.

5 If X was reduced in (1), set $RESULT = Pi/2 - R$, otherwise $RESULT = R$.

6 Set the sign of the $RESULT$ according to the sign of the argument.

Notes

1) For $|X| > ATmax$, $|\tan^{-1}(X)|$ is indistinguishable from $\pi/2$ in the floating-point format. For single-length, $ATmax = 1.68 * 10^7$, and for double-length $ATmax = 9 * 10^{15}$, approximately.

2) This function is numerically very stable, despite the complicated argument reduction. The worst errors occur just above $2 - \sqrt{3}$, but are no more than 3.2 ulp.

3) It is also very well behaved with respect to errors in the argument, i.e. the error amplification factors are always small.

4) The argument reduction scheme ensures that the identities $\tan^{-1}(X) = \pi/2 - \tan^{-1}(1/X)$, and $\tan^{-1}(X) = \pi/6 + \tan^{-1}((\sqrt{3} * X - 1)/(\sqrt{3} + X))$ are preserved.

ATAN2

REAL32 FUNCTION ATAN2 (VAL REAL32 X, VAL REAL32 Y)

REAL64 FUNCTION DATAN2 (VAL REAL64 X, VAL REAL64 Y)

These compute the angular co-ordinate $\tan^{-1}(Y/X)$ (in radians) of a point whose X and Y co-ordinates are given.

Domain: [−Inf, Inf] x [−Inf, Inf]
Range: $(-Pi, Pi]$
Primary Domain: See note 2.

Exceptions

(0, 0) and (±Inf,±Inf) give **undefined.NaN**.

Propagated Error

$$A = X(1 \pm Y)/(X^2 + Y^2), \quad R = X(1 \pm Y)/(\tan^{-1}(Y/X)(X^2 + Y^2)) \qquad \text{(See note 3)}$$

Generated Error   (See note 2)

The Algorithm

    1 If X, the first argument, is zero, set the result to $\pm\pi/2$, according to the sign of Y, the second argument.

    2 Otherwise set $Rwork:=$ **ATAN** (Y/X). Then if $Y < 0$ set $RESULT = Rwork - Pi$, otherwise set $RESULT = Pi - Rwork$.

Notes

1) This two-argument function is designed to perform rectangular-to-polar co-ordinate conversion.

2) See the notes for **ATAN** for the primary domain and estimates of the generated error.

3) The error amplification factors were derived on the assumption that the relative error in Y is $\pm$ that in X, otherwise there would be separate factors for X and Y. They are small except near the origin, where the polar co-ordinate system is singular.

SINH

    REAL32 FUNCTION SINH (VAL REAL32 X)

    REAL64 FUNCTION DSINH (VAL REAL64 X)

These compute: $\sinh(X)$

Domain:	$[-Hmax, Hmax]$	$= [-89.4, 89.4]$S, $[-710.5, 710.5]$D
Range:	$(-$**Inf, Inf**$)$	
Primary Domain:	$(-1.0, 1.0)$	

Exceptions

$X < -Hmax$ gives $-$**Inf**, and $X > Hmax$ gives **Inf**.

Propagated Error

$$A = X\cosh(X), \quad R = X\coth(X) \quad \text{(See note 3)}$$

Generated Error

	Primary Domain		$[1.0, XBig]$		(See note 2)
	MRE	RMSRE	MRE	RMSRE	
Single Length:	0.91 ulp	0.26 ulp	1.41 ulp	0.34 ulp	
Double Length:	0.67 ulp	0.22 ulp	1.31 ulp	0.33 ulp	

The Algorithm

    1 If $|X| > XBig$, set $Rwork:=$ **EXP** $(|X| - \ln(2))$.

    2 If $XBig > |X| > 1.0$, set temp:= **EXP** $(|X|)$, and set $Rwork = (temp - 1/temp)/2$.

    3 Otherwise compute $\sinh(|X|)$ with a floating-point rational approximation.

    4 In all cases, set $RESULT = \pm Rwork$ according to the sign of X.

Notes

1) $Hmax$ is the point at which sinh(X) becomes too large to be represented in the floating-point format.

2) $XBig$ is the point at which $e^{-|X|}$ becomes insignificant compared with $e^{|X|}$, (in floating-point). For single-length it is 8.32, and for double-length it is 18.37.

3) This function is quite stable with respect to errors in the argument. Relative error is magnified near zero, but the absolute error is a better measure near the zero of the function and it is diminished there. For large arguments absolute errors are magnified, but since the function is itself large, relative error is a better criterion, and relative errors are not magnified unduly for any argument in the domain, although the output does become less reliable near the ends of the range.

COSH

REAL32 FUNCTION COSH (VAL REAL32 X)

REAL64 FUNCTION DCOSH (VAL REAL64 X)

These compute: cosh(X)

Domain:	$[-Hmax, Hmax]$	$= [-89.4, 89.4]$S, $[-710.5, 710.5]$D
Range:	$[1.0, \textbf{Inf})$	
Primary Domain:	$[-XBig, XBig]$	$= [-8.32, 8.32]$S $[-18.37, 18.37]$D

Exceptions

$|X| > Hmax$ gives **Inf**.

Propagated Error

$A = X \sinh(X)$, $R = X \tanh(X)$ (See note 3)

Generated Error

Primary Domain Error:	**MRE**	**RMS**
Single Length:	1.24 ulp	0.32 ulp
Double Length:	1.24 ulp	0.33 ulp

The Algorithm

1 If $|X| > XBig$, set $result:=$ **EXP** $(|X| - \ln(2))$.

2 Otherwise, set $temp:=$ **EXP** $(|X|)$, and set $result = (temp + 1/temp)/2$.

Notes

1) $Hmax$ is the point at which cosh(X) becomes too large to be represented in the floating-point format.

2) $XBig$ is the point at which $e^{-|X|}$ becomes insignificant compared with $e^{|X|}$ (in floating-point).

3) Errors in the argument are not seriously magnified by this function, although the output does become less reliable near the ends of the range.

TANH

> **REAL32 FUNCTION TANH (VAL REAL32 X)**

> **REAL64 FUNCTION DTANH (VAL REAL64 X)**

These compute: $\tanh(X)$

Domain: [−Inf, Inf]
Range: [1.0, 1.0]
Primary Domain: [−Log(3)/2, Log(3)/2] = [−0.549, 0.549]

Exceptions

None.

Propagated Error

$$A = X/\cosh^2(X), \qquad R = X/\sinh(X)\cosh(X)$$

Generated Error

Primary Domain Error:	MRE	RMS
Single Length:	0.53 ulp	0.2 ulp
Double Length:	0.53 ulp	0.2 ulp

The Algorithm

1 If $|X| > \ln(3)/2$, set $temp := $ **EXP** $(|X|/2)$. Then set $Rwork = 1 - 2/(1 + temp)$.

2 Otherwise compute $Rwork = \tanh(|X|)$ with a floating point rational approximation.

3 In both cases, set $RESULT = \pm Rwork$ according to the sign of X.

Notes

1) As a floating-point number, $\tanh(X)$ becomes indistinguishable from its asymptotic values of ± 1.0 for $|X| > HTmax$, where $HTmax$ is 8.4 for single-length, and 19.06 for double-length. Thus the output of **TANH** is equal to ± 1.0 for such X.

2) This function is very stable and well-behaved, and errors in the argument are always diminished by it.

RAN

> **REAL32,INT32 FUNCTION RAN (VAL INT32 X)**

> **REAL64,INT64 FUNCTION DRAN (VAL INT64 X)**

These produce a pseudo-random sequence of integers, and a corresponding sequence of floating-point numbers between zero and one.

Domain: Integers (see note 1)
Range: [0.0, 1.0] x Integers

Exceptions

None.

The Algorithm

1 Produce the next integer in the sequence: $N_{k+1} = (aN_k + 1)_{mod\,M}$

2 Treat N_{k+1} as a fixed-point fraction in [0,1), and convert it to floating point.

3 Output the floating point result and the new integer.

Notes

1) This function has two results, the first a real, and the second an integer (both 32 bits for single-length, and 64 bits for double-length). The integer is used as the argument for the next call to **RAN**, i.e. it 'carries' the pseudo-random linear congruential sequence N_k, and it should be kept in scope for as long as **RAN** is used. It should be initialised before the first call to **RAN** but not modified thereafter except by the function itself.

2) If the integer parameter is initialised to the same value, the same sequence (both floating-point and integer) will be produced. If a different sequence is required for each run of a program it should be initialised to some 'random' value, such as the output of a timer.

3) The integer parameter can be copied to another variable or used in expressions requiring random integers. The topmost bits are the most random. A random integer in the range $[0, L]$ can conveniently be produced by taking the remainder by $(L + 1)$ of the integer parameter shifted right by one bit. If the shift is not done an integer in the range $[-L, L]$ will be produced.

4) The modulus M is 2^{32} for single-length and 2^{64} for double-length, and the multipliers, a, have been chosen so that all M integers will be produced before the sequence repeats. However several different integers can produce the same floating-point value and so a floating-point output may be repeated, although the *sequence* of such will not be repeated until M calls have been made.

5) The floating-point result is uniformly distributed over the output range, and the sequence passes various tests of randomness, such as the 'run test', the 'maximum of 5 test' and the 'spectral test'.

6) The double-length version is slower to execute, but 'more random' than the single-length version. If a highly-random sequence of single-length numbers is required, this could be produced by converting the output of **DRAN** to single-length. Conversely if only a relatively crude sequence of double-length numbers is required, **RAN** could be used for higher speed and its output converted to double-length.

14.3.2   IMS T414 elementary function library

The version of the library described by this section has been written for 32-bit processors without hardware for floating-point arithmetic. Functions from it will give results very close, but not identical to, those produced by the corresponding functions from the previous library.

This is the version specifically intended to derive maximum performance from the IMS T414. The single-precision functions make use of the **FMUL** instruction available on the B revision of that processor and successor 32-bit processors without floating-point hardware. The library is compiled for transputer class **TB**.

The tables and notes at the beginning of the previous library section apply equally here. However all the functions are contained in one library. To use this library a program header must include the line:

> #USE t4math

ALOG

> **REAL32 FUNCTION ALOG (VAL REAL32 X)**
>
> **REAL64 FUNCTION DALOG (VAL REAL64 X)**

These compute: $\log_e(X)$

Domain:	$(0, XMAX]$
Range:	$[MinLog, MaxLog]$ (See Note 2)
Primary Domain:	$[\sqrt{2}/2, \sqrt{2}) = [0.7071, 1.4142)$

Exceptions

All arguments outside the domain generate an **undefined.NaN**.

Propagated Error

$$A \equiv 1, \qquad R = 1/\log_e(X)$$

Generated Error

Primary Domain Error:	MRE	RMSRE
Single Length:	1.19 ulp	0.36 ulp
Double Length:	2.4 ulp	1.0 ulp

The Algorithm

1 Split X into its exponent N and fraction F.

2 Find the natural log of F with a fixed-point rational approximation, and convert it into a floating-point number LnF.

3 Compute ln(2) $*$ N with extended precision and add it to LnF to get the result.

Notes

1) The term ln(2) $*$ N is much easier to compute (and more accurate) than LnF, and it is larger provided N is not 0 (i.e. for arguments outside the primary domain). Thus the accuracy of the result improves as the modulus of log(X) increases.

2) The minimum value that can be produced, $MinLog$, is the logarithm of the smallest denormalised floating-point number. For single length $Minlog$ is −103.28, and for double length it is −745.2. The maximum value $MaxLog$ is the logarithm of $XMAX$. For single length it is 88.72, and for double-length it is 709.78.

3) Since **Inf** is used to represent *all* values greater than $XMAX$ its logarithm cannot be defined.

4) This function is well-behaved and does not seriously magnify errors in the argument.

ALOG10

> **REAL32 FUNCTION ALOG10 (VAL REAL32 X)**
>
> **REAL64 FUNCTION DALOG10 (VAL REAL64 X)**

These compute: $\log_{10}(X)$

Domain:	$(0, XMAX]$
Range:	$[MinL10, MaxL10]$ (See Note 2)
Primary Domain:	$[\sqrt{2}/2, \sqrt{2}) = [0.7071, 1.4142)$

Exceptions

All arguments outside the domain generate an **undefined.NaN**.

Propagated Error

$$A \equiv \log_{10}(e), \qquad R = \log_{10}(e)/\log_e(X)$$

Generated Error

Primary Domain Error:	MRE	RMSRE
Single Length:	1.43 ulp	0.39 ulp
Double Length:	2.64 ulp	0.96 ulp

The Algorithm

    1 Set $temp := $ **ALOG (X)** .

    2 If $temp$ is a **NaN**, copy it to the output, otherwise set result = $\log(e) * temp$

Notes

1) See note 1 for **ALOG**.

2) The minimum value that can be produced, $MinL10$, is the base-10 logarithm of the smallest denormalised floating-point number. For single length $MinL10$ is -44.85, and for double length it is -323.6. The maximum value $MaxL10$ is the base-10 logarithm of $XMAX$. For single length $MaxL10$ is 38.53, and for double-length it is 308.26.

3) Since **Inf** is used to represent *all* values greater than $XMAX$ its logarithm cannot be defined.

4) This function is well-behaved and does not seriously magnify errors in the argument.

EXP

        REAL32 FUNCTION EXP (VAL REAL32 X)

        REAL64 FUNCTION DEXP (VAL REAL64 X)

These compute: e^X

Domain:	$[-\text{Inf}, MaxLog)$	$= [-\text{Inf}, 88.72)\text{S},$	$[-\text{Inf}, 709.78)\text{D}$
Range:	$[0, \text{Inf})$	(See note 4)	
Primary Domain:	$[-Ln2/2, Ln2/2)$	$= [-0.3466, 0.3466)$	

Exceptions

All arguments outside the domain generate an **Inf**.

Propagated Error

$A = Xe^X, \qquad R = X$

Generated Error

Primary Domain Error:	**MRE**	**RMSRE**
Single Length:	0.51 ulp	0.21 ulp
Double Length:	0.5 ulp	0.21 ulp

The Algorithm

    1 Set N = integer part of $X/\ln(2)$.

    2 Compute the remainder of X by $\ln(2)$, using extended precision arithmetic.

    3 Convert the remainder to fixed-point, compute its exponential using a fixed-point rational function, and convert the result back to floating point.

    4 Increase the exponent of the result by N. If N is sufficiently negative the result must be denormalised.

Notes

1) $MaxLog$ is $\log_e(XMAX)$.

2) The analytical properties of e^x make the relative error of the result proportional to the absolute error of the argument. Thus the accuracy of step 2, which prepares the argument for the rational

approximation, is crucial to the performance of the subroutine. It is completely accurate when N = 0, i.e. in the primary domain, and becomes less accurate as the magnitude of N increases. Since N can attain larger negative values than positive ones, **EXP** is least accurate for large, negative arguments.

3) For sufficiently negative arguments (below −87.34 for single-length and below −708.4 for double-length) the output is denormalised, and so the floating-point number contains progressively fewer significant digits, which degrades the accuracy. In such cases the error can theoretically be a factor of two.

4) Although the true exponential function is never zero, for large negative arguments the true result becomes too small to be represented as a floating-point number, and **EXP** underflows to zero. This occurs for arguments below −103.9 for single-length, and below −745.2 for double-length.

5) The propagated error is considerably magnified for large positive arguments, but diminished for large negative arguments.

POWER

> **REAL32 FUNCTION POWER (VAL REAL32 X, VAL REAL32 Y)**
>
> **REAL32 FUNCTION DPOWER (VAL REAL64 X, VAL REAL64 Y)**

These compute: X^Y

Domain:	[0, Inf] x [−Inf, Inf]
Range:	(Inf, Inf)
Primary Domain:	See note 3.

Exceptions

If the first argument is outside its domain, **undefined.NaN** is returned. If the true value of X^Y exceeds $XMAX$, **Inf** is returned. In certain other cases other **NaN**s are produced: See note 2.

Propagated Error

$A = Y X^Y (1 \pm \log_e(X)), \quad R = Y(1 \pm \log_e(X))$ (See note 4)

Generated Error

Example Range Error:	**MRE**	**RMSRE**	(See note 3)
Single Length:	1.0 ulp	0.24 ulp	
Double Length:	13.2 ulp	1.73 ulp	

The Algorithm

Deal with special cases: either argument = 1, 0, +**Inf** or −**Inf** (see note 2). Otherwise:

(a) For single precision:

    1 Compute $L = \log_2(X)$ in fixed point, where X is the first argument.

    2 Compute $W = Y \times L$ in double precision, where Y is the second argument.

    3 Compute 2^W in fixed point and convert to floating-point result.

(b) For double precision:

    1 Compute $L = \log_2(X)$ in extended precision, where X is the first argument.

    2 Compute $W = Y \times L$ in extended precision, where Y is the second argument.

    3 Compute $RESULT = 2^W$ in extended precision.

Notes

1) This subroutine implements the mathematical function x^y to a much greater accuracy than can be attained using the **ALOG** and **EXP** functions, by performing each step in higher precision.

2) Results for special cases are as follows:

First Input (X)	Second Input (Y)	Result
< 0	ANY	undefined.NaN
0	≤ 0	undefined.NaN
0	$0 < Y \leq XMAX$	0
0	Inf	unstable.NaN
$0 < X < 1$	Inf	0
$0 < X < 1$	−Inf	Inf
1	$-XMAX \leq Y \leq XMAX$	1
1	± Inf	unstable.NaN
$1 < X \leq XMAX$	Inf	Inf
$1 < X \leq XMAX$	−Inf	0
Inf	$1 \leq Y \leq Inf$	Inf
Inf	$-Inf \leq Y \leq -1$	0
Inf	$-1 < Y < 1$	undefined.NaN
otherwise	0	1
otherwise	1	X

3) Performing all the calculations in extended precision makes the double-precision algorithm very complex in detail, and having two arguments makes a primary domain difficult to specify. As an indication of accuracy, the functions were evaluated at 100 000 points logarithmically distributed over (0.1, 10.0), with the exponent linearly distributed over $(-35.0, 35.0)$ (single-length), and $(-300.0, 300.0)$ (double-length), producing the errors given above. The errors are much smaller if the exponent range is reduced.

4) The error amplification factors are calculated on the assumption that the relative error in Y is ± that in X, otherwise there would be separate factors for both X and Y. It can be seen that the propagated error will be greatly amplified whenever $\log_e(X)$ or Y is large.

SIN

REAL32 FUNCTION SIN (VAL REAL32 X)

REAL64 FUNCTION DSIN (VAL REAL64 X)

These compute: sine(X) (where X is in radians)

Domain: $[-Smax, Smax]$ $= [-12868.0, 12868.0]$S, $[-2.1*10^8, 2.1*10^8]$D
Range: $[-1.0, 1.0]$
Primary Domain: $[-Pi/2, Pi/2]$ $= [-1.57, 1.57]$

Exceptions

All arguments outside the domain generate an **inexact.NaN**, except ±**Inf**, which generates an **undefined.NaN**.

Propagated Error

$A = X \cos(X)$, $R = X \cot(X)$

Generated Error (See note 3)

Range:

	Primary Domain		$[0, 2Pi]$	
	MRE	**RMSRE**	**MAE**	**RMSAE**
Single Length:	0.65 ulp	0.22 ulp	0.74 ulp	0.18 ulp
Double Length:	0.56 ulp	0.21 ulp	0.64 ulp	0.16 ulp

The Algorithm

1 Set N = integer part of $|X|/Pi$.

2 Compute the remainder of $|X|$ by Pi, using extended precision arithmetic.

3 Convert the remainder to fixed-point, compute its sine using a fixed-point rational function, and convert the result back to floating point.

4 Adjust the sign of the result according to the sign of the argument and the evenness of N.

Notes

1) For arguments outside the primary domain the accuracy of the result depends crucially on step 2. The extended precision corresponds to K extra bits in the representation of π ($K = 8$ for single-length and 12 for double-length). If the argument agrees with an integer multiple of π to more than K bits there is a loss of significant bits in the remainder, equal to the number of extra bits of agreement, and this causes a loss of accuracy in the result.

2) The extra precision of step 2 is lost if N becomes too large, and the cut-off $Smax$ is chosen to prevent this. In any case for large arguments the 'granularity' of floating-point numbers becomes a significant factor. For arguments larger than $Smax$ a change in the argument of 1 ulp would change more than half of the significant bits of the result, and so the result is considered to be essentially indeterminate.

3) The propagated error has a complex behaviour. The propagated relative error becomes large near each zero of the function (outside the primary range), but the propagated absolute error only becomes large for large arguments. In effect, the error is seriously amplified only in an interval about each irrational zero, and the width of this interval increases roughly in proportion to the size of the argument.

4) Since only the remainder of X by Pi is used in step 3, the symmetry $\sin(x + n\pi) = \pm\sin(x)$ is preserved, although there is a complication due to differing precision representations of π.

5) The output range is not exceeded. Thus the output of **SIN** is always a valid argument for **ASIN**.

COS

REAL32 FUNCTION COS (VAL REAL32 X)

REAL64 FUNCTION DCOS (VAL REAL64 X)

These compute: cosine (X) (where X is in radians)

Domain: $[-Smax, Smax]$ = $[-12868.0, 12868.0]$S, $[-2.1 * 10^8, 2.1 * 10^8]$D
Range: $[-1.0, 1.0]$
Primary Domain: See note 1.

Exceptions

All arguments outside the domain generate an **inexact.NaN**, except $\pm$**Inf**, which generates an **undefined.NaN**.

Propagated Error

$$A = -X\sin(X), \qquad R = -X\tan(X) \quad \text{(See note 4)}$$

Generated Error

Range:

	[0, Pi/4)		[0, 2Pi]	
	MRE	RMSRE	MAE	RMSAE
Single Length:	1.0 ulp	0.28 ulp	0.81 ulp	0.17 ulp
Double Length:	0.93 ulp	0.26 ulp	0.76 ulp	0.18 ulp

The Algorithm

1 Set N = integer part of $(|X| + Pi/2)/Pi$.

2 Compute the remainder of $(|X| + Pi/2)$ by Pi, using extended precision arithmetic.

3 Compute the remainder to fixed-point, compute its sine using a fixed-point rational function, and convert the result back to floating point.

4 Adjust the sign of the result according to the evenness of N.

Notes

1) Inspection of the algorithm shows that argument reduction always occurs, thus there is no 'primary domain' for **COS**. So for all arguments the acuracy of the result depends crucially on step 2. The extended precision corresponds to K extra bits in the representation of π ($K = 8$ for single-length and 12 for double length). If the argument agrees with an odd integer multiple of $\pi/2$ to more than K bits there is a loss of significant bits in the remainder, equal to the number of extra bits of agreement, and this causes a loss of accuracy in the result.

2) The extra precision of step 2 is lost if N becomes too large, and the cut-off $Smax$ is chosen to prevent this. In any case for large arguments the 'granularity' of floating-point numbers becomes a significant factor. For arguments larger than $Smax$ a change in the argument of 1 ulp would change more than half of the significant bits of the result, and so the result is considered to be essentially indeterminate.

3) For small arguments the errors are not evenly distributed. As the argument becomes smaller there is an increasing bias towards negative errors (which is to be expected from the form of the Taylor series). For the single-length version and X in $[-0.1, 0.1]$, 62% of the errors are negative, whilst for X in $[-0.01, 0.01]$, 70% of them are.

4) The propagated error has a complex behaviour. The propagated relative error becomes large near each zero of the function, but the propagated absolute error only becomes large for large arguments. In effect, the error is seriously amplified only in an interval about each irrational zero, and the width of this interval increases roughly in proportion to the size of the argument.

5) Since only the remainder of $(|X|+Pi/2)$ by Pi is used in step 3, the symmetry $\cos(x+n\pi) = \pm\cos(x)$ is preserved. Moreover, since the same rational approximation is used as in **SIN**, the relation $\cos(x) = \sin(x + \pi/2)$ is also preserved. However, in each case there is a complication due to the different precision representations of π.

6) The output range is not exceeded. Thus the output of **COS** is always a valid argument for **ACOS**.

TAN

```
REAL32 FUNCTION TAN (VAL REAL32 X)

REAL64 FUNCTION DTAN (VAL REAL64 X)
```

These compute: $\tan(X)$ (where X is in radians)

Domain:	$[-Tmax, Tmax]$	$= [-6434.0, 6434.0]$S	$[-1.05 * 10^8, 1.05 * 10^8]$D
Range:	$(-$Inf, Inf$)$		
Primary Domain:	$[-Pi/4, Pi/4]$	$= [-0.785, 0.785]$	

Exceptions

All arguments outside the domain generate an **inexact.NaN**, except $\pm$Inf, which generate an **unde-fined.NaN**. Odd integer multiples of $\pi/2$ may produce **unstable.NaN**.

Propagated Error

$$A = X(1 + \tan^2(X)), \quad R = X(1 + \tan^2(X))/\tan(X) \quad \text{(See note 4)}$$

Generated Error

Primary Domain Error:	**MRE**	**RMSRE**
Single Length:	3.5 ulp	0.23 ulp
Double Length:	0.69 ulp	0.23 ulp

The Algorithm

1 Set N – integer part of $X/(Pi/2)$.

2 Compute the remainder of X by $Pi/2$, using extended precision arithmetic.

3 Convert the remainder to fixed point, compute its tangent using a fixed-point rational func-tion, and convert the result back to floating point.

4 If N is odd, take the reciprocal.

5 Set the sign of the result according to the sign of the argument.

Notes

1) R is large whenever X is near to an integer multiple of $\pi/2$, and so tan is very sensitive to small errors near its zeros and singularities. Thus for arguments outside the primary domain the acuracy of the result depends crucially on step 2. The extended precision corresponds to K extra bits in the representation of $\pi/2$ ($K = 8$ for single-length and 12 for double-length). If the argument agrees with an integer multiple of $\pi/2$ to more than K bits there is a loss of significant bits in the remainder, approximately equal to the number of extra bits of agreement, and this causes a loss of accuracy in the result.

2) The extra precision of step 2 is lost if N becomes too large, and the cut-off $Tmax$ is chosen to prevent this. In any case for large arguments the 'granularity' of floating-point numbers becomes a significant factor. For arguments larger than $Tmax$ a change in the argument of 1 ulp would change more than half of the significant bits of the result, and so the result is considered to be essentially indeterminate.

3) Step 3 of the algorithm has been slightly modified in the double-precision version from that given in Cody & Waite to avoid fixed-point underflow in the polynomial evaluation for small arguments.

4) Tan is quite badly behaved with respect to errors in the argument. Near its zeros outside the primary domain the relative error is greatly magnified, though the absolute error is only proportional to the size of the argument. In effect, the error is seriously amplified in an interval about each irrational zero, whose width increases roughly in proportion to the size of the argument. Near its singularities both absolute and relative errors become large, so any large output from this function is liable to be seriously contaminated with error, and the larger the argument, the smaller the maximum output which can be trusted. If step 4 of the algorithm requires division by zero, an **unstable.NaN** is produced instead.

5) Since only the remainder of X by $Pi/2$ is used in step 3, the symmetry $\tan(x + n\pi) = \tan(x)$ is preserved, although there is a complication due to the differing precision representations of π. Moreover, by step 4 the symmetry $\tan(x) = 1/\tan(\pi/2 - x)$ is also preserved.

ASIN

> **REAL32 FUNCTION ASIN (VAL REAL32 X)**
>
> **REAL64 FUNCTION DASIN (VAL REAL64 X)**

These compute: $\mathrm{sine}^{-1}(X)$ (in radians)

Domain: $[-1.0, 1.0]$
Range: $[-Pi/2, Pi/2]$
Primary Domain: $[-0.5, 0.5]$

Exceptions

All arguments outside the domain generate an **undefined.NaN**.

Propagated Error

$A = X/\sqrt{1 - X^2}, \quad R = X/(\sin^{-1}(X)\sqrt{1 - X^2})$

Generated Error

	Primary Domain		$[-1.0, 1.0]$	
	MRE	RMSRE	MAE	RMSAE
Single Length:	0.53 ulp	0.21 ulp	1.35 ulp	0.33 ulp
Double Length:	2.8 ulp	1.4 ulp	2.34 ulp	0.64 ulp

The Algorithm

1 If $|X| > 0.5$, set $Xwork := $ **SQRT** $((1 - |X|)/2)$.
   Compute $Rwork = \mathrm{arcsine}(-2 * Xwork)$ with a floating-point rational approximation, and set the result $= Rwork + Pi/2$.

2 Otherwise compute the result directly using the rational approximation.

3 In either case set the sign of the result according to the sign of the argument.

Notes

1) The error amplification factors are large only near the ends of the domain. Thus there is a small interval at each end of the domain in which the result is liable to be contaminated with error: however since both domain and range are bounded the *absolute* error in the result cannot be large.

2) By step 1, the identity $\sin^{-1}(x) = \pi/2 - 2\sin^{-1}(\sqrt{(1 - x)/2})$ is preserved.

ACOS

> **REAL32 FUNCTION ACOS (VAL REAL32 X)**
>
> **REAL64 FUNCTION DACOS (VAL REAL64 X)**

These compute: $\mathrm{cosine}^{-1}(X)$ (in radians)

Domain: $[-1.0, 1.0]$
Range: $[0, Pi]$
Primary Domain: $[-0.5, 0.5]$

Exceptions

All arguments outside the domain generate an **undefined.NaN**.

Propagated Error

$$A = -X/\sqrt{1 - X^2}, \quad R = -X/(\sin^{-1}(X)\sqrt{1 - X^2})$$

Generated Error

	Primary Domain	[−1.0, 1.0]		
	MRE	RMSRE	MAE	RMSAE
Single Length:	1.1 ulp	0.38 ulp	2.4 ulp	0.61 ulp
Double Length:	1.3 ulp	0.34 ulp	2.9 ulp	0.78 ulp

The Algorithm

1 If $|X| > 0.5$, set $Xwork := \textbf{SQRT}((1 - |X|)/2)$. Compute $Rwork = \text{arcsine}\ (2 * Xwork)$ with a floating-point rational approximation. If the argument was positive, this is the result, otherwise set the result $= Pi - Rwork$.

2 Otherwise compute $Rwork$ directly using the rational approximation. If the argument was positive, set result $= Pi/2 - Rwork$, otherwise result $= Pi/2 + Rwork$.

Notes

1) The error amplification factors are large only near the ends of the domain. Thus there is a small interval at each end of the domain in which the result is liable to be contaminated with error, although this interval is larger near 1 than near −1, since the function goes to zero with an infinite derivative there. However since both the domain and range are bounded the *absolute* error in the result cannot be large.

2) Since the rational approximation is the same as that in **ASIN**, the relation $\cos^{-1}(x) = \pi/2 - \sin^{-1}(x)$ is preserved.

ATAN

REAL32 FUNCTION ATAN (VAL REAL32 X)

REAL64 FUNCTION DATAN (VAL REAL64 X)

These compute: $\tan^{-1}(X)$ (in radians)

Domain: [−Inf, Inf]
Range: $[-Pi/2, Pi/2]$
Primary Domain: $[-z, z], \quad z = 2 - \sqrt{3} = 0.2679$

Exceptions

None.

Propagated Error

$$A = X/(1 + X^2), \quad R = X/(\tan^{-1}(X)(1 + X^2))$$

Generated Error

Primary Domain Error:	MRE	RMSRE
Single Length:	0.53 ulp	0.21 ulp
Double Length:	1.27 ulp	0.52 ulp

The Algorithm

1 If $|X| > 1.0$, set $Xwork = 1/|X|$, otherwise $Xwork = |X|$.

2 If $Xwork > 2 - \sqrt{3}$, set $F = (Xwork * \sqrt{3} - 1)/(Xwork + \sqrt{3})$, otherwise $F = Xwork$.

3 Compute $Rwork = \arctan(F)$ with a floating-point rational approximation.

4 If $Xwork$ was reduced in (2), set $R = Pi/6 + Rwork$, otherwise $R = Rwork$.

5 If X was reduced in (1), set $RESULT = Pi/2 - R$, otherwise $RESULT = R$.

6 Set the sign of the $RESULT$ according to the sign of the argument.

Notes

1) For $|X| > ATmax$, $|\tan^{-1}(X)|$ is indistinguishable from $\pi/2$ in the floating-point format. For single-length, $ATmax = 1.68 * 10^7$, and for double-length $ATmax = 9 * 10^{15}$, approximately.

2) This function is numerically very stable, despite the complicated argument reduction. The worst errors occur just above $2 - \sqrt{3}$, but are no more than 1.8 ulp.

3) It is also very well behaved with respect to errors in the argument, i.e. the error amplification factors are always small.

4) The argument reduction scheme ensures that the identities $\tan^{-1}(X) = \pi/2 - \tan^{-1}(1/X)$, and $\tan^{-1}(X) = \pi/6 + \tan^{-1}((\sqrt{3} * X - 1)/(\sqrt{3} + X))$ are preserved.

ATAN2

REAL32 FUNCTION ATAN2 (VAL REAL32 X, VAL REAL32 Y)

REAL64 FUNCTION DATAN2 (VAL REAL64 X, VAL REAL64 Y)

These compute the angular co-ordinate $\tan^{-1}(Y/X)$ (in radians) of a point whose X and Y co-ordinates are given.

Domain: [−Inf, Inf] x [−Inf, Inf]
Range: $(-Pi, Pi]$
Primary Domain: See note 2.

Exceptions

(0, 0) and ($\pm$Inf,$\pm$Inf) give **undefined.NaN**.

Propagated Error

$A = X(1 \pm Y)/(X^2 + Y^2)$, $R = X(1 \pm Y)/(\tan^{-1}(Y/X)(X^2 + Y^2))$ (See note 3)

Generated Error

See note 2.

The Algorithm

1 If X, the first argument, is zero, set the result to $\pm\pi/2$, according to the sign of Y, the second argument.

2 Otherwise set $Rwork := \textbf{ATAN}(Y/X)$. Then if $Y < 0$ set $RESULT = Rwork - Pi$, otherwise set $RESULT = Pi - Rwork$.

Notes

1) This two-argument function is designed to perform rectangular-to-polar co-ordinate conversion.

2) See the notes for **ATAN** for the primary domain and estimates of the generated error.

3) The error amplification factors were derived on the assumption that the relative error in Y is $\pm$ that in X, otherwise there would be separate factors for X and Y. They are small except near the origin, where the polar co-ordinate system is singular.

SINH

> **REAL32 FUNCTION SINH (VAL REAL32 X)**
>
> **REAL64 FUNCTION DSINH (VAL REAL64 X)**

These compute: $\sinh(X)$

Domain: $[-Hmax, Hmax]$ $= [-89.4, 89.4]$S, $[-710.5, 710.5]$D
Range: $(-\text{Inf}, \text{Inf})$
Primary Domain: $(-1.0, 1.0)$

Exceptions

$X < -Hmax$ gives $-\text{Inf}$, and $X > Hmax$ gives Inf.

Propagated Error

$A = X\cosh(X)$, $\quad R = X\coth(X)$ (See note 3)

Generated Error

	Primary Domain		$[1.0, XBig]$		(See note 2)
	MRE	RMSRE	MRE	RMSRE	
Single Length:	0.89 ulp	0.3 ulp	0.98 ulp	0.31 ulp	
Double Length:	1.3 ulp	0.51 ulp	1.0 ulp	0.3 ulp	

The Algorithm

1 If $|X| > XBig$, set $Rwork := \textbf{EXP}\,(|X| - ln(2))$.

2 If $XBig \geq |X| \geq 1.0$, set $temp := \textbf{EXP}\,(|X|)$, and set $Rwork = (temp - 1/temp)/2$.

3 Otherwise compute $Rwork = \sinh(|X|)$ with a fixed-point rational approximation.

4 In all cases, set $RESULT = \pm Rwork$ according to the sign of X.

Notes

1) $Hmax$ is the point at which $\sinh(X)$ becomes too large to be represented in the floating-point format.

2) $XBig$ is the point at which $e^{-|X|}$ becomes insignificant compared with $e^{|X|}$, (in floating-point). For single-length it is 8.32, and for double-length it is 18.37.

3) This function is quite stable with respect to errors in the argument. Relative error is magnified near zero, but the absolute error is a better measure near the zero of the function and it is diminished there. For large arguments absolute errors are magnified, but since the function is itself large, relative error is a better criterion, and relative errors are not magnified unduly for any argument in the domain, although the output does become less reliable near the ends of the range.

COSH

> **REAL32 FUNCTION COSH (VAL REAL32 X)**
>
> **REAL64 FUNCTION DCOSH (VAL REAL64 X)**

These compute: $\cosh(X)$

Domain: $[-Hmax, Hmax]$ $= [-89.4, 89.4]$S, $[-710.5, 710.5]$D
Range: $[1.0,$ **Inf**$)$
Primary Domain: $[-XBig, XBig]$ $= [-8.32, 8.32]$S $[-18.37, 18.37]$D

Exceptions

$|X| > Hmax$ gives **Inf**.

Propagated Error

$A = X \sinh(X),$ $R = X \tanh(X)$ (See note 3)

Generated Error

Primary Domain Error: **MRE**   **RMS**
Single Length: 0.99 ulp   0.3 ulp
Double Length: 1.23 ulp   0.3 ulp

The Algorithm

1 If $|X| > XBig$, set $result := $ **EXP** $(|X| - \ln(2))$.

2 Otherwise, set $temp := $ **EXP** $(|X|)$, and set $result = (temp + 1/temp)/2$.

Notes

1) $Hmax$ is the point at which $\cosh(X)$ becomes too large to be represented in the floating-point format.

2) $XBig$ is the point at which $e^{-|X|}$ becomes insignificant compared with $e^{|X|}$ (in floating-point).

3) Errors in the argument are not seriously magnified by this function, although the output does become less reliable near the ends of the range.

TANH

> **REAL32 FUNCTION TANH (VAL REAL32 X)**
>
> **REAL64 FUNCTION DTANH (VAL REAL64 X)**

These compute: $\tanh(X)$

Domain: $[-$**Inf**, **Inf**$]$
Range: $[-1.0, 1.0]$
Primary Domain: $[-Log(3)/2, Log(3)/2] =$ $[-0.549, 0.549]$

Exceptions

None.

Propagated Error

$A = X/\cosh^2(X),$ $R = X/\sinh(X)\cosh(X)$

Generated Error

Primary Domain Error: MRE   RMS
Single Length: 0.52 ulp   0.2 ulp
Double Length: 4.6 ulp   2.6 ulp

The Algorithm

1 If $|X| > \ln(3)/2$, set $temp := \textbf{EXP}\,(|X|/2)$. Then set $Rwork = 1 - 2/(1 + temp)$.

2 Otherwise compute $Rwork = \tanh(|X|)$ with a floating-point rational approximation.

3 In both cases, set $RESULT = \pm Rwork$ according to the sign of X.

Notes

1) As a floating-point number, $\tanh(X)$ becomes indistinguishable from its asymptotic values of ± 1.0 for $|X| > HTmax$, where $HTmax$ is 8.4 for single-length, and 19.06 for double-length. Thus the output of **TANH** is equal to ± 1.0 for such X.

2) This function is very stable and well-behaved, and errors in the argument are always diminished by it.

RAN

REAL32,INT32 FUNCTION RAN (VAL INT32 X)

REAL64,INT64 FUNCTION DRAN (VAL INT64 X)

These produce a pseudo-random sequence of integers, and a corresponding sequence of floating-point numbers between zero and one

Domain: Integers (see note 1)
Range: [0.0, 1.0] x Integers

Exceptions

None.

The Algorithm

1 Produce the next integer in the sequence: $N_{k+1} = (aN_k + 1)_{mod\,M}$

2 Treat N_{k+1} as a fixed-point fraction in [0,1), and convert it to floating point.

3 Output the floating point result and the new integer.

Notes

1) This function has two results, the first a real, and the second an integer (both 32 bits for single-length, and 64 bits for double-length). The integer is used as the argument for the next call to **RAN**, i.e. it 'carries' the pseudo-random linear congruential sequence N_k, and it should be kept in scope for as long as **RAN** is used. It should be initialised before the first call to **RAN** but not modified thereafter except by the function itself.

2) If the integer parameter is initialised to the same value, the same sequence (both floating-point and integer) will be produced. If a different sequence is required for each run of a program it should be initialised to some 'random' value, such as the output of a timer.

3) The integer parameter can be copied to another variable or used in expressions requiring random integers. The topmost bits are the most random. A random integer in the range [0, L] can conveniently be produced by taking the remainder by $(L + 1)$ of the integer parameter shifted right by one bit. If the shift is not done an integer in the range $[-L, L]$ will be produced.

4) The modulus M is 2^{32} for single-length and 2^{64} for double-length, and the multipliers, a, have been chosen so that all M integers will be produced before the sequence repeats. However several different integers can produce the same floating-point value and so a floating-point output may be repeated, although the *sequence* of such will not be repeated until M calls have been made.

5) The floating-point result is uniformly distributed over the output range, and the sequence passes various tests of randomness, such as the 'run test', the 'maximum of 5 test' and the 'spectral test'.

6) The double-length version is slower to execute, but 'more random' than the single-length version. If a highly-random sequence of single-length numbers is required, this could be produced by converting the output of **DRAN** to single-length. Conversely if only a relatively crude sequence of double-length numbers is required, **RAN** could be used for higher speed and its output converted to double-length.

14.4   Introduction to input/output libraries (`hostlibs`, `iolibs`)

These libraries support a wide variety of operations commonly classified as input/output. This includes conversions of representation in both directions between occam types and text strings, general text string manipulation and procedures for communicating values in various styles between processes using the occam channel model of communication.

An occam program may use procedures drawn from any combination of these libraries if an appropriate **#USE** directive is inserted in place of a declaration for each library referenced. If a program uses named constants used by the libraries it will be necessary also to include **#USE** directives for the necessary constant libraries.

If library procedures are used whose formal parameters include channels with named protocols, then the library containing the declaration of these protocols must be **#USE**d before the library declaring the procedures. For convenience all protocols used by libraries included in the TDS are declared in a single library **strmhdr**. A **#USE** for **strmhdr** is supplied automatically whenever an **EXE** or **UTIL** is compiled, but must be supplied by the user if required in a **PROGRAM** or an **SC**.

The structure of the remainder of this chapter is as follows:

This section continues with a general discussion of the models of input and output supported.

Tables of all the procedures in the libraries are then presented.

Then the individual procedures are grouped into 7 principal subject groups for each of which is given a general introduction and then detailed specifications of all the procedures.

The groups are:

- Environment enquiries
- Representation conversions and string handling
- Terminals and text streams
- Buffers, multiplexors and protocol converters
- Access to host filing system
- Access to the TDS's folded file store
- Access to transputer board peripherals

The name of the library in which the procedure may be found appears at the top right hand side of each specification.

The actions performed by **hostlibs** procedures corresponding directly to **iserver** commands are described in further detail in section 16.5.

14.5   Tables of contents of the input/output libraries (`hostlibs`, `iolibs`)

The input output libraries are in two main groups **hostlibs** and **iolibs** which are held in separate directories. These groupings are mainly historical but reflect in particular the addition of the **iserver** interface using **SP** protocol in the current version of the TDS. Procedures in **iolibs** do not use SP protocol. Procedures in **hostlibs** either use this protocol or are designed for use in conjunction with procedures which do.

A compilation unit using any of the protocols **SP**, **SS** or **KS** must include (explicitly or implicitly) the line:

>   **#USE strmhdr**

above the **#USE** for any library which declares one or more procedures using one of the protocols.

This line is supplied implicitly by the compiler for **EXE**s and **UTIL**s which are compiled with the parameter **tds2.style.exe** FALSE. In such **EXE**s and **UTIL**s the user must not include a **#USE** for **strmhdr**.

The name of the group to which the library belongs appears at the top right hand side of each table.

14.5.1   Basic type i/o conversion library `ioconv`   `iolibs`

Procedure	Parameter Specifiers
STRINGTOINT	BOOL Error, INT n, VAL []BYTE string
INTTOSTRING	INT len, []BYTE string, VAL INT n
STRINGTOHEX	BOOL Error, INT n, VAL []BYTE string
HEXTOSTRING	INT len, []BYTE string, VAL INT n
STRINGTOBOOL	BOOL Error, b, VAL []BYTE string
BOOLTOSTRING	INT len, []BYTE string, VAL BOOL b

To use this library a program header (all targets) must include the line:

>   **#USE ioconv**

The number to string conversion procedures are defined in the occam 2 reference manual. Input conversion procedures return two results, a boolean error indication and the converted value. Output conversions all return an integer which is the number of significant characters written into the string.

14.5.2   Extra type i/o conversion library `extrio`   `iolibs`

Procedure	Parameter Specifiers
STRINGTOINT16	BOOL Error, INT16 n, VAL []BYTE string
INT16TOSTRING	INT len, []BYTE string, VAL INT16 n
STRINGTOINT32	BOOL Error, INT32 n, VAL []BYTE string
INT32TOSTRING	INT len, []BYTE string, VAL INT32 n
STRINGTOINT64	BOOL Error, INT64 n, VAL []BYTE string
INT64TOSTRING	INT len, []BYTE string, VAL INT64 n
STRINGTOHEX16	BOOL Error, INT16 n, VAL []BYTE string
HEX16TOSTRING	INT len, []BYTE string, VAL INT16 n
STRINGTOHEX32	BOOL Error, INT32 n, VAL []BYTE string
HEX32TOSTRING	INT len, []BYTE string, VAL INT32 n
STRINGTOHEX64	BOOL Error, INT64 n, VAL []BYTE string
HEX64TOSTRING	INT len, []BYTE string, VAL INT64 n
STRINGTOREAL32	BOOL Error, REAL32 X, VAL []BYTE string
REAL32TOSTRING	INT len, []BYTE string, VAL REAL32 X, VAL INT Ip, Dp
STRINGTOREAL64	BOOL Error, REAL64 X, VAL []BYTE string
REAL64TOSTRING	INT len, []BYTE string, VAL REAL64 X, VAL INT Ip, Dp

To use this library a program header (all targets) must include the line:

> #USE extrio

For further information on the procedures provided by this library see the occam 2 Reference Manual.

14.5.3   String handling library `strings`   `iolibs`

Result	Function	Parameter Specifiers
BOOL	is.in.range	VAL BYTE char, bottom, top
BOOL	is.upper	VAL BYTE char
BOOL	is.lower	VAL BYTE char
BOOL	is.digit	VAL BYTE char
BOOL	is.hex.digit	VAL BYTE char
BOOL	is.id.char	VAL BYTE char
	to.upper.case	[]BYTE str
	to.lower.case	[]BYTE str
INT	compare.strings	VAL []BYTE str1, str2
BOOL	eqstr	VAL []BYTE s1,s2
	str.shift	[]BYTE str, VAL INT start, len, shift, BOOL not.done
	delete.string	INT len, []BYTE str, VAL INT start, size, BOOL not.done
	insert.string	VAL []BYTE new.str, INT len, []BYTE str, VAL INT start, BOOL not.done
INT	string.pos	VAL []BYTE search, str
INT	char.pos	VAL BYTE search, VAL []BYTE str
INT, BYTE	search.match	VAL []BYTE possibles, str
INT, BYTE	search.no.match	VAL []BYTE possibles, str

Procedure	Parameter Specifiers
add.char	INT len, []BYTE str, VAL BYTE char
append.char	INT len, []BYTE str, VAL BYTE char
add.text	INT len, []BYTE str, VAL []BYTE text
append.text	INT len, []BYTE str, VAL []BYTE text
add.int	INT len, []BYTE str, VAL INT number, width
append.int	INT len, []BYTE str, VAL INT number, width
add.hex.int	INT len, []BYTE str, VAL INT number, width
append.hex.int	INT len, []BYTE str, VAL INT number, width
add.real32	INT len, []BYTE str, VAL REAL32 number, VAL INT Ip,Dp
append.real32	INT len, []BYTE str, VAL REAL32 number, VAL INT Ip,Dp
add.real64	INT len, []BYTE str, VAL REAL64 number, VAL INT Ip,Dp
append.real64	INT len, []BYTE str, VAL REAL64 number, VAL INT Ip,Dp
add.int32	INT len, []BYTE str, VAL INT32 number, VAL INT width
append.int32	INT len, []BYTE str, VAL INT32 number, VAL INT width
add.int64	INT len, []BYTE str, VAL INT64 number, VAL INT width
append.int64	INT len, []BYTE str, VAL INT64 number, VAL INT width
add.hex.int32	INT len, []BYTE str, VAL INT32 number, VAL INT width
append.hex.int32	INT len, []BYTE str, VAL INT32 number, VAL INT width
add.hex.int64	INT len, []BYTE str, VAL INT64 number, VAL INT width
append.hex.int64	INT len, []BYTE str, VAL INT64 number, VAL INT width
next.word.from.line	VAL []BYTE line, INT ptr,len, []BYTE word,BOOL ok
next.int.from.line	VAL []BYTE line, INT ptr, number, BOOL ok

To use this library a program header (all targets) must include the line:

    #USE strings

The procedures and functions in this group provide the basis for string handling in occam. They are consistent with the absence of dynamic space allocation, insofar as they work in terms of a declared array and a used part of that array defined by an upper bound.

These functions and procedures facilitate simple manipulation of names, commands, replies, etc.

See section 14.8

14.5.4 Host i/o basic procedure library splib   hostlibs

The **SP** protocol used on channels **fs**, **ts** of these procedures is independent of word length and is implemented for all transputer target types. These procedures correspond directly to **iserver** commands.

Procedure	Parameter Specifiers
so.open	CHAN OF SP fs, ts, VAL[]BYTE name, VAL BYTE type, mode INT32 streamid, BYTE result
so.close	CHAN OF SP fs, ts, VAL INT32 streamid, BYTE result
so.read	CHAN OF SP fs, ts, VAL INT32 streamid, INT bytes.read, []BYTE data
so.write	CHAN OF SP fs, ts, VAL INT32 streamid, VAL[] BYTE data, INT length
so.gets	CHAN OF SP fs, ts, VAL INT32 streamid, INT bytes.read, []BYTE data, BYTE result
so.puts	CHAN OF SP fs, ts, VAL INT32 streamid, VAL[]BYTE data, BYTE result
so.flush	CHAN OF SP fs, ts, VAL INT32 streamid, BYTE result
so.seek	CHAN OF SP fs, ts, VAL INT32 streamid, offset, origin, BYTE result
so.tell	CHAN OF SP fs, ts, VAL INT32 streamid, INT32 position, BYTE result
so.eof	CHAN OF SP fs, ts, VAL INT32 streamid, BYTE result
so.ferror	CHAN OF SP fs, ts, VAL INT32 streamid, INT32 error.no, INT length, []BYTE message, BYTE result
so.remove	CHAN OF SP fs, ts, VAL[]BYTE name, BYTE result
so.rename	CHAN OF SP fs, ts, VAL[]BYTE oldname, newname, BYTE result
so.getkey	CHAN OF SP fs, ts, BYTE key, result
so.pollkey	CHAN OF SP fs, ts, BYTE key, result
so.getenv	CHAN OF SP fs, ts, VAL[]BYTE name, INT length, []BYTE value, BYTE result
so.time	CHAN OF SP fs, ts, INT32 localtime, UTCtime
so.system	CHAN OF SP fs, ts, VAL[]BYTE command, INT32 status, BYTE result
so.exit	CHAN OF SP fs, ts, VAL INT32 status
so.commandline	CHAN OF SP fs, ts, VAL BYTE all, INT length, []BYTE string, BYTE result
so.core	CHAN OF SP fs, ts, VAL INT32 offset, INT bytes.read, []BYTE data, BYTE result
so.version	CHAN OF SP fs, ts, BYTE version, host, os, board

To use this library a program header (all targets) must include the line:

```
#USE splib
```

14.5.5   Hostio general and screen output procedure library `solib`   `hostlibs`

Procedure	Parameter Specifiers
so.open.temp	CHAN OF SP fs, ts, VAL BYTE type, [so.temp.filename.length]BYTE filename, INT32 streamid, BYTE result
so.test.exists	CHAN OF SP fs, ts, VAL[]BYTE filename, BOOL exists
so.popen.read	CHAN OF SP fs, ts, VAL[]BYTE filename, path.variable.name, VAL BYTE open.type, INT full.len, []BYTE full.name, INT32 stream.id, BYTE result
so.parse.command.line	CHAN OF SP fs, ts, VAL[][]BYTE option.strings, VAL[]INT option.parameters.required, []BOOL option.exists, [][2]INT option.parameters, INT error.len, []BYTE line
so.write.string	CHAN OF SP fs, ts, VAL[]BYTE string
so.fwrite.string	CHAN OF SP fs, ts, VAL INT32 streamid VAL[]BYTE string, BYTE result
so.write.char	CHAN OF SP fs, ts, VAL BYTE char
so.fwrite.char	CHAN OF SP fs, ts, VAL INT32 streamid, VA1 BYTE char, BYTE result
so.write.string.nl	CHAN OF SP fs, ts, VAL[]BYTE string
so.fwrite.string.nl	CHAN OF SP fs, ts, VAL INT32 streamid, VAL[]BYTE string, BYTE result
so.write.nl	CHAN OF SP fs, ts
so.fwrite.nl	CHAN OF SP fs, ts, VAL INT32 streamid, BYTE result
so.write.int	CHAN OF SP fs, ts, VAL INT n, width
so.fwrite.int	CHAN OF SP fs, ts, VAL INT32 streamid, VAL INT n, width, BYTE result
so.write.hex.int	CHAN OF SP fs, ts, VAL INT n, width
so.fwrite.hex.int	CHAN OF SP fs, ts, VAL INT32 streamid, VAL INT n, width, BYTE result
so.write.int64	CHAN OF SP fs, ts, VAL INT64 n, VAL INT width
so.fwrite.int64	CHAN OF SP fs, ts, VAL INT32 streamid, VAL INT64 n, VAL INT width, BYTE result
so.write.hex.int64	CHAN OF SP fs, ts, VAL INT64 n, VAL INT width
so.fwrite.hex.int64	CHAN OF SP fs, ts, VAL INT32 streamid, VAL INT64 n, VAL INT width, BYTE result

Procedure	Parameter Specifiers
`so.write.real32`	`CHAN OF SP fs, ts, VAL REAL32 r, VAL INT Ip, Dp`
`so.fwrite.real32`	`CHAN OF SP fs, ts, VAL INT32 streamid,` `VAL real32 r, VAL INT Ip, Dp, BYTE result`
`so.write.real64`	`CHAN OF SP fs, ts, VAL REAL64 r, VAL INT Ip, Dp`
`so.fwrite.real64`	`CHAN OF SP fs, ts, VAL INT32 streamid,` `VAL real64 r, VAL INT Ip, Dp, BYTE result`
`so.read.line`	`CHAN OF SP fs, ts, INT len, []BYTE line,` `BYTE result`
`so.read.echo.line`	`CHAN OF SP fs, ts, INT len, []BYTE line,` `BYTE result`
`so.time.to.date`	`VAL INT32 input.time, [so.date.len]INT date`
`so.date.to.ascii`	`VAL[so.date.len]INT date, VAL BOOL long.years,` `days.first, [so.time.string.len] BYTE string`
`so.time.to.ascii`	`VAL INT32 time, VAL BOOL long.years, days.first,` `[so.time.string.len] BYTE string`
`so.today.date`	`CHAN OF SP fs, ts, [so.date.len]INT date`
`so.today.ascii`	`CHAN OF SP fs,ts, VAL BOOL long.years, days.first,` `[so.time.string.len] BYTE string`

To use this library a program header (all targets) must include the line:

```
#USE solib
```

14.5.6 Keyboard input library `sklib`

Procedure	Parameter Specifiers

`so.ask`	`CHAN OF SP fs,ts, VAL[]BYTE prompt, replies,`
	`VAL BOOL display.possible.replies, echo.reply,`
	`INT reply.number`

`so.read.echo.int`	`CHAN OF SP fs, ts, INT n, BOOL error`
`so.read.echo.hex.int`	`CHAN OF SP fs, ts, INT n, BOOL error`
`so.read.echo.any.int`	`CHAN OF SP fs, ts, INT n, BOOL error`
`so.read.echo.int64`	`CHAN OF SP fs, ts, INT64 n, BOOL error`
`so.read.echo.hex.int64`	`CHAN OF SP fs, ts, INT64 n, BOOL error`
`so.read.echo.real32`	`CHAN OF SP fs, ts, REAL32 n, BOOL error`
`so.read.echo.real64`	`CHAN OF SP fs, ts, REAL64 n, BOOL error`

To use this library a program header (all targets) must include the line:

```
#USE sklib
```

14.5.7   Host and stream i/o interface library `spinterf`   `hostlibs`

Procedure	Parameter Specifiers
`so.buffer`	`CHAN OF SP fs, ts, from.user, to.user,` `CHAN OF BOOL stopper`

`so.multiplexor`	`CHAN OF SP fs, ts,`
	`[]CHAN OF SP from.user, to.user,`
	`CHAN OF BOOL stopper`

`so.overlapped.buffer`	`CHAN OF SP fs, ts, from.user, to.user,` `CHAN OF BOOL stopper`

`so.overlapped.multiplexor`	`CHAN OF SP fs, ts,`
	`[]CHAN OF SP from.user, to.user,`
	`CHAN OF BOOL stopper, []INT queue`

`so.keystream.from.kbd`	`CHAN OF SP fs, ts, CHAN OF KS keys.out,` `CHAN OF BOOL stopper, VAL INT ticks.per.poll`
`so.keystream.from.file`	`CHAN OF SP fs, ts, CHAN OF KS keys.out,` `VAL[]BYTE filename, BYTE result`
`so.keystream.from.stdin`	`CHAN OF SP fs, ts,` `CHAN OF KS keys.out, BYTE result`
`so.scrstream.to.file`	`CHAN OF SP fs, ts, CHAN OF SS scrn,` `VAL[]BYTE filename, BYTE result`
`so.scrstream.to.stdout`	`CHAN OF SP fs, ts,` `CHAN OF SS scrn, BYTE result`
`so.scrstream.to.ANSI`	`CHAN OF SP fs, ts, CHAN OF SS scrn`
`so.scrstream.to.TVI920`	`CHAN OF SP fs, ts, CHAN OF SS scrn`

To use this library a program header must include the line:

   `#USE spinterf`

Note that although these procedures are supplied compiled for all targets the use of **KS** protocol which is word-length dependent may require careful system building on **T2** transputers.

14.5.8   Protocol conversion library `afsp`   `hostlibs`

Procedure	Parameter Specifiers

`af.to.sp`	`CHAN OF SP fs, ts,`
	`CHAN OF ANY from.user, to.user,`
	`VAL BOOL passthrough.Terminate.Cmd`

To use this library a program header must include the line:

   `#USE afsp`

14.5.9   Keystream and screenstream library streamio   iolibs

Procedure	Parameter Specifiers
ss.write.char	CHAN OF SS sink, VAL BYTE char
ss.write.string	CHAN OF SS sink, VAL []BYTE str
ss.write.nl	CHAN OF SS sink
ss.write.int	CHAN OF SS sink, VAL INT number, field
ss.write.hex.int	CHAN OF SS sink, VAL INT number, field
ss.write.text.line	CHAN OF SS sink, VAL []BYTE str
ss.write.endstream	CHAN OF SS sink
ss.goto.xy	CHAN OF SS sink, VAL INT x, y
ss.clear.eol	CHAN OF SS sink
ss.clear.eos	CHAN OF SS sink
ss.beep	CHAN OF SS sink
ss.up	CHAN OF SS sink
ss.down	CHAN OF SS sink
ss.left	CHAN OF SS sink
ss.right	CHAN OF SS sink
ss.insert.char	CHAN OF SS sink, VAL BYTE char
ss.delete.chl	CHAN OF SS sink
ss.delete.chr	CHAN OF SS sink
ss.ins.line	CHAN OF SS sink
ss.del.line	CHAN OF SS sink
ks.read.echo.char	CHAN OF KS source, CHAN OF SS echo, INT char
ks.read.echo.hex.int	CHAN OF KS source, CHAN OF SS echo, INT number, char
ks.read.echo.int	CHAN OF KS source, CHAN OF SS echo, INT number, char
ks.read.echo.text.line	CHAN OF KS source, CHAN OF SS echo, INT len, []BYTE line, INT char
ks.read.char	CHAN OF KS source, INT char
ks.read.hex.int	CHAN OF KS source, INT number, char
ks.read.int	CHAN OF KS source, INT number, char
ks.read.text.line	CHAN OF KS source, INT len, []BYTE line, INT char
ss.write.int64	CHAN OF SS sink, VAL INT64 number, VAL INT field
ss.write.hex.int64	CHAN OF SS sink, VAL INT64 number, VAL INT field

Procedure	Parameter Specifiers
`ks.read.echo.int64`	`CHAN OF KS source, CHAN OF SS echo,` `INT64 number, INT char`
`ks.read.echo.hex.int64`	`CHAN OF KS source, CHAN OF SS echo,` `INT64 number, INT char`
`ks.read.int64`	`CHAN OF KS source, INT64 number, INT char`
`ks.read.hex.int64`	`CHAN OF KS source, INT64 number, INT char`
`ss.write.real32`	`CHAN OF SS sink, VAL REAL32 number,` `VAL INT Ip, Dp`
`ss.write.real64`	`CHAN OF SS sink, VAL REAL64 number,` `VAL INT Ip, Dp`
`ks.get.real.with.del`	`CHAN OF KS in, CHAN OF SS echo, INT len,` `[]BYTE str, INT char`
`ks.read.echo.real32`	`CHAN OF KS source, CHAN OF SS echo,` `REAL32 number, INT char`
`ks.read.echo.real64`	`CHAN OF KS source, CHAN OF SS echo,` `REAL64 number, INT char`
`ks.get.real.string`	`CHAN OF KS in, INT len, []BYTE str, INT char`
`ks.read.real32`	`CHAN OF KS source, REAL32 number, INT char`
`ks.read.real64`	`CHAN OF KS source, REAL64 number, INT char`
`ss.write.int32`	`CHAN OF SS sink, VAL INT32 number,` `VAL INT field`
`ss.write.hex.int32`	`CHAN OF SS sink, VAL INT32 number,` `VAL INT field`
`ks.read.echo.int32`	`CHAN OF KS source, CHAN OF SS echo,` `INT32 number, INT char`
`ks.read.echo.hex.int32`	`CHAN OF KS source, CHAN OF SS echo,` `INT32 number, INT char`
`ks.read.int32`	`CHAN OF KS source, INT32 number, INT char`
`ks.read.hex.int32`	`CHAN OF KS source, INT32 number, INT char`

To use this library a program header must include the line:

> `#USE streamio`

The procedures in this library should be used in preference to the corresponding ones in `userio` in all new occam programs.

14.5.10   Screenstream interface procedure library `ssinterf`   `iolibs`

Procedure	Parameter Specifiers
`ss.scrstream.to.array`	`CHAN OF SS scrn, []BYTE buffer`
`ss.scrstream.from.array`	`CHAN OF SS scrn, VAL[] BYTE buffer`
`ss.scrstream.fan.out`	`CHAN OF SS scrn, screen.out1, screen.out2`
`ss.scrstream.sink`	`CHAN OF SS scrn`
`ss.scrstream.copy`	`CHAN OF SS scrn, scrn.out`
`ss.scrstream.to.ANSI.bytes`	`CHAN OF SS scrn, CHAN OF BYTE ansi`
`ss.scrstream.to.TVI920.bytes`	`CHAN OF SS scrn, CHAN OF BYTE ansi`
`ss.scrstream.to.fold`	`CHAN OF SS scrn, CHAN OF ANY` `from.uf,to.uf`
`ks.keystream.sink`	`CHAN OF KS keys`
`ks.keystream.to.screen`	`CHAN OF KS keyboard, CHAN OF SS screen`
`ks.keystream.from.fold`	`CHAN OF ANY from.uf,to.uf,CHAN OF KS kbd,` `VAL INT fold.number, INT result`

To use this library a program header must include the line:

```
#USE ssinterf
```

The procedures in this library should be used in preference to the corresponding ones in `interf` in all new occam programs.

14.5.11   General purpose i/o procedure library `userio`   `iolibs`

Procedure	Parameter Specifiers
`write.char`	`CHAN OF ANY sink, VAL BYTE char`
`write.len.string`	`CHAN OF ANY sink, VAL INT len, VAL []BYTE str`
`write.full.string`	`CHAN OF ANY sink, VAL []BYTE str`
`newline`	`CHAN OF ANY sink`
`write.int`	`CHAN OF ANY sink, VAL INT number, field`
`write.hex.int`	`CHAN OF ANY sink, VAL INT number, field`
`write.text.line`	`CHAN OF ANY sink, VAL []BYTE str`
`write.endstream`	`CHAN OF ANY sink`
`goto.xy`	`CHAN OF ANY sink, VAL INT x, y`
`clear.eol`	`CHAN OF ANY sink`
`clear.eos`	`CHAN OF ANY sink`
`beep`	`CHAN OF ANY sink`
`up`	`CHAN OF ANY sink`
`down`	`CHAN OF ANY sink`
`left`	`CHAN OF ANY sink`
`right`	`CHAN OF ANY sink`
`insert.char`	`CHAN OF ANY sink, VAL BYTE char`
`delete.chl`	`CHAN OF ANY sink`
`delete.chr`	`CHAN OF ANY sink`
`ins.line`	`CHAN OF ANY sink`
`del.line`	`CHAN OF ANY sink`

Procedure	Parameter Specifiers
read.echo.char	CHAN OF INT source, CHAN OF ANY echo, INT char
read.echo.hex.int	CHAN OF INT source, CHAN OF ANY echo, INT number, char
read.echo.int	CHAN OF INT source, CHAN OF ANY echo, INT number, char
read.echo.text.line	CHAN OF INT source, CHAN OF ANY echo, INT len, []BYTE line, INT char
read.char	CHAN OF INT source, INT char
read.hex.int	CHAN OF INT source, INT number, char
read.int	CHAN OF INT source, INT number, char
read.text.line	CHAN OF INT source, INT len, []BYTE line, INT char
write.int64	CHAN OF ANY sink, VAL INT64 number, VAL INT field
write.hex.int64	CHAN OF ANY sink, VAL INT64 number, VAL INT field
read.echo.int64	CHAN OF INT source, CHAN OF ANY echo, INT64 number, INT char
read.echo.hex.int64	CHAN OF INT source, CHAN OF ANY echo, INT64 number, INT char
read.int64	CHAN OF INT source, INT64 number, INT char
read.hex.int64	CHAN OF INT source, INT64 number, INT char
write.real32	CHAN OF ANY sink, VAL REAL32 number, VAL INT Ip, Dp
write.real64	CHAN OF ANY sink, VAL REAL64 number, VAL INT Ip, Dp
get.real.with.del	CHAN OF INT in, CHAN OF ANY echo, INT len, []BYTE str, INT char
read.echo.real32	CHAN OF INT source, CHAN OF ANY echo, REAL32 number, INT char
read.echo.real64	CHAN OF INT source, CHAN OF ANY echo, REAL64 number, INT char
get.real.string	CHAN OF INT in, INT len, []BYTE str, INT char
read.real32	CHAN OF INT source, REAL32 number, INT char
read.real64	CHAN OF INT source, REAL64 number, INT char

In all new occam programs the corresponding procedures in streamio should be used in preference to those tabulated above.

Procedure	Parameter Specifiers
create.new.fold	CHAN OF ANY from.ws, to.ws, INT fold.number, VAL []BYTE comment, VAL []INT attributes, VAL []BYTE fileid, INT errornum
write.record.item	CHAN OF ANY from.ws, to.ws, VAL []BYTE record, INT errornum
write.number.item	CHAN OF ANY from.ws, to.ws, VAL INT number, INT errornum
write.top.crease	CHAN OF ANY from.ws, to.ws, VAL []BYTE comment, VAL []INT attributes, VAL BYTE file.or.fold, VAL []BYTE fileid, INT errornum
write.fold.top.crease	CHAN OF ANY from.ws, to.ws, VAL []BYTE comment, VAL []INT attributes, INT errornum
write.filed.top.crease	CHAN OF ANY from.ws, to.ws, VAL []BYTE comment, VAL []INT attributes, VAL []BYTE fileid, INT errornum
write.bottom.crease	CHAN OF ANY from.ws, to.ws, INT errornum
finish.new.fold	CHAN OF ANY from.ws, to.ws, VAL INT fold.number, VAL BOOL must.unfile, INT errornum
read.fold.heading	CHAN OF ANY from.rs, to.rs, VAL INT fold.number, INT len.comment, []BYTE comment, []INT attributes, INT errornum
read.file.name	CHAN OF ANY from.rs, to.rs, VAL INT fold.number, INT len.file.id, []BYTE file.id, INT errornum
open.folded.stream	CHAN OF ANY from.rs, to.rs, VAL INT fold.number, BYTE first.item, BOOL not.filed, INT errornum
read.record.item	CHAN OF ANY from.rs, to.rs, INT len, []BYTE record, BYTE next.item
read.number.item	CHAN OF ANY from.rs, to.rs, INT number, BYTE next.item
read.error.item	CHAN OF ANY from.rs, to.rs, INT status, BYTE next.item
read.fold.top.crease	CHAN OF ANY from.rs, to.rs, INT len.comment, []BYTE comment, []INT attributes, BYTE next.item
read.filed.top.crease	CHAN OF ANY from.rs, to.rs, INT len.comment, []BYTE comment, []INT attributes, INT len.fileid, []BYTE fileid, BYTE next.item
read.bottom.crease	CHAN OF ANY from.rs, to.rs, []INT attributes, BYTE next.item
input.record.item	CHAN OF ANY from.rs, INT len, []BYTE record, VAL BYTE next.item
input.number.item	CHAN OF ANY from.rs, INT number, VAL BYTE next.item
input.error.item	CHAN OF ANY from.rs, INT status, VAL BYTE next.item
input.top.crease	CHAN OF ANY from.rs, to.rs, INT len.comment, []BYTE comment, []INT attributes, INT len.fileid, []BYTE fileid, VAL BYTE next.item

Procedure	Parameter Specifiers
skip.item	CHAN OF ANY from.rs, to.rs, BYTE next.item
enter.fold	CHAN OF ANY from.rs, to.rs, BYTE next.item
exit.fold	CHAN OF ANY from.rs, to.rs, BYTE next.item
repeat.fold	CHAN OF ANY from.rs, to.rs, BYTE next.item
close.folded.stream	CHAN OF ANY from.rs, to.rs, VAL INT fold.number, VAL BOOL must.unfile, INT errornum

To use this library a program header must include the line:

> #USE userio

14.5.12   Low level user filer interface support library `ufiler`   `iolibs`

Procedure	Parameter Specifiers
get.stream.result	CHAN OF ANY fs, INT result
clean.string	INT len, []BYTE str
truncate.file.id	INT len, VAL[]BYTE id
number.of.folds	CHAN OF ANY from.uf, to.uf, INT n, result
write.fold.string	CHAN OF ANY from.uf, to.uf, VAL INT seq.no, VAL INT len, VAL []BYTE data, INT result
create.fold	CHAN OF ANY from.uf, to.uf, INT new.fold.number, VAL []INT attributes, INT result
send.command	CHAN OF ANY from.uf, to.uf, VAL BYTE op, VAL INT seq.no, INT result
make.filed	CHAN OF ANY from.uf, to.uf, VAL INT seq.no, VAL INT id.len, VAL []BYTE file.id, INT result
open.stream	CHAN OF ANY fs, ts, VAL BYTE op, VAL INT fold.no, INT result
read.fold.string	CHAN OF ANY from.uf, to.uf, VAL INT seq.no, INT len, []BYTE data, INT result
read.fold.attr	CHAN OF ANY from.uf, to.uf, VAL INT seq.no, []INT attributes, INT result
open.data.stream	CHAN OF ANY from.rs, to.rs VAL INT fold.number, BYTE first.item, BOOL not.filed, INT errornum
close.uf.stream	CHAN OF ANY from.rs, to.rs, VAL INT fold.number, VAL BOOL must.unfile, INT errornum
read.data.record	CHAN OF ANY from.rs, to.rs INT len, []BYTE record, BYTE next.item

To use this library a program header must include the line:

> #USE ufiler

All the procedures are contained in one SC, which makes use of the library `filerhdr`.

Procedure	Parameter Specifiers
`scrstream.to.array`	`CHAN OF ANY scrn, []BYTE buffer`
`scrstream.from.array`	`VAL[] BYTE buffer, CHAN OF ANY scrn`

`scrstream.to.file`	`CHAN OF ANY scrn, CHAN OF ANY from.uf,`
	`to.uf, VAL[]BYTE fold.title, INT fold.number,`
	`INT result`

`scrstream.multiplexor`	`[]CHAN OF ANY screen.in, CHAN OF ANY`
	`screen.out,`
	`CHAN OF INT stopper`

`scrstream.fan.out`	`CHAN OF ANY scrn, screen.out1, screen.out2`
`scrstream.sink`	`CHAN OF ANY scrn`
`scrstream.copy`	`CHAN OF ANY scrn, scrn.out`
`keystream.from.file`	`CHAN OF ANY from.uf, to.uf, CHAN OF INT kbd,` `VAL INT fold.number, INT result`
`keystream.sink`	`CHAN OF INT keys`
`keystream.to.screen`	`CHAN OF INT keyboard, CHAN OF ANY screen`

Additional procedures `scrstream.to.server`, `keystream.from.server`, `scrstream.to.ANSI` and `scrstream.to.TVI920` are also included in this library for compatibility with earlier versions of the TDS. Where possible new programs should use `ssinterf` in preference to this library.

To use this library a program header must include the line:

   `#USE interf`

Procedure	Parameter Specifiers

`B00x.term.p.driver`	`CHAN OF ANY from.user.scrn,`
	`CHAN OF INT to.user.kbd, VAL INT board.type,`
	`port, baud.rate, screen.type`

To use this library a program header must include the line:

   `#USE t4board`   or
   `#USE t2board`

14.5.15   Other libraries

The other libraries supplied are for compatibility with earlier versions of the TDS only and should not be used in new programs. Their specifications remain unchanged.

14.6   Protocols and formal parameter conventions

Three named protocols are used in the input/output libraries in the TDS. These are all declared in the library `strmhdr`, see appendix D.

`SP` protocol is the protocol used on communications to the `iserver`, including those buffered inside the TDS and passed to an `EXE` or `UTIL`. This protocol is discussed in full in section 16.5. The protocol assumes bidirectional communication using a pair of channels.

SS protocol is the so-called TDS screenstream protocol, which is used for unidirectional communication of ASCII text, with optional inclusion of screen control operations in a terminal type independent representation. Although originally designed for communication with the screen of the terminal on which the TDS is running, this protocol is of general applicability and may be used in conjunction with multiplexors, protocol converters, etc. for any communication of sequential text streams between occam processes on one or more processors.

KS protocol is an integer protocol used for communicating individual keystrokes from a real or simulated keyboard. Various meanings have been given to values which can be derived by the TDS from sequences of physical keystrokes. These are primarily for the internal purposes of the TDS editor, debugger, etc. In user programs it is probably advisable to restrict use of this protocol to the usual ASCII character set.

SS and **KS** protocols are formalisations of anarchic **CHAN OF ANY** protocols used in earlier versions of the TDS. To minimise disturbance to users who have used these earlier versions, two versions of procedures using these protocols are now provided. The versions in the libraries **userio** and **interf** use the old protocols and the old procedure names. The versions in the library **streamio** and **ssinterf** use the new protocols and the corresponding new naming conventions discussed below. The **streamio** library corresponds to a similarly named library in INMOS occam toolset products, and it is recommended that procedures from this library be used in all new programs. With care it is possible to mix procedures from the two groups of libraries but this is not recommended.

IMPORTANT NOTE: On 16-bit transputers the **CHAN OF ANY** version of screenstream protocol is not the same as **SS** and they should not be mixed in T2 programs.

The following conventions have been adopted in naming procedures and constructing formal parameter lists:

Wherever a pair of channels are used together for bi-directional communication, the channels will appear as adjacent formal parameters with the one on which inputs to the current process will occur first.

Any procedure which communicates using **SP** protocol will have as its first two formal parameters a pair of channels **from.isv** and **to.isv** (not necesarily with these names) used respectively for communication from the server to the program and from the program to the server. All procedures with names beginning **so.** are of this kind.

Any procedure which only communicates outwards on a channel with TDS screenstream (**SS**) protocol will have this channel as its first formal parameter. All procedures with names beginning **ss.** are of this kind.

Any procedure which only receives input on a channel with TDS keystream (**KS**) protocol will have this channel as its first formal parameter. All procedures with names beginning **ks.** are of this kind.

14.7   Environment enquiries

The procedures in this group use the **iserver** interface to ask for information available on the host computer. They are in the libraries **splib** and **solib**. They are designed for use in hosted applications only. Some details of the specifications of these procedures are host type dependent.

so.getenv   **splib**

```
PROC so.getenv (CHAN OF SP fs, ts, VAL[]BYTE name,
                INT length, []BYTE value, BYTE result)
```

Returns the string defined for the host environment variable **name**. If **name** is not defined on the system **result** takes the value **spr.operation.failed**.

The result returned can take any of the following values:

spr.ok	The operation was successful.
spr.bad.name	The specified name is a null string.
spr.operation.failed	Could not read environment string.

      spr.bad.packet.size   **SIZE name** is too large
                                   (> **sp.max.getenvname.size**).
      spr.buffer.overflow   Environment string too large for **value**.

If the buffer overflows **length** is set to the buffer size.

so.time   oplib

    PROC so.time (CHAN OF SP fs, ts, INT32 localtime, UTCtime)

Returns the Coordinated Universal Time and local time if they are available on the system. Both times are expressed as the number of seconds that have elapsed since midnight on 1st January, 1970. If either time is unavailable then it will have a value of zero.

so.system   splib

    PROC so.system (CHAN OF SP fs, ts, VAL[]BYTE command,
                      INT32 status, BYTE result)

Passes the string **command** to the host command processor for execution. If the command string is of zero length **result** takes the value **spr.ok** if there is a host command processor, otherwise an error is returned. If **command** is non-zero in length then **status** contains the host-specified value of the command, otherwise it is undefined.

The result returned can take any of the following values:

      spr.ok   Host command processor exists.
      spr.bad.packet.size   The array **command** is too large
                                   (> **sp.max.systemcommand.size**).

so.commandline   splib

    PROC so.commandline (CHAN OF SP fs, ts, VAL BYTE all,
                          INT length, []BYTE string, BYTE result

Returns the command line passed to the server when it was invoked.
If **all** has the value **sp.short.commandline** then all options are stripped from the command line. If **all** is **sp.whole.commandline** then the command line is returned exactly as it was invoked. See also section 16.5.7.

The result returned can take any of the following values:

      spr.ok   The operation was successful.
      spr.buffer.overflow   Command line too long for **string**.

If the buffer overflows **length** is set to the buffer size.

so.core   splib

    PROC so.core (CHAN OF SP fs, ts, VAL INT32 offset,
                  INT bytes.read, []BYTE data, BYTE result)

Returns the contents of the root transputer's memory as peeked from the transputer when **iserver** is invoked with the analyse ('**SA**') option. The start of the memory segment is given by **offset**, and the number of bytes by the size of the **data** vector. An error is returned if **offset** is larger than the total amount of peeked memory.

The result returned can take any of the following values:

spr.ok	The operation was successful.
spr.operation.failed	The operation failed.
spr.bad.packet.size	The array **data** is too large (> **sp.max.corerequest.size**).

This procedure can also be used to determine whether the memory was peeked (whether the server was invoked with the 'SA' option), by specifying a size of zero for **data** and **offset**. If the routine fails the memory was not peeked.

so.version   **splib**

```
PROC so.version (CHAN OF SP fs, ts,
                 BYTE version, host, os, board)
```

Returns version information about the server and the host on which it is running. A value of zero for any of the items indicates that the information is unavailable.

The version of the server is given by **version**. The value should be divided by ten to yield the true version number. For example, a value of 15 means version 1.5.

The host machine type is given by **host**, and can take any of the following values:

sph.PC	IBM PC
sph.NECPC	NEC PC
sph.VAX	DEC VAX
sph.SUN3	SUN Microsystems Sun-3
sph.SUN4	SUN Microsystems Sun-4

Values up to 127 are reserved for use by INMOS.

The host operating system is given by **os**, and can take any of the following values:

spo.DOS	DOS
spo.HELIOS	HELIOS
spo.VMS	VMS
spo.SunOS	SunOS

Values up to 127 are reserved for use by INMOS.

The interface board type is given by **board**, and can take any of the following values:

spb.B004	IMS B004
spb.B008	IMS B008
spb.B010	IMS B010
spb.B011	IMS B011
spb.B014	IMS B014
spb.DRX11	DRX-11
spb.QT0	Caplin QT0

Values up to 127 are reserved for use by INMOS.

so.exit   **splib**

```
PROC so.exit (CHAN OF SP fs, ts, VAL INT32 status)
```

Terminates the server, which returns the value of **status** to its caller. If **status** has the special value **sps.success** then the server will terminate with a host specific 'success' result. If **status** has the special value **sps.failure** then the server will terminate with a host specific 'failure' result. See also section 16.5.7.

so.parse.command.line   **solib**

     PROC so.parse.command.line (CHAN OF SP fs, ts,
                        VAL[][]BYTE option.strings,
                        VAL[]INT option.parameters.required,
                        []BOOL option.exists,
                        [][2]INT option.parameters,
                        INT error.len, []BYTE line

This procedure reads the command line and parses it for specified options and associated parameters.

The parameter **option.strings** contains a list of all the possible options. Options may be any length up to 255 bytes and are *not* case sensitive. To read a parameter that has no preceding option (such as a file name) then the first option string should be empty (contain only spaces). The optionstrings must be filled out with spaces so that they are all the same length. For example, consider a program can be supplied with a file name, and any of three options 'A', 'B' and 'C'. The array **option.strings** would look like this:

VAL option.strings IS [" ", "A", "B", "C"]:

The parameter **option.parameters.required** indicates if the corresponding option (in **option.strings**) requires a parameter. The values it may take are:

spopt.never   Never takes a parameter.
spopt.maybe   Optionally takes a parameter.
spopt.always   Must take a parameter.

Continuing the above example, the file name must be supplied and none of the options take parameters, except for 'C', which may or may not have a parameter, then **option.parameters.required** would look like this:

      VAL option.parameters.required IS
        [sopt.always, sopt.never,
         sopt.never, sopt.maybe]:

If an option was present on the command line **option.exists** is set to TRUE, otherwise it is set to FALSE.

If an option was followed by a parameter then the position in the array **line** where the parameter starts and the length of the parameter are given by the first and second elements respectively of the parameter element in **option.parameters**.

If an error occurs whilst the command line is being parsed then
error.len will be greater than zero and **line** will contain an error message of the given length. If no error occurs then **line** will contain the command line as supplied by the host file server.

so.today.date   **solib**

     PROC so.today.date (CHAN OF SP fs, ts, [so.date.len]INT date)

Gives today's date as six integers, stored in the **date** array. The format of the array is the same as for **so.time.to.date**. If the date is unavailable all elements in **date** are set to zero.

so.today.ascii   **solib**

     PROC so.today.ascii (CHAN OF SP fs,ts,
                    VAL BOOL long.years, days.first,
                    [so.time.string.len] BYTE string)

Gives today's date as an ASCII string, in the same format as procedure **so.date.to.ascii**. If the date is unavailable **string** is filled with spaces.

14.8   Representation conversions and string handling

These procedures do not have channels as parameters and are potentially usable in any occam programs. The set of procedures provided should not be considered as complete. In particular application environments additional or alternative procedures may usefully be written to supersede or supplement these.

For further information on the procedures provided by the libraries **ioconv** and **extrio** see the occam 2 Reference Manual.

Time and date functions

so.time.to.date   solib

> PROC so.time.to.date (VAL INT32 input.time, [so.date.len]INT date)

Converts time (as supplied by **so.time**) to six integers, stored in the **date** array. The elements of the array are as follows:

Element of array	Data
0	Seconds
1	Minutes
2	Hour (24 hour clock)
3	Day
4	Month
5	Year

so.date.to.ascii   solib

> PROC so.date.to.ascii (VAL[so.date.len]INT date,
>   VAL BOOL long.years, days.first,
>   [so.time.string.len] BYTE string)

Converts an array of six integers containing the date (as supplied by **so.time.to.date**) into an ASCII string of the form:

> *HH:MM:SS DD/MM/YYYY*

If **long.years** is FALSE then year field is reduced to two characters, the last two characters being padded with spaces. If **days.first** is FALSE then the ordering of day and month is changed (to the U.S. standard).

so.time.to.ascii   solib

> PROC so.time.to.ascii (VAL INT32 time,
>   VAL BOOL long.years, days.first,
>   [so.time.string.len] BYTE string)

Converts time (as supplied by **so.time**) into an ASCII string, as described for **so.date.to.ascii**.

Character handling functions

is.in.range   strings

> BOOL FUNCTION is.in.range (VAL BYTE char, bottom, top)

Returns **TRUE** if the value of **char** is in the range defined by **bottom** and **top** inclusive.

is.upper   strings

    BOOL FUNCTION is.upper (VAL BYTE char)

    Returns **TRUE** if **char** is an ASCII upper case letter.

is.lower   strings

    BOOL FUNCTION is.lower (VAL BYTE char)

    Returns **TRUE** if **char** is an ASCII lower case letter.

is.digit   strings

    BOOL FUNCTION is.digit (VAL BYTE char)

    Returns **TRUE** if **char** is an ASCII decimal digit.

is.hex.digit   strings

    BOOL FUNCTION is.hex.digit (VAL BYTE char)

    Returns **TRUE** if **char** is an ASCII hexadecimal digit. Upper or lower case letters A–F are allowed

is.id.char   strings

    BOOL FUNCTION is.id.char (VAL BYTE char)

    Returns **TRUE** if **char** is an ASCII character which can be part of an occam name.

to.upper.case   strings

    PROC to.upper.case ([]BYTE str)

    Converts all alphabetic characters in **str** to upper case.

to.lower.case   strings

    PROC to.lower.case ([]BYTE str)

    Converts all alphabetic characters in **str** to lower case.

String comparison functions

Strings may be compared for order or for equality.

compare.strings   strings

    INT FUNCTION compare.strings (VAL []BYTE str1, str2)

    This general purpose lexicographic ordering function compares two strings. (lexicographic ordering is the ordering used in dictionaries etc., using the ASCII values of the bytes). It returns one of the 5 results 0, 1, −1, 2, −2 as follows.

        0 The strings are exactly the same in length and content.

        1 **str2** is a leading sub-string of **str1**

     −1 **str1** is a leading sub-string of **str2**

        2 **str1** is lexicographically later than **str2**

     −2 **str2** is lexicographically later than **str1**

So if **s** is **"abcd"**:

```
compare.strings ("abc", [s FROM 0 FOR 3])  =  0
compare.strings ("abc", [s FROM 0 FOR 2])  =  1
compare.strings ("abc", s)                 = -1
compare.strings ("bc", s)                  =  2
compare.strings ("a4", s)                  = -2
```

eqstr   strings

> **BOOL FUNCTION eqstr (VAL []BYTE s1,s2)**

This is an optimised test for string equality. It returns **TRUE** if the two strings are the same size and have the same contents.

String editing procedures

A string to be edited will be stored in an array which may have some unused space at its end. The editing operations supported are deletion of a number of charaters, with the closing up of the gap created, and insertion of a new string starting at any position within a string, at which a gap of the necessary size is created.

These two operations are supported by a lower level procedure for shifting a consecutive substring left or right within the array. This lower level procedure does exhaustive tests against overflow.

str.shift   strings

> **PROC str.shift ([]BYTE str, VAL INT start,**
> **len, shift, BOOL not.done)**

Takes a substring **[str FROM start FOR len]**, and copies it to a position **shift** places to the right. Any implied actions involving bytes outside the string are not performed and cause the error flag **not.done** to be set **TRUE**.

delete.string   strings

> **PROC delete.string (INT len, []BYTE str,**
> **VAL INT start, size, BOOL not.done)**

Deletes **size** bytes from the string **str** starting an **str[start]**. There are initially **len** significant characters in **str** and it is decremented appropriately. If **start** is outside the string, or **size** is negative or greater than the string length, then no action occurs and **not.done** is set **TRUE**.

insert.string   strings

> **PROC insert.string (VAL []BYTE new.str, INT len,**
> **[]BYTE str, VAL INT start, BOOL not.done)**

Creates a gap in **str** before **str[start]** and copies the string **new.str** into it. There are initially **len** significant characters in **str** and **len** is incremented by the length of **new.str** inserted. Any overflow of the declared SIZE of **str** results in truncation at the right and setting **not.done** to **TRUE**. This procedure may be used for simple concatenation on the right by setting **start = len** or on the left by setting **start = 0**. This method of concatenation differs from that using the **append.** procedures in that it can never cause the program to stop.

String searching functions

These procedures allow a string to be searched for a match with a single byte or a string of bytes, or for a byte which is one of a set of possible bytes, or for a byte which is not one of a set of bytes. Searches insensitive to alphabetic case should use **to.upper.case** or **to.lower.case** on both operands before using these procedures.

string.pos   strings

> INT FUNCTION string.pos (VAL []BYTE search, str)

> Returns the position in **str** of the first occurrence of a sub-string which exactly matches **search**. Returns −1 if there is no such match.

char.pos   strings

> INT FUNCTION char.pos (VAL BYTE search, VAL []BYTE str)

> Returns the position in **str** of the first occurrence of the byte **search**. Returns −1 if there is no such byte.

search.match   strings

> INT, BYTE FUNCTION search.match (VAL []BYTE possibles, str)

> Searches **str** for any one of the bytes in the array **possibles**. If one is found its index and identity are returned as results. If none is found then −1, 255(BYTE) are returned.

search.no.match   strings

> INT, BYTE FUNCTION search.no.match (VAL []BYTE possibles, str)

> Searches **str** for a byte which does not match any one of the bytes in the array **possibles**. If one is found its index and identity are returned as results. If none is found then −1, 255(BYTE) are returned.

String add/append functions

The **add** and **append** procedures produce identical results provided that the array does not overflow. If the array overflows **add** truncates the data and continues processing, **append** behaves like **STOP**.

add.char   strings
append.char   strings

> PROC append.char (INT len, []BYTE str, VAL BYTE char)

> Writes a byte **char** into the array **str** at **str[len]**. **len** is incremented by 1.

add.text   strings
append.text   strings

> PROC append.text (INT len, []BYTE str, VAL []BYTE text)

> Writes a string **text** into the array **str**, starting at **str[len]** and computing a new value for **len**.

add.int   *strings*
append.int   *strings*

>   PROC append.int (INT len, []BYTE str, VAL INT number, width)

Converts **number** into a sequence of ASCII decimal digits padded out with leading spaces and an optional sign to the specified **width** width if necessary. If the number cannot be represented in **width** characters it is widened as necessary. A zero value for **width** will give minimum width. The converted number is written into the array **str** starting at **str[len]** and **len** is incremented.

add.hex.int   *strings*
append.hex.int   *strings*

>   PROC append.hex.int (INT len, []BYTE str, VAL INT number, width)

Converts **number** into a sequence of ASCII hexadecimal digits, using upper case letters, preceded by **#**. The total number of characters sent is always **width+1**, padding out with **0** or **F** on the left if necessary. The number is truncated at the left if the field is too narrow, thereby allowing the less significant part of any number to be printed. The converted number is written into the array **str** starting at **str[len]** and **len** is incremented.

add.real32   *strings*
append.real32   *strings*

>   PROC append.real32 (INT len, []BYTE str,
>   VAL REAL32 number, VAL INT Ip, Dp)

Converts **number** into a sequence of ASCII decimal digits padded out with leading spaces and an optional sign to the specified number of digits **Ip** before and **Dp** after the decimal point. The converted number is written into the array **str** starting at **str[len]** and **len** is incremented.

The total added width will be **Ip + Dp + 2** except in the following special cases:

If the value will not fit, an exponential form is used.

If **Ip** is zero, an exponential form with **Dp** significant digits is used, giving a field width of **Dp + 6**.

If **Ip** and **Dp** are zero, a minimum field width free format is used.

Numbers which correspond to the IEEE standard concepts of 'Infinity' and 'NotaNumber' produce the texts **Inf** and **NaN**, respectively.

In exponential forms a number in the range [1.0, 10.0) is followed by **E**, a **+** or **–** sign, and a 2 digit decimal exponent.

add.real64   *strings*
append.real64   *strings*

>   PROC append.real64 (INT len, []BYTE str,
>   VAL REAL64 number, VAL INT Ip, Dp)

Converts **number** into a sequence of ASCII decimal digits padded out with leading spaces and an optional sign to the specified number of digits **Ip** before and **Dp** after the decimal point.

Details as for **REAL32** but allowing 3 digits for the exponent.

add.int32   strings
append.int32   strings

      PROC append.int32 (INT len, []BYTE str,
                    VAL INT32 number, VAL INT width)

    As **append.int** but for 32-bit integers

add.int64   strings
append.int64   strings

      PROC append.int64 (INT len, []BYTE str,
                    VAL INT64 number, VAL INT width)

    As **append.int** but for 64-bit integers

add.hex.int32   strings
append.hex.int32   strings

      PROC append.hex.int32 (INT len, []BYTE str,
                    VAL INT32 number, VAL INT width)

    As **append.hex.int** but for 32-bit integers

add.hex.int64   strings
append.hex.int64   strings

      PROC append.hex.int64 (INT len, []BYTE str,
                    VAL INT64 number, VAL INT width)

    As **append.hex.int** but for 64-bit integers

Line parsing

next.word.from.line   strings

      PROC next.word.from.line (VAL []BYTE line, INT ptr, len,
                    []BYTE word, BOOL ok)

    If **ok** is FALSE on entry then no action is taken.

    Skips leading spaces and tabs and reads the next word from the string **line**. The value of **ptr** is
the starting point for the search.

    If the first non-space/non-tab character found is not a printable ASCII character, or if the end of the
string is reached, the boolean **ok** is set to FALSE. If the word is followed by a space, tab or the end
of the string, **ok** is set to TRUE, otherwise it is set to FALSE.

    The pointer **ptr** is updated either to the position beyond the last character read or to the end of the
string.

next.int.from.line   strings

      PROC next.int.from.line (VAL []BYTE line,
                    INT ptr, number, BOOL ok)

    If **ok** is FALSE on entry then no action is taken.

    Skips leading spaces and tabs and reads the next integer from the string **line**. The value of **ptr**
is the starting point for the search.

If the first non-space/non-tab character found is not a digit or sign (+ or −), or if the end of the string is reached, the boolean **ok** is set to FALSE. If the integer is followed by a space, tab or the end of the string, **ok** is set to TRUE and the integer is returned in **number**, otherwise **ok** is set to FALSE.

The pointer **ptr** is updated either to the position beyond the last character read or to the end of the string.

14.9   Terminals and text streams

A terminal may be treated in one of two distict ways, supported by procedures in the library groups **hostlibs** and **iolibs** respectively.

The iserver model of the terminal is the set of standard streams, standard input, standard output and standard error. The behaviour of these streams is defined to be that supplied by the C library in the implementation of C used to compile the **iserver**. Such streams are sequences of ASCII characters and have no abstractions of control operations. Such operations must be coded as appropriate escape sequences for the actual terminal or low-level driving software being used.

The alternative model of the terminal is as a pair of occam channels, the input channel using TDS keystream **KS** protocol, and the output channel using TDS screenstream **SS** protocol. Both these protocols include abstractions: on the keystream certain combinations of values, generated by one or more physical key depressions, may be treated as single 'cooked' keystrokes; on the screenstream the usual screen control operations of cursor positioning and movement, deletion and insertion of lines and characters, clearing, etc are encoded with their own special representation. Another important feature of the TDS keystream model is that it is possible to perform an occam **ALT** operation on the keyboard channel of an **EXE** using this model, and so to write interruptable programs. As the **iserver** is not capable of asynchronous keyboard handling, the TDS polls the keyboard at intervals determined by the system or the user in an EXE. See section 16.6.2.

The library includes collections of procedures for communicating values along channels using either of these models. It also includes protocol conversion procedures for conversion between the two.

Procedures in the libraries **splib**, **solib** and **sklib** use the standard input/standard output model.

Procedures in the libraries **streamio**, **ssinterf**, **userio** and **interf** use the TDS stream models.

Whereas combinations of the models may be used for communications between the processes of an application, it is desirable that accesses to a particular terminal, such as that of the host in a hosted application use either one approach or the other but not a mixture. The choice will be based on such matters as interruptability, need for abstraction from terminal type dependencies, adequacy of the line-based standard input view of the keyboard, etc.

14.9.1   The simple input and output procedures (TDS stream models)

Two TDS stream models of input/output are supported by appropriate sets of procedures.

A simple model of input and output which is applicable both to an interactive terminal and to sequential text files is based on a sequence of lines of text, separated by carriage return characters. This model is also appropriate for communication between the processes of an occam program, if the information being sent is essentially a sequential text stream.

The second model is the folded stream model, which allows a hierarchical data structure to be traversed, with the option to omit parts of the structure, to repeat parts of the traverse, etc.

The simple user procedures provide access to a stream of characters as input, and a stream of characters as output. The characters are received from, or transmitted to, the environment as ASCII values, represented as **INT** values on input and **BYTE** values on output. The procedures enable the programmer to think also, if desirable, in terms of higher level concepts such as numbers and strings. The set of procedures provided is not exhaustive, and users should feel free to add to the set for their own purposes.

The user does not need, in the first instance, to be aware of the protocol used on the channels used by these procedures. A simple program containing only sequential code, or doing screen output only from one branch of its parallel structure can do all its terminal input and output through procedures using screenstream output.

The output procedures are of two kinds. On the one hand there are procedures to output characters, numbers and strings at the current cursor position, and an explicit newline procedure. On the other hand there is a procedure for outputting a complete line of text which the user has built up in an array of bytes. Procedures are provided in the libraries **ioconv**, **extrio** and **strings** for converting numbers into strings.

The user may decide whether or not to echo keyboard input to the screen as it is input. Each procedure has a version with and without the ability to echo. It is normally preferable to use the versions with echo, as these are coded to respond to the delete key for simple corrections of keyboard errors. Note that the effect of the delete key is restricted to the current procedure call and cannot influence the result of previous calls of procedures on the same line of input.

Screenstream output procedures

These all have a first parameter **sink** which is the channel on which output commands are sent to cause an appropriate sequence of text characters to be generated. The actual parameter corresponding to this should be a channel using the screen stream (output) protocol. Such a channel is the **screen** channel of an executable procedure (**EXE**). It may alternatively be an input channel to one of the interface procedures described in section 14.10 on the libraries **interf** and **ssinterf**. The procedures are designed for streams of ASCII characters and may not be appropriate for binary byte streams.

Each line of text may be terminated either by an explicit call of the procedure **ss.write.text.line** or **ss.write.nl**, or by including the character pair "*c*n" in a string. Both these characters should normally be sent as one or other may be ignored in some circumstances, and both are usually required on real terminals. The preferred convention is to terminate each line with a **ss.write.nl** (rather than preceding it with one).

Procedures for sending terminal control codes down a screen stream channel should only be used when the receiving process is controlling a terminal or is forwarding commands to one which does. They can not sensibly be used if the receiver is creating a file. The control codes are terminal-independent codes for a number of common screen handling operations. They are described in detail in section 16.6.

Keystream input procedures

Values from a keyboard channel are expressed as integers. Positive integers are key codes, negative ones are error numbers. Error numbers are only likely if the source is a filing system interface procedure. Key codes may be either simple ASCII codes, or they may be encoded representations of function keys as used by the TDS editor interface. These are discussed in detail in section 16.6. End of line is normally represented by the value **INT'*c'** alone.

Each procedure is provided in two forms, with and without echo. Echo is the return of an ASCII code to an output channel using screen protocol. A procedure with echo also allows characters to be deleted (as long as a terminator has not yet been encountered).

The versions without echo perform input only and are otherwise identical to those which perform echo and handle deletions.

The reading procedures for numeric values all obey the same conventions with respect to leading characters and terminators. The first character which may be part of the number is assumed to have been already read and must be provided by the caller as the initial value of the **INT** parameter **char**. This requires that the actual parameter corresponding to **char** must be a variable and must be initialised before using any of the read number procedures. The initial value may be obtained by a call of **ks.read.char** or may be a dummy value such as **INT' '**.

The procedure then reads characters, ignoring everything before the first valid character (a digit, or for decimal numbers, a + or − sign). A number in the appropriate syntax is then read, terminated by the first character which cannot validly continue the number, which is returned in the parameter **char**. If an error occurs an error number is returned in **char**.

These conventions are adopted to facilitate the coding of input from text streams where numbers are embedded within text and may have arbitrary terminators.

Although, strictly, hexadecimal digits greater than **9** and the decimal exponent symbol **E** are defined to be upper case characters these procedures will accept the corresponding lower case characters, but will echo them in upper case. The character **'$'** is NOT accepted as an alternative to **'#'**.

There is one common error value which any of these procedures may generate if a number which is out of range or otherwise invalid is encountered. This value has the name **ft.number.error** and is defined in the library **uservals**. If such a value is returned the value of the parameter **number** is undefined.

The procedures with echo copy the actual characters input to the echo channel, converting lower case hexadecimal digits or **e** in a **REAL** number to upper case and acting on delete characters as they are received. The first character (passed in as **char**) is assumed to have been already echoed, and an immediately subsequent delete will delete it. The terminating character is echoed even when it is subsequently converted into an error indication.

14.9.2   Procedures supporting screenstream output

ss.write.char   streamio

      PROC ss.write.char (CHAN OF SS sink, VAL BYTE char)

write.char   userio

      PROC write.char (CHAN OF ANY sink, VAL BYTE char)

    Sends the ASCII value **char** down **sink** to the current position in the output line.

ss.write.string   streamio

      PROC ss.write.string (CHAN OF SS sink, VAL []BYTE str)

write.full.string   userio

      PROC write.full.string (CHAN OF ANY sink, VAL []BYTE str)

    str is any string all of whose characters are sent to **sink**. This procedure should be used for constant text strings, and other strings of known length (probably expressed as an array segment). The maximum length of **str** is **st.max.string.size**. Strings longer than this will cause a STOP.

write.len.string   userio

      PROC write.len.string (CHAN OF ANY sink,
                        VAL INT len, VAL[]BYTE str)

    str is any string with **len** or more characters, the first **len** of these are sent to **sink**. This procedure should be used for text passed as an array segment whose size is computed at run-time. The maximum length of **str** is **st.max.string.size**. Strings longer than this will cause a STOP.

ss.write.nl   streamio

      PROC ss.write.nl (CHAN OF SS sink)

newline   userio

      PROC newline (CHAN OF ANY sink)

    Sends "*c*n" to **sink**.

ss.write.int   streamio
      PROC ss.write.int (CHAN OF SS sink, VAL INT number, field)
write.int   userio
      PROC write.int (CHAN OF ANY sink, VAL INT number, field)

      Converts **number** into a sequence of ASCII decimal digits padded out with leading spaces and an optional sign to the specified **field** width if necessary. If the number cannot be represented in **field** characters it is widened as necessary. A zero value for **field** will give minimum width. The converted number is sent to **sink**.

ss.write.hex.int   streamio
      PROC ss.write.hex.int (CHAN OF SS sink, VAL INT number, field)
write.hex.int   userio
      PROC write.hex.int (CHAN OF ANY sink, VAL INT number, field)

      Converts **number** into a sequence of ASCII hexadecimal digits, using upper case letters, preceded by **#**. The total number of characters sent is always **field + 1**, padding out with **0** or **F** on the left if necessary. The number is truncated at the left if the field is too narrow, thereby allowing the less significant part of any number to be printed. The converted number is sent to **sink**.

ss.write.text.line   streamio
      PROC ss.write.text.line (CHAN OF SS sink, VAL []BYTE str)
write.text.line   userio
      PROC write.text.line (CHAN OF ANY sink, VAL []BYTE str)

      A line of characters from **str**, optionally terminated by a '***c**' is sent to **sink** followed by a newline. This procedure should be used for text which the programmer organises into complete lines.

ss.write.int32   streamio
      PROC ss.write.int32 (CHAN OF SS sink,
                       VAL INT32 number, VAL INT field)

      As **ss.write.int** but for 32-bit integer values.

ss.write.hex.int32   streamio
      PROC ss.write.hex.int32 (CHAN OF SS sink,
                       VAL INT32 number, VAL INT field)

      As **ss.write.hex.int** but for 32-bit integer values.

ss.write.int64   streamio
      PROC ss.write.int64 (CHAN OF SS sink,
                       VAL INT64 number, VAL INT field)
write.int64   userio
      PROC write.int64 (CHAN OF ANY sink,
                       VAL INT64 number, VAL INT field)

      As **write.int** but for 64-bit integer values.

`ss.write.hex.int64`   **streamio**

> PROC ss.write.hex.int64 (CHAN OF SS sink,
>   VAL INT64 number, VAL INT field)

`write.hex.int64`   **userio**

> PROC write.hex.int64 (CHAN OF ANY sink,
>   VAL INT64 number, VAL INT field)

As **write.hex.int** but for 64-bit integer values.

`ss.write.real32`   **streamio**

> PROC ss.write.real32 (CHAN OF SS sink,
>   VAL REAL32 number, VAL INT Ip, Dp)

`write.real32`   **userio**

> PROC write.real32 (CHAN OF ANY sink,
>   VAL REAL32 number, VAL INT Ip, Dp)

Converts **number** into a sequence of ASCII decimal digits padded out with leading spaces and an optional sign to the specified number of digits **Ip** before and **Dp** after the decimal point. The converted number is sent to **sink**.

The total width will be **Ip + Dp + 2** except in the following special cases:

If the value will not fit, an exponential form is used.

If **Ip** is zero, an exponential form with **Dp** significant digits is used, giving a field width of **Dp + 6**.

If **Ip** and **Dp** are zero, a minimum field width free format is used.

Numbers which correspond to the IEEE standard concepts of 'Infinity' and 'NotaNumber' produce the texts **Inf** and **NaN**, respectively.

In exponential forms a number in the range [1.0, 10.0) is followed by **E**, a + or − sign, and a 2 digit decimal exponent.

`ss.write.real64`   **streamio**

> PROC ss.write.real64 (CHAN OF SS sink,
>   VAL REAL64 number, VAL INT Ip, Dp)

`write.real64`   **userio**

> PROC write.real64 (CHAN OF ANY sink,
>   VAL REAL64 number, VAL INT Ip, Dp)

Converts **number** into a sequence of ASCII decimal digits padded out with leading spaces and an optional sign to the specified number of digits **Ip** before and **Dp** after the decimal point.

Details as for **REAL32** but allowing 3 digits for the exponent.

`ss.write.endstream`   **streamio**

> PROC ss.write.endstream (CHAN OF SS sink)

`write.endstream`   **userio**

> PROC write.endstream (CHAN OF ANY sink)

Sends a special stream terminator value to **sink**. A call of this is needed if **sink** is a file interface, or other interface procedure without an explicit stoppping channel, but not if it is a real screen channel (parameter of the **EXE**).

```
ss.goto.xy                                                        streamio
      PROC ss.goto.xy (CHAN OF SS sink, VAL INT x, y)
goto.xy                                                           userio
      PROC goto.xy (CHAN OF ANY sink, VAL INT x, y)
```

Sends the cursor to screen position (x,y). The origin (0,0) is at the top left corner of the screen.

```
ss.clear.eol                                                      streamio
      PROC ss.clear.eol (CHAN OF SS sink)
clear.eol                                                         userio
      PROC clear.eol (CHAN OF ANY sink)
```

Clears from the cursor position to the end of the current screen line.

```
ss.clear.eos                                                      streamio
      PROC ss.clear.eos (CHAN OF SS sink)
clear.eos                                                         userio
      PROC clear.eos (CHAN OF ANY sink)
```

Clears from the cursor position to the end of the current line and all lines below.

```
ss.beep                                                           streamio
      PROC ss.beep (CHAN OF SS sink)
beep                                                              userio
      PROC beep (CHAN OF ANY sink)
```

Sends a bell code to the terminal.

```
ss.up                                                             streamio
      PROC ss.up (CHAN OF SS sink)
up                                                                userio
      PROC up (CHAN OF ANY sink)
```

Moves the cursor one line up the screen.

```
ss.down                                                           streamio
      PROC ss.down (CHAN OF SS sink)
down                                                              userio
      PROC down (CHAN OF ANY sink)
```

Moves the cursor one line down the screen.

```
ss.left                                                           streamio
      PROC ss.left (CHAN OF SS sink)
left                                                              userio
      PROC left (CHAN OF ANY sink)
```

Moves the cursor one place left.

ss.right   `streamio`

    PROC ss.right (CHAN OF SS sink)

right   `userio`

    PROC right (CHAN OF ANY sink)

    Moves the cursor one place right.

The next five pairs procedures are not guaranteed to be fully implemented on all terminal types. They should only be used when the terminal being used can perform the required effect.

ss.insert.char   `streamio`

    PROC ss.insert.char (CHAN OF SS sink, VAL BYTE char)

insert.char   `userio`

    PROC insert.char (CHAN OF ANY sink, VAL BYTE char)

    The character at the cursor and those to the right of it are moved one place to the right and the character **char** is inserted at the cursor. The cursor moves one place right.

ss.delete.chl   `streamio`

    PROC ss.delete.chl (CHAN OF SS sink)

delete.chl   `userio`

    PROC delete.chl (CHAN OF ANY sink)

    The character to the left of the cursor is deleted. The character at the cursor and those to the right of it are moved one place left. The cursor moves one place left.

ss.delete.chr   `streamio`

    PROC ss.delete.chr (CHAN OF SS sink)

delete.chr   `userio`

    PROC delete.chr (CHAN OF ANY sink)

    The character at the cursor is deleted and all following characters on the line are moved one place left. The cursor does not move.

ss.ins.line   `streamio`

    PROC ss.ins.line (CHAN OF SS sink)

ins.line   `userio`

    PROC ins.line (CHAN OF ANY sink)

    The current line and those lines below it are moved down one line on the screen, losing the bottom line. The current line becomes blank.

ss.del.line   `streamio`

    PROC del.line (CHAN OF SS sink)

del.line   `userio`

    PROC del.line (CHAN OF ANY sink)

    The current line is deleted and all lines below it are moved up one line. The bottom line becomes blank.

14.9.3   Procedures supporting keystream input

These are in two groups: those which only perform input and those which echo the input to the screen and allow in-line deletion of characters input.

ks.read.char   streamio

     PROC ks.read.char (CHAN OF KS source, INT char)

read.char   userio

     PROC read.char (CHAN OF INT source, INT char)

     Returns ASCII value of next **char** from source, (If input is from a file end of line is signified by the value **INT' *c'**).

ks.read.int   streamio

     PROC ks.read.int (CHAN OF KS source, INT number,char)

read.int   userio

     PROC read.int (CHAN OF INT source, INT number,char)

     char must be initialised to the first input character previously removed from the input stream with, for example, **read.char**. Skips input up to a digit, **#**, **+** or **−**, then reads a sequence of digits to the first non-digit and converts the digits to an integer which is returned in **number**. If the first significant character is a '**#**' then a hexadecimal number is input, thereby allowing the user the option of which number base to use.

     If the input is invalid **ft.number.error** is returned in **char**; otherwise the terminating character is returned in **char**.

ks.read.hex.int   streamio

     PROC ks.read.hex.int (CHAN OF KS source, INT number, char)

read.hex.int   userio

     PROC read.hex.int (CHAN OF INT source, INT number, char)

     char must be initialised to the first input character previously removed from the input stream with, for example, **read.char**. Skips input up to a valid hexadecimal digit, then reads a sequence of hex digits to the first non-digit, returned as **char**, and converts the digits to an integer which is returned in **number**.

     If the input is invalid **ft.number.error** is returned in **char**; otherwise the terminating character is returned in **char**.

ks.read.text.line   streamio

     PROC ks.read.text.line (CHAN OF KS source, INT len,
                     []BYTE line, INT char)

read.text.line   userio

     PROC read.text.line (CHAN OF INT source, INT len,
                     []BYTE line, INT char)

     Reads text into the array **line** up to and including '***c'**, or up to and excluding any error code. Any '***n'** encountered is thrown away. **len** is the length of the line. A terminating '***c'** is always stored in the array. If there is an error its code is returned in **char**, otherwise the value of **char** will be **INT** '***c'**. If the array is filled before a '***c'** is encountered all further characters are ignored. Note that some TDS function codes have values which exceed 255 (see appendix D), this procedure will ignore such values completely.

`ks.read.int32`   streamio

     PROC ks.read.int32 (CHAN OF KS source, INT32 number, INT char)

     As **ks.read.int**, but for 32-bit integers.

`ks.read.hex.int32`   streamio

     PROC ks.read.hex.int32 (CHAN OF KS source, INT32 number, INT char)

     As **read.hex.int**, but for 32-bit integers.

`ks.read.int64`   streamio

     PROC read.int64 (CHAN OF KS source, INT64 number, INT char)

`read.int64`   userio

     PROC read.int64 (CHAN OF INT source, INT64 number, INT char)

     As **read.int**, but for 64-bit integers.

`ks.read.hex.int64`   streamio

     PROC ks.read.hex.int64 (CHAN OF KS source, INT64 number, INT char)

`read.hex.int64`   userio

     PROC read.hex.int64 (CHAN OF INT source, INT64 number, INT char)

     As **read.hex.int**, but for 64-bit integers.

`ks.get.real.string`   streamio

     PROC ks.get.real.string (CHAN OF KS in, INT len,
                  []BYTE str, INT char)

`get.real.string`   userio

     PROC get.real.string (CHAN OF INT in, INT len,
                  []BYTE str, INT char)

     For internal use only by the following two procedures.

`ks.read.real32`   streamio

     PROC ks.read.real32 (CHAN OF KS source, REAL32 number, INT char)

`read.real32`   userio

     PROC read.real32 (CHAN OF INT source, REAL32 number, INT char)

char must be initialised to the first input character previously removed from the input stream with, for example, **read.char**. Skips input up to a digit, **+** or **−**, then reads a sequence of digits (with decimal point and optional exponent) up to the first invalid character, returned as **char**. Converts the digits to a floating point value in **number**.

If the input is invalid ft.number.error is returned in **char**, otherwise the terminating character is returned in **char**.

Only numbers conforming to the syntax of occam REAL constants (without the type symbol), are accepted as valid REAL numbers. This means in particular that there must be at least one digit before the decimal point and there must be a sign before the optional exponent part. The only relaxation of the rule is the allowing of **e** as an alternative to **E** to introduce the decimal exponent.

```
ks.read.real64                                                   streamio
      PROC ks.read.real64 (CHAN OF KS source, REAL64 number, INT char)
read.real64                                                      userio
      PROC read.real64 (CHAN OF INT source, REAL64 number, INT char)
```

As **read.real.32**, but for 64-bit real numbers.

```
ks.read.echo.char                                               streamio
      PROC read.echo.char (CHAN OF KS source,
                           CHAN OF SS echo, INT char)
read.echo.char                                                  userio
      PROC read.echo.char (CHAN OF INT source,
                           CHAN OF ANY echo, INT char)
```

Returns ASCII value of next **char** from source, (if input is from a file interface procedure (see library **interf** section 14.10) end of line is normally signified by the value **INT'*c'**). No deletions are allowed. A **'*c'** is echoed with the character pair **"*c*n"**. All other ASCII control codes and TDS function codes are not echoed.

```
ks.read.echo.int                                                streamio
      PROC ks.read.echo.int (CHAN OF KS source,
                             CHAN OF SS echo, INT number, char)
read.echo.int                                                   userio
      PROC read.echo.int (CHAN OF INT source,
                          CHAN OF ANY echo, INT number, char)
```

char must be initialised to the first input character previously removed from the input stream with, for example, **read.echo.char**. Skips input up to a digit, **#**, **+** or **−**, then reads a sequence of digits to the first non-digit and converts the digits to an integer in **number**. If the first significant character is a **#** then a hexadecimal number is input, thereby allowing the user the option of which number base to use.

If the input is invalid **ft.number.error** is returned in **char**; otherwise the terminating character is returned in **char**.

```
ks.read.echo.hex.int                                            streamio
      PROC ks.read.echo.hex.int (CHAN OF KS source,
                                 CHAN OF SS echo, INT number, char)
read.echo.hex.int                                               userio
      PROC read.echo.hex.int (CHAN OF INT source,
                              CHAN OF ANY echo, INT number, char)
```

char must be initialised to the first input character previously removed from the input stream with, for example, **read.char**. Skips input up to a valid hexadecimal digit, then reads a sequence of hex digits to the first non-digit and converts the digits to an integer in **number**.

If the input is invalid **ft.number.error** is returned in **char**; otherwise the terminating character is returned in **char**.

ks.read.echo.text.line   streamio
     PROC **ks.read.echo.text.line** (CHAN OF KS source, CHAN OF SS echo,
                              INT len, []BYTE line, INT char)
read.echo.text.line   userio
     PROC **read.echo.text.line** (CHAN OF INT source, CHAN OF ANY echo,
                             INT len, []BYTE line, INT char)

Reads text into the array **line** up to and including '***c**', or up to and excluding any error code. A final '***c**' is always stored in the array. Any '***n**' encountered is thrown away. **len** is the length of the line including the terminator. If there is an error its code is returned as **char**, otherwise the value of **char** will be **INT'*c'**. If the array is filled before a '***c**' is encountered all further characters are ignored. Note that some TDS function codes have values which exceed 255 (see appendix D), this procedure will ignore such values completely.

ks.read.echo.int32 (T2 only)   streamio
     PROC **ks.read.echo.int32** (CHAN OF KS source,
                        CHAN OF SS echo,
                        INT32 number, INT char)

As **ks.read.echo.int** but reads 32-bit numbers.

ks.read.echo.hex.int32 (T2 only)   streamio
     PROC **ks.read.echo.hex.int32** (CHAN OF KS source,
                            CHAN OF SS echo,
                            INT32 number, INT char)

As **ks.read.echo.hex.int** but reads 32-bit numbers.

ks.get.real.with.del   streamio
     PROC **ks.get.real.with.del** (CHAN OF KS in, CHAN OF SS echo,
                          INT len, []BYTE str, INT char)
get.real.with.del   userio
     PROC **get.real.with.del** (CHAN OF INT in, CHAN OF ANY echo,
                        INT len, []BYTE str, INT char)

For internal use only by the following two procedures.

ks.read.echo.real32   streamio
     PROC **ks.read.echo.real32** (CHAN OF KS source, CHAN OF SS echo,
                        REAL32 number, INT char)
read.echo.real32   userio
     PROC **read.echo.real32** (CHAN OF INT source, CHAN OF ANY echo,
                      REAL32 number, INT char)

char must be initialised to the first input character previously removed from the input stream with, for example, **read.char**. Skips input up to a digit, + or −, then reads a sequence of digits (with decimal point and optional exponent) up to the first invalid character, returned as **char**. Converts the digits to a floating point value in **number**.

If the input is invalid **ft.number.error** is returned in **char**, otherwise the terminating character is returned in **char**.

`ks.read.echo.real64`   streamio

      PROC ks.read.echo.real64 (CHAN OF KS source, CHAN OF SS echo,
                        REAL64 number, INT char)

`read.echo.real64`   userio

      PROC read.echo.real64 (CHAN OF INT source, CHAN OF ANY echo,
                      REAL64 number, INT char)

As **read.echo.real32**, but for 64-bit real numbers.

14.9.4 Procedures supporting the standard input model of the keyboard

`so.getkey`   splib

      PROC so.getkey (CHAN OF SP fs, ts, BYTE key, result)

As **so.pollkey** but waits for a key if none is available.

See also section 16.5.7. This is the preferred procedure for individual character input in EXEs and hosted programs.

The key is not echoed on the screen.

The results can take the same values as **so.pollkey**.

`so.pollkey`   splib

      PROC so.pollkey (CHAN OF SP fs, ts, BYTE key, result)

Reads a single character from the keyboard. If no key is available then it returns immediately with **spr.operation.failed**. The key is not echoed on the screen.

See also section 16.5.7. This procedure is used by the TDS in its implementation of the **keyboard** channel of an EXE.

The result returned can take any of the following values:

  spr.ok   The read was successful.
  spr.operation.failed   The read failed.

`so.ask`   sklib

      PROC so.ask (CHAN OF SP fs,ts, VAL[]BYTE prompt, replies,
              VAL BOOL display.possible.replies, echo.reply,
              INT reply.number)

Prompts on the screen for a user response on the keyboard. The prompt is specified by the string **prompt**, and the list of permitted replies by the string **replies**. Only single character responses are permitted, and alphabetic characters are *not* case sensitive. For example if the permitted responses are 'Y', 'N' and 'Q' then the **replies** string would contain the characters **"YNQ"**, and 'y', 'n' and 'q' would also be accepted. **reply.number** indicates which response was typed, numbered from zero.

If **display.possible.replies** is TRUE the permitted replies are displayed on the screen. If **echo.reply** is TRUE the user's response is displayed.

The procedure will not return until a valid response has been typed.

The following procedures are based on a simple view of the keyboard as a device on which all input communications consist of a whole line of input terminated by a RETURN. Application code may be easier to write, but there is less flexibility of user interface design than in the TDS keystream approach.

so.read.line   solib

> **PROC so.read.line (CHAN OF SP fs, ts, INT len,**
> **[]BYTE line, BYTE result)**

Reads a line of text from the keyboard, without echoing it on the screen. The line is read until RETURN is pressed at the keyboard.

The result returned can take any of the following values:

> **spr.ok**   The read was successful.
> **spr.operation.failed**   The read failed.

so.read.echo.line   solib

> **PROC so.read.echo.line (CHAN OF SP fs, ts, INT len,**
> **[]BYTE line, BYTE result)**

As **so.read.line**, but user input is echoed on the screen.

so.read.echo.int   sklib

> **PROC so.read.echo.int (CHAN OF SP fs, ts,**
> **INT n, BOOL error)**

Reads a decimal integer typed at the keyboard and displays it on the screen. The number must be terminated by 'RETURN'. The boolean **error** is set to TRUE if an invalid integer is typed.

so.read.echo.hex.int   sklib

> **PROC so.read.echo.hex.int (CHAN OF SP fs, ts, INT n, BOOL error)**

As **so.read.echo.int** but reads a number in hexadecimal format. The number must be prefixed with either '#', which directly indicates a hexadecimal number, or '%', which assumes the prefix **#8000....**

so.read.echo.any.int   sklib

> **PROC so.read.echo.any.int (CHAN OF SP fs, ts, INT n, BOOL error)**

As **so.read.echo.int** but accepts numbers in either decimal or hexadecimal format. Hexadecimal numbers must be prefixed with either '#', which specifies the number directly, or '%', which assumes the prefix **#8000....** .

so.read.echo.int64   sklib

> **PROC so.read.echo.int64 (CHAN OF SP fs, ts, INT64 n, BOOL error)**

As **so.read.echo.int** but reads 64-bit numbers.

so.read.echo.hex.int64   sklib

> **PROC so.read.echo.hex.int64 (CHAN OF SP fs, ts,**
> **INT64 n, BOOL error)**

As **so.read.echo.hex.int** but reads 64-bit numbers.

so.read.echo.real32   **sklib**

>   PROC so.read.echo.real32 (CHAN OF SP fs, ts, REAL32 n, BOOL error)

Reads a real number typed at the keyboard and displays it on the screen. The number must be terminated by 'RETURN'. The boolean variable **error** is set to TRUE if an invalid number is typed.

so.read.echo.real64   **sklib**

>   PROC so.read.echo.real64 (CHAN OF SP fs, ts, REAL64 n, BOOL error)

As **so.read.echo.real32**, but for 64-bit real numbers.

14.9.5 Procedures supporting the standard output model of the screen

These procedures are closely related to corresponding procedures which provide access to arbitrary output files on the host and so for convenience of documentation are described in section 14.11.

14.10 Buffers, multiplexors and protocol converters

These interface procedures are designed to be called in parallel with an application process using the TDS screenstream and keystream input output procedures. They enable such processes to be interfaced to the TDS folded file system and host filing systems. An example of the use of interface procedures appears in section 6.7.2 and further examples are included with the software.

The TDS provides a set of system channels as parameters to an executable program (**EXE**) which support a versatile virtual terminal interface and access to a generalised filing system accessible through the idea of filed folds. Direct access to host files is also provided.

One end of each of these channels is available to the user, and the other end is in the run-time system. These channels have well-defined protocols, and communications must conform to these protocols.

The main body of a user's application may either be written to interface directly to the system channels or may be written to run in parallel with interface procedures which use these channels.

The simple user procedures may be used to access files if suitable interface procedures are used.

Two distinct classes of file may be accessed. Folded files which are part of the fold structure of the development system, and host files which are not. Note that although in a hosted implementation all files are accessible as host files, those which are folded files have a particular internal structure defined in appendix G of this manual, and should be accessed through the TDS user filer interface, or through other software which can decode the special TDS folded file representation.

Access to host files may be obtained by means of communications on the system channel pair **from.filer** and **to.filer** using the obsolete **tkf** protocol, or preferably by the **iserver** channel pair **from.isv** and **to.isv** using **SP** protocol.

A pair of interface procedures are described which allow an existing host text file to be read sequentially as if from a keyboard, and a new host text file to be created and written to as if it were a screen. The implementations of these procedures may be host-dependent because of the different ways of handling the ends of text lines in different operating systems. The interface procedures hide these possible difficulties and treat all text files as sequences of ASCII character strings separated by **'*c'** characters.

In chapter 16 further details of the communications across channels to the filing system are described. There are several different ways in which such communication can be organised according to whether the program is running inside the TDS, loaded directly by a server, either written by the user or, for example, the host file server **iserver** supplied with the TDS.

iserver allows non-sequential access, read/write access, block access, etc.

The interface to the server is defined in chapter 16, 'System interfaces'.

One or more filed folds which are themselves members of a single fold bundle may be accessed by an executable program (**EXE**) called with the cursor on the fold line of the bundle. Each of these folds is potentially the root of a nested tree structure.

Folded files to be read may be created as folds by the editor, or by another program using this library. New filed folds may be created within the bundle and written into. Such files will be readable by the editor.

The principal limitation is that no more than four root folds may be simultaneously in use, either for input or for output. All access to these filed folds is sequential, and the procedures below are designed to facilitate the reading of existing files as if they were a source of characters like a keyboard, and the writing of new files as if they were a simple screen or printer. Folds used for input may include nested folds, but such structure will not be visible.

Access to a filed fold is obtained by a pair of channels which must be corresponding elements of the channel arrays **from.user.filer** and **to.user.filer** passed to the **EXE** as parameters. Interface procedures take parameters which represent these channels, terminal channels, fold numbers and fold comment text. The members of the bundle of folds are identified by a fold number which starts at 1 for the top fold of the bundle. The whole bundle, or just a simple fold, may be accessed as fold number 0.

There are also some generally applicable multiplexing and channel consuming procedures which are useful for organising the plumbing of the channels in a program. It is particularly important when building these procedures into a program to ensure that the proper termination of each interface procedure is assured. If this is not done it will not be possible to return cleanly to the development system after calling the program.

so.buffer   *spinterf*

```
PROC so.buffer (CHAN OF SP fs, ts, from.user, to.user,
                CHAN OF BOOL stopper)
```

This procedure buffers data between the user and the host or any other pair of processes using **SP** protocol. It can be used by processes on a network to pass data to the host across intervening processes. It is terminated by sending a **FALSE** value on the channel **stopper**.

so.multiplexor   *spinterf*

```
PROC so.multiplexor (CHAN OF SP fs, ts,
                     []CHAN OF SP from.user, to.user,
                     CHAN OF BOOL stopper)
```

This procedure multiplexes any number of pairs of **SP** protocol channels onto a single pair of **SP** protocol channels, which may go to the file server or another **SP** protocol multiplexor (or buffer). It is terminated by sending a **FALSE** value on the channel **stopper**.

so.overlapped.buffer   *spinterf*

```
PROC so.overlapped.buffer (CHAN OF SP fs, ts, from.user, to.user,
                           CHAN OF BOOL stopper)
```

Similar to **so.buffer**, but contains built-in knowledge of host file server commands and allows many host communications to occur simultaneously through a train of processes. This can improve efficiency if the communications pass through many processes before reaching the server. It is terminated by a **FALSE** value on the channel **stopper**.

so.overlapped.multiplexor   **spinterf**

```
PROC so.overlapped.multiplexor (CHAN OF SP fs, ts,
                          []CHAN OF SP from.user, to.user,
                          CHAN OF BOOL stopper, []INT queue)
```

Similar to **so.multiplexor**, but can pipeline server requests. The number of requests that can be pipelined is determined by the size of **queue**, which must provide one word for each request that can be pipelined. Pipelining improves efficiency if the server requests have to pass through many processes on the way to and from the server. It is terminated by a **FALSE** value on the channel **stopper**.

so.keystream.from.kbd   **spinterf**

```
PROC so.keystream.from.kbd (CHAN OF SP fs, ts,
                          CHAN OF KS keys.out,
                          CHAN OF BOOL stopper,
                          VAL INT ticks.per.poll)
```

Reads characters from the keyboard using **so.pollkey** and outputs them one at a time as integers on the channel **keys.out**. It is terminated by sending **FALSE** on the boolean channel **stopper**. The procedure polls the keyboard at an interval determined by the value of **ticks.per.poll**, in transputer clock cycles. **ticks.per.poll** must not be zero.

After **FALSE** is sent on the channel **stopper** the procedure sends the negative value **ft.terminated** on **keys.out** to mark the end of the file.

so.keystream.from.file   **spinterf**

```
PROC so.keystream.from.file (CHAN OF SP fs, ts,
                          CHAN OF KS keys.out,
                          VAL[]BYTE filename, BYTE result)
```

As **so.keystream.from.kbd**, but reads characters from the specified file. Terminates automatically when it has reached the end of the file and all the characters have been read from the **keys.out** channel. The negative value **ft.terminated** is sent on the channel **keys.out** to mark the end of the file. The result value returned will be one of those returned by **so.gets**.

so.keystream.from.stdin   **spinterf**

```
PROC so.keystream.from.stdin (CHAN OF SP fs, ts,
                          CHAN OF KS keys.out, BYTE result)
```

As **so.keystream.from.kbd**, but reads from the standard input stream. The standard input stream is normally assigned to the keyboard, but can be redirected by the host operating system.

so.scrstream.to.file   **spinterf**

```
PROC so.scrstream.to.file (CHAN OF SP fs, ts, CHAN OF SS scrn,
                          VAL[]BYTE filename, BYTE result)
```

Creates a new file with the specified name and writes the data sent on channel **scrn** to it. The **scrn** channel uses the screen stream protocol which is used by all the stream output library routines. It terminates on receipt of the stream terminator from **ss.write.endstream**, or on an error condition. The error code returned by **result** can be any result returned by **so.puts**, which is called by this procedure.

If used in conjunction with **ss.scrstream.fan.out** it may be used to file a copy of everything sent to the screen.

so.scrstream.to.stdout   **spinterf**

    PROC so.scrstream.to.stdout (CHAN OF SP fs, ts,
                                CHAN OF SS scrn, BYTE result)

Performs the same operation as **so.scrstream.to.file**, but writes to the standard output stream. The standard output stream goes to the screen, but can be redirected to a file by the host operating system.

ss.scrstream.to.ANSI.bytes   **ssinterf**

    PROC ss.scrstream.to.ANSI.bytes (CHAN OF SS scrn, CHAN OF BYTE ansi)

scrstream.to.ANSI   **interf**

    PROC scrstream.to.ANSI (CHAN OF ANY scrn, CHAN OF BYTE ansi)

so.scrstream.to.ANSI   **spinterf**

    PROC so.scrstream.to.ANSI (CHAN OF SP fs, ts, CHAN OF SS scrn)

Converts screen stream protocol into a stream of **BYTE**s according to the requirements of ANSI terminal screen protocol. Not all of the screen stream commands are supported, as some are not straightforward to implement. Refer to the source of the procedure to determine which commands are supported. The procedure terminates on receipt of the stream terminator from **ss.write.endstream**.

The **spinterf** version of this procedure sends the bytes to the standard output host stream. The other versions send the BYTE on the channel **ansi**.

ss.scrstream.to.TVI920.bytes   **ssinterf**

    PROC ss.scrstream.to.TVI920.bytes(CHAN OF SS scrn,CHAN OF BYTE tvi)

scrstream.to.TVI920   **interf**

    PROC scrstream.to.TVI920 (CHAN OF ANY scrn, CHAN OF BYTE tvi)

so.scrstream.to.TVI920   **spinterf**

    PROC so.scrstream.to.TVI920 (CHAN OF SP fs, ts, CHAN OF SS scrn)

Converts screen stream protocol into a stream of **BYTE**s according to the requirements of TVI920 (and compatible) terminals. Not all of the screen stream commands are supported, as some are not straightforward to implement. Refer to the source of the procedure to determine which commands are supported. The procedure terminates on receipt of the stream terminator from **ss.write.endstream**.

The **spinterf** version of this procedure sends the bytes to the standard output host stream. The other versions send the BYTE on the channel **tvi**.

ss.scrstream.to.array   **ssinterf**

    PROC ss.scrstream.to.array (CHAN OF SS scrn, []BYTE buffer)

scrstream.to.array   **interf**

    PROC scrstream.to.array (CHAN OF ANY scrn, []BYTE buffer)

A screen stream whose total size does not exceed the capacity of **buffer** may be buffered by this procedure, for subsequent onward transmission using **scrstream.from.array**.

`ss.scrstream.from.array`   ssinterf

> PROC ss.scrstream.from.array (CHAN OF SS scrn, VAL[] BYTE buffer)

`scrstream.from.array`   interf

> PROC scrstream.from.array (VAL[] BYTE buffer, CHAN OF ANY scrn)

Regenerates a screen stream buffered in **buffer** by a previous call of **scrstream.to.array**.

`ss.scrstream.to.fold`   ssinterf

> PROC ss.scrstream.to.fold (CHAN OF SS scrn,
>   CHAN OF ANY from.uf, to.uf
>   VAL[]BYTE fold.title,
>   INT fold.number, INT result)

`scrstream.to.file`   interf

> PROC scrstream.to.file (CHAN OF ANY scrn, from.uf, to.uf
>   VAL[]BYTE fold.title,
>   INT fold.number, INT result)

This procedure may be used to file a text stream, generated in screen stream protocol, in a new filed fold. If used in conjunction with **scrstream.fan.out** it may be used to file a copy of everything a program sends to the screen.

A new filed fold is created at the end of the current bundle, and its position is returned as **fold.number**. The filed fold has attributes **ft.opstext** and **fc.comment.text**. The string **fold.title** is written as its fold comment, **fold.title** is truncated at the first space or '.' to generate a file name. If it is empty a name will be created by the TDS.

Text to be filed is received on channel **scrn** in screen stream protocol as generated by simple user output procedures. The procedure terminates on receipt of the code generated by **write.endstream**. If any filing system error condition occurs the input screen stream is consumed as usual but an error is signalled in **result** when the procedure terminates.

`scrstream.multiplexor`   interf

> PROC scrstream.multiplexor ([]CHAN OF ANY screen.in, screen.out,
>   CHAN OF INT stopper)

Multiplexes a collection of channels using screen protocol into a single such channel. The input channels must be an array **screen.in**, and the output channel is **screen.out**. Each change of input channel directs output to a new line on the screen, tagged by the channel index. Any integer input on **stopper** terminates the multiplexor. The endstream command generated by **write.endstream** is ignored.

`ss.scrstream.fan.out`   ssinterf

> PROC ss.scrstream.fan.out (CHAN OF SS scrn, screen.out1, screen.out2)

`scrstream.fan.out`   interf

> PROC scrstream.fan.out (CHAN OF ANY scrn, screen.out1, screen.out2)

Sends copies of everything received on its input channel **scrn** to both of the output channels. Uses screen protocol. Terminated by calling **write.endstream** on the input channel.

ss.scrstream.sink   ssinterf

    PROC **ss.scrstream.sink** (CHAN OF SS scrn)

scrstream.sink   interf

    PROC **scrstream.sink** (CHAN OF ANY scrn)

Reads characters preceded by **tt.out.byte** and ignores them; also ignores all other **tt.** commands except **tt.endstream** (generated by **write.endstream**) which terminates the process.

ss.scrstream.copy   ssinterf

    PROC **ss.scrstream.copy** (CHAN OF SS scrn, scrn.out)

scrstream.copy   interf

    PROC **scrstream.copy** (CHAN OF ANY scrn, scrn.out)

Sends a screen stream received on **scrn** out again on **scrn.out**. Terminates on receipt of **tt.endstream** which is not sent on. This procedure is sometimes needed as a buffer (e.g. between a link and a multiplexor whose inputs are specified as an array).

ss.keystream.from.fold   ssinterf

    PROC **ss.keystream.from.fold** (CHAN OF ANY from.uf, to.uf,
                              CHAN OF KS kbd, VAL INT fold.number,
                              INT result)

keystream.from.file   interf

    PROC **keystream.from.file** (CHAN OF ANY from.uf, to.uf,
                              CHAN OF INT kbd, VAL INT fold.number,
                              INT result)

This procedure may be used to generate a stream of characters from a fold. If the fold is not filed then it will be filed for the duration of this procedure and then unfiled again.

The file in member **fold.number** of the current fold bundle (0 = whole bundle, 1 = first fold inside it ...) is opened. Its contents are output on channel **kbd** as if from a keyboard, with ***c** as line terminator between lines.

The file is closed on any error condition or when its last character has been read, followed by outputting **ft.terminated**. The procedure then terminates with an error number in **result**.

ks.keystream.sink   ssinterf

    PROC **ks.keystream.sink** (CHAN OF KS keys)

keystream.sink   interf

    PROC **keystream.sink** (CHAN OF INT keys)

Reads integers until the value **ft.terminated**, then terminates.

ks.keystream.to.screen   ssinterf

    PROC ks.keystream.to.screen (CHAN OF KS keyboard,CHAN OF SS screen)

keystream.to.screen   interf

    PROC keystream.to.screen (CHAN OF INT keyboard, CHAN OF ANY screen)

This procedure converts from key stream protocol to screen stream protocol. On its input channel **keyboard** it receives a sequence of integers which may be ASCII values, or coded function keys. ASCII values are passed through unchanged, except for '***c**' which is followed by '***n**' Those function keys which have a corresponding screen function (simple cursor moves, etc) are converted into this screen function, others ring the bell.

The procedure may also be used in programs which were originally written for earlier implementations of occam which required text for the screen to be output as a sequence of integer values. All negative values received, other than **ft.terminated**, are ignored.

The procedure terminates on receipt of **ft.terminated**.

af.to.sp   afsp

    PROC af.to.sp (CHAN OF SP fs, ts,
                CHAN OF ANY from.user, to.user,
                VAL BOOL passthrough.Terminate.Cmd)

Converts channels from the AFSERVER protocol to the SP protocol. **fs** and **ts** are the SP channels from and to the host file server, and **from.user** and **to.user** are the channels between the application program and the AFSERVER. The boolean **passthrough.Terminate.Cmd** controls the action of the AFSERVER terminate command; if set to **TRUE** it will terminate the server, if set to **FALSE** it will not.

14.11 Access to host filing system

Filing systems on the host computer are made accessible by way of the **iserver** and its associated SP protocol channels. The protocols used on these channels are described in detail in section 16.5. The library procedures which support this interface include low level procedures which exactly match each of the alternatives of the protocol, and higher level procedures which call the lower level ones. The user need not be concerned with which of the procedures described here are the low level ones.

The **iserver** can support an arbitrary number of simultaneously open files. Each **iserver** transaction consists of a pair of communications, one from the program to the server, followed by one from the server to the program. The server responds to each transaction in turn and does not support any kind of parallelism on the host. The same applies to the use of the SP protocol channel pair in an **EXE**.

If an occam application requires access to the server from more than one concurrent process then the user must provide any necessary multiplexing of access using one of the procedures described above. Use of library procedures for server transactions will usually ensure adequate sequencing of communications.

The filing system interface supports the opening of files in a variety of modes, text/binary, input/output/update, etc. It is the responsibility of the user to ensure that files are properly used. In some implementations the C library used in **iserver** may return error codes in cases of misuse but this is not in general guaranteed. When a file is opened a value called a stream identifier is returned. As this identifier must be quoted for all subsequent operations on that file it is possible for many files to be simultaneously open.

The procedures described below include the **so.write.** procedures which allow output to be sent to the standard output stream (terminal screen). These are closely related to the corresponding **so.fwrite.** procedures and are described here because of the similarity.

Output to a printer may be coded by opening a file with the appropriate reserved DOS name (e.g. **PRN**). The contents of such a file should be written using the text line oriented procedure **so.puts**, allowing the server to insert the appropriate line separators.

The library procedures allow arbitrarily large arrays to be passed for output or for the receipt of input. Such operations are broken down into an appropriate sequence of smaller transfers in the code of the library procedures. The relatively small size of these lower level blocks enables general purpose multiplexors and buffers to be written in occam which requires array sizes to be known at compile time.

so.open   **splib**

```
PROC so.open (CHAN OF SP fs, ts, VAL[]BYTE name,
              VAL BYTE type, mode, INT32 streamid, BYTE result)
```

Opens the file given by **name** and returns a stream identifier **streamid** for all future operations on the file until it is closed. File type is specified by **type** and the mode of opening by **mode**.

type can take the following values:

spt.binary	File contains raw bytes only.
spt.text	File contains text records separated by newline sequences.

mode can take the following values:

spm.input	Open existing file for reading.
spm.output	Open new file, or truncate an existing one, for writing.
spm.append	Open a new file, or append to an existing one, for writing.
spm.existing.update	Open an existing file for update (reading and writing), starting at beginning of the file.
spm.new.update	Open new file, or truncate existing one, for update.
spm.append.update	Open new file, or append to an existing one, for update.

result can take the following values:

spr.ok	The open was successful.
spr.operation.failed	The open failed.
spr.bad.name	Invalid file name.
spr.bad.type	Invalid file type.
spr.bad.mode	Invalid open mode.
spr.bad.packet.size	File name too large.

so.close   **splib**

```
PROC so.close (CHAN OF SP fs, ts, VAL INT32 streamid, BYTE result)
```

Closes the stream identified by **streamid**.

The result returned can take any of the following values:

spr.ok	The close was successful.
spr.operation.failed	The close failed.

so.read   **splib**

```
PROC so.read (CHAN OF SP fs, ts, VAL INT32 streamid,
              INT bytes.read, []BYTE data)
```

Reads a block of bytes from the specified stream up to a maximum given by the size of the array **data**. If **bytes.read** returned is not the same as the size of **data** then the end of the file has been reached or an error has occurred.

so.write   splib

> PROC so.write (CHAN OF SP fs, ts, VAL INT32 streamid,
>   VAL[] BYTE data, INT length)

Writes a block of data to the specified stream. If **length** is less than the size of **data** then an error has occurred.

so.gets   splib

> PROC so.gets (CHAN OF SP fs, ts, VAL INT32 streamid,
>   INT bytes.read, []BYTE data, BYTE result)

Reads a line from the specified input stream. Characters are read until a newline sequence is found, the end of the file is reached, or all characters in **data** have been read. The newline sequence is not included in the returned array. If the read fails then either the end of file has been reached or an error has occurred.

The result returned can take any of the following values:

spr.ok	The read was successful.
spr.operation.failed	The read failed.
spr.bad.packet.size	**data** is too large (> **sp.max.readbuffer.size**).
spr.buffer.overflow	The line was larger than the buffer **data**. **bytes.read** contains the size of the buffer.

so.puts   splib

> PROC so.puts (CHAN OF SP fs, ts, VAL INT32 streamid,
>   VAL[]BYTE data, BYTE result)

Writes a line to the specified output stream. A newline sequence is added to the end of the line. The size of **data** must be less than or equal to the hostio constant **sp.max.writebuffer.size**.

The result returned can take any of the following values:

spr.ok	The write was successful.
spr.operation.failed	The write failed.
spr.bad.packet.size	SIZE **data** is too large (> **sp.max.writebuffer.size**).

so.flush   splib

> PROC so.flush (CHAN OF SP fs, ts,
>   VAL INT32 streamid, BYTE result)

Flushes the specified output stream. All internally buffered data is written to the stream. Write and put operations that are directed to standard output are flushed automatically.

The result returned can take any of the following values:

spr.ok	The flush was successful.
spr.operation.failed	The flush failed.

so.seek   splib

> PROC so.seek (CHAN OF SP fs, ts,
>   VAL INT32 streamid, offset, origin, BYTE result)

Sets the file position for the specified stream. A subsequent read or write will access data at the new position.

For a binary file the new position will be **offset** bytes from the position defined by **origin**. For a text file **offset** must be zero or a value returned by **so.tell**, in which case **origin** must be **spo.start**.

origin may take the following values:

spo.start   The start of the file.
spo.current   The current position in the file.
spo.end   The end of the file.

The result returned can take any of the following values:

spr.ok   The operation was successful.
spr.operation.failed   The seek failed.
spr.bad.origin   Invalid origin.

---

so.tell   **splib**

    PROC so.tell (CHAN OF SP fs, ts, VAL INT32 streamid,
                INT32 position, BYTE result)

Returns the current file position for the specified stream.

The result returned can take any of the following values:

spr.ok   The operation was successful.
spr.operation.failed   The tell failed.

---

so.eof   **splib**

    PROC so.eof (CHAN OF SP fs, ts,
              VAL INT32 streamid, BYTE result)

Tests whether the specified stream has reached the end of a file.

The result returned can take any of the following values:

spr.ok   End of file has been reached.
spr.operation.failed   The end of file has not been reached.

---

so.ferror   **splib**

    PROC so.ferror (CHAN OF SP fs, ts,
              VAL INT32 streamid, INT32 error.no, INT length,
              []BYTE message, BYTE result)

Indicates whether an error has occurred on the specified stream. The integer **error.no** is a host defined error number. The message will have **length** zero if no message can be provided.

The result returned can take any of the following values:

spr.ok   An error has occurred.
spr.operation.failed   No error has occurred.
spr.buffer.overflow   An error has occurred but the message is too
                       large for the buffer.

If the buffer overflows **length** is set to the buffer size.

so.remove splib

> **PROC so.remove (CHAN OF SP fs, ts,**
> **   VAL[]BYTE name, BYTE result)**

Deletes the specified file.

The result returned can take any of the following values:

spr.ok	The delete was successful.
spr.operation.failed	The delete failed.
spr.bad.name	Null name supplied.
spr.bad.packet.size	SIZE name is too large (> **sp.max.removename.size**).

so.rename splib

> **PROC so.rename (CHAN OF SP fs, ts,**
> **   VAL[]BYTE oldname, newname, BYTE result)**

Renames the specified file.

The result returned can take any of the following values:

spr.ok	The operation was successful.
spr.operation.failed	The rename failed.
spr.bad.name	Null name supplied.
spr.bad.packet.size	File names are too large (SIZE name1 + SIZE name2 > **sp.max.renamename.size**).

so.open.temp solib

> **PROC so.open.temp (CHAN OF SP fs, ts, VAL BYTE type,**
> **   [so.temp.filename.length]BYTE filename,**
> **   INT32 streamid, BYTE result**

Opens a temporary file in **spm.update** mode. Temporary files are created with names of the form **temp***nn*. The *nn* suffix on the name **temp***nn* is incremented up to a maximum of 9999 until an unused number is found. If the number exceeds 2 digits the last character of **temp** is overwritten. For example: if the number exceeds 99 the **p** is overwritten , as in **tem999**; if the number exceeds 999, the **m** is overwritten, as in **te9999**. File type can be **spt.binary** or **spt.text**, as with **so.open**. The name of the file actually opened is returned in **filename**.

The result returned can take any of the following values:

> **spr.ok**   The open was successful.

so.test.exists solib

> **PROC so.test.exists (CHAN OF SP fs, ts,**
> **   VAL[]BYTE filename, BOOL exists)**

Tests if the specified file exists. The value of **exists** is TRUE if the file exists, otherwise it is FALSE.

so.popen.read   solib

> PROC so.popen.read (CHAN OF SP fs, ts,
>   VAL[]BYTE filename, path.variable.name,
>   VAL BYTE open.type, INT full.len,
>   []BYTE full.name, INT32 stream.id,
>   BYTE result

As **so.open** but if the file is not found in the current directory, and the filename does not include a directory name, uses the directory path string associated with the environment variable given as a string in **path.variable.name** and performs a search in each directory in the path in turn. Directory names in this path string must include a terminating directory separator and be separated by semicolons.

Example for a DOS system:

> **\tds3\;\tds3\system**

File type can be **spt.binary** or **spt.text**, as with **so.open**. The mode of opening is always **spm.input**.

The name of the file opened is returned in **full.name**, and the length of the file name is returned in **full.len**. If no file is opened **full.len** is set to zero.

The result returned can take any of the following values:

spr.ok	The open was successful.
spr.operation.failed	The open failed.
spr.bad.name	Null name supplied.
spr.bad.type	Invalid file type specified.
spr.bad.packet.size	File name too large.

so.write.string   solib

> PROC so.write.string (CHAN OF SP fs, ts, VAL[]BYTE string)

Writes the string **string** to the screen. The size of **string** must not exceed **sp.max.writebuffer.size** (see appendix D.3).

so.fwrite.string   solib

> PROC so.fwrite.string (CHAN OF SP fs, ts, VAL INT32 streamid
>   VAL[]BYTE string, BYTE result)

Writes a string to the specified stream.

so.write.char   solib

> PROC so.write.char (CHAN OF SP fs, ts, VAL BYTE char)

Writes the single byte **char** to the screen.

so.fwrite.char   solib

> PROC so.fwrite.char (CHAN OF SP fs, ts, VAL INT32 streamid,
>   VAL BYTE char, BYTE result)

Writes a single character to the specified stream.

so.write.string.nl solib

>   PROC so.write.string.nl (CHAN OF SP fs, ts, VAL[]BYTE string)

>   As **so.write.string**, but appends a newline sequence to the end of the string. The size of **string** must not exceed **sp.max.writebuffer.size**−2 (see appendix D.3).

so.fwrite.string.nl solib

>   PROC so.fwrite.string.nl (CHAN OF SP fs, ts, VAL INT32 streamid,
>   VAL[]BYTE string, BYTE result)

>   As **so.fwrite.string**, but appends a newline sequence to the end of the string.

so.write.nl solib

>   PROC so.write.nl (CHAN OF SP fs, ts)

>   Writes a newline sequence to the screen.

so.fwrite.nl solib

>   PROC so.fwrite.nl (CHAN OF SP fs, ts,
>   VAL INT32 streamid, BYTE result)

>   Writes a newline sequence to the specified stream.

so.write.int solib

>   PROC so.write.int (CHAN OF SP fs, ts, VAL INT n, width)

>   Writes the value **n** (of type **INT**) to the screen as decimal ASCII digits, padded out with leading spaces and an optional sign to the specified **width**. If the **width** is too small for the number it is widened as necessary; a zero value for **width** specifies minimum width.

so.fwrite.int solib

>   PROC so.fwrite.int (CHAN OF SP fs, ts, VAL INT32 streamid,
>   VAL INT n, width, BYTE result)

>   Writes the value **n** (of type **INT**) to the specified stream as decimal ASCII digits, padded out with leading spaces and an optional sign to the specified **width**. If the **width** is too small for the number it is widened as necessary; a zero value for **width** will give minimum width.

so.write.hex.int solib

>   PROC so.write.hex.int (CHAN OF SP fs, ts, VAL INT n, width)

>   Writes the value **n** (of type **INT**) to the screen as hexadecimal ASCII digits, preceded by the '#' character. The number of characters to be printed is specified by **width**. If **width** is larger than the size of the number then the number is padded with leading '0's or 'F's as appropriate. If **width** is smaller than the size of the number, the number is truncated to **width** digits.

so.fwrite.hex.int solib

>   PROC so.fwrite.hex.int (CHAN OF SP fs, ts, VAL INT32 streamid,
>   VAL INT n, width, BYTE result)

>   Writes the value **n** (of type **INT**) to the specified stream as hexadecimal ASCII digits preceded by the '#' character. The number of characters to be printed is specified by **width**. If **width** is larger than the size of the number then the number is padded with leading '0's or 'F's as appropriate. If **width** is smaller than the size of the number, then the number is truncated to **width** digits.

so.write.int64   **solib**

    PROC so.write.int64 (CHAN OF SP fs, ts, VAL INT64 n, VAL INT width)

    As **so.write.int** but for 64-bit integers. The **field** parameter behaves as in **so.write.int**.

so.fwrite.int64   **solib**

    **PROC so.fwrite.int64 (CHAN OF SP fs, ts, VAL INT32 streamid,
                         VAL INT64 n, VAL INT width, BYTE result)**

    As **so.fwrite.int** but for 64-bit integers.
    The **field** parameter behaves as in **so.fwrite.int**.

so.write.hex.int64   **solib**

    **PROC so.write.hex.int64 (CHAN OF SP fs, ts, VAL INT64 n,
                         VAL INT width)**

    As **so.write.hex.int** but for 64-bit integers.
    The **width** parameter behaves as in **so.write.hex.int**.

so.fwrite.hex.int64   **solib**

    **PROC so.fwrite.hex.int64 (CHAN OF SP fs, ts, VAL INT32 streamid,
                         VAL INT64 n, VAL INT width, BYTE result)**

    As **so.fwrite.hex.int** but for 64-bit integers.
    The **width** parameter behaves as in **so.fwrite.hex.int**.

so.write.real32   **solib**

    **PROC so.write.real32 (CHAN OF SP fs, ts, VAL REAL32 r,
                       VAL INT Ip, Dp)**

Writes the value **r** (of type **REAL32**) to the screen as decimal ASCII digits. The number is padded out with leading spaces and an optional sign bit to the number of digits specified by **m** before and **n** after the decimal point. The total width of the number is **m** + **n** + 2, except in the cases described under **REAL32TOSTRING** (see the 'occam 2 reference manual').

so.fwrite.real32   **solib**

    **PROC so.fwrite.real32 (CHAN OF SP fs, ts, VAL INT32 streamid,
                        VAL real32 r, VAL INT Ip, Dp, BYTE result)**

Writes the value **r** (of type **REAL32**) to the specified stream as decimal ASCII digits. The number is padded out with leading spaces and an optional sign bit to the number of digits specified by **m** before and **n** after the decimal point. The total width of the number is **m** + **n** + 2, except in the cases described under **REAL32TOSTRING** (see the 'occam 2 reference manual').

so.write.real64   **solib**

    **PROC so.write.real64 (CHAN OF SP fs, ts, VAL REAL64 r,
                       VAL INT Ip, Dp)**

As **so.write.real32** but for 64-bit real numbers. Allows 3 digits for the exponent.

so.fwrite.real64   solib

       PROC so.fwrite.real64 (CHAN OF SP fs, ts, VAL INT32 streamid,
                              VAL real64 r, VAL INT Ip, Dp, BYTE result)

     As **so.fwrite.real32** but for 64-bit real numbers. Allows 3 digits for the exponent.

14.12 Access to the TDS's folded file store

Output to and input from the folded file store

The folded stream access procedures are in the library **userio**, with subsidiary procedures in the library
ufiler.

This group of procedures enable the user to write and read hierarchically structured data mapped on to the
folding system of the TDS. They are therefore only suitable for inclusion in executable procedures (**EXE**s) or
in **PROGRAM**s which will run in communication with an **EXE** at run time. The process within the TDS which
handles these communications is called the user filer.

The procedures provided do not exercise all the facilities available across the folded stream interface, but
support a subset. Programmers may wish to extend the set of procedures provided (in a similar style if this is
appropriate) to give the facilities they require. The full facilities of the interface are described in section 16.7.

All these procedures have a pair of channels as their first two parameters, the first of these is a channel
from the environment into the current process and the second is a channel from the current process to the
environment.

Communications across these channels are of three kinds (from the point of view of a user program communicating with the TDS):

- user filer control mode,

- file stream input modes (folded and data),

- file stream output modes (folded and data).

In any one sequence of communications using a pair of these channels the channel pair is first used in user
filer control mode sending commands or queries on the output channel and receiving data or responses on
the input channel. Actual data transfers are then carried out either in a file stream input mode or in a file
stream output mode. These modes are defined to be data stream modes or folded stream modes according
to the particular open command used to switch from command mode.

Data stream modes are for reading operations where any internal fold structure is to be ignored or for writing
operations where no internal fold structure is to be created in the new fold. They are used in the interface procedures **ks.keystream.from.fold**, **keystream.from.file**, **ss.scrstream.to.fold** and
scrstream.to.file described in section 14.10.

Folded stream modes give the user the potential to navigate the fold structure of an existing fold and to
create a nested fold structure in a new fold. The procedures described here support a significant subset of
the possible operations in these modes.

On exhaustion of a stream the channel pair reverts to user filer control mode.

The procedures are presented in two groups, those for writing new folded streams and those for reading
existing folded streams. They all make use of lower level procedures from the library **ufiler** described at
the end of this section.

The descriptions of the procedures assume familiarity with the structure of folded data as described in
appendix G of this manual.

The following conventions apply to the parameters of the procedures:

1 The first two parameters are channels; in a call where these procedures are being used to communicate directly with the TDS, the first will be an element from the array **from.user.filer** and the second must be the corresponding element (with the same subscript) from the array **to.user.filer**.

2 A parameter called **fold.number** is an integer defining the position of the root fold being used in the operation with respect to the closed fold bundle on which the cursor is positioned when the **EXE** program is run. The following example shows how folds are numbered:

```
{{{   fold.number = 0 - cursor here
...   fold.number = 1
any text or blank line not counted
...   fold.number = 2

...   fold.number = 3
and so on
}}}
```

Fold number 0 cannot be accessed concurrently with any other fold.

3 **[]BYTE** parameters are used to pass text as lines of data or as fold line comment and/or file names. A terminating '***c**' character is always permitted in strings passed to the environment. This is removed when necessary by the procedure.

4 A **[]BYTE** parameter used to return a string read by the procedure is always preceded by an **INT** parameter whose computed value defines the length of the string read. The array must be big enough for the expected data (256 for text, 512 for arbitrary data).

5 A parameter called **attributes** is used to communicate an array of three integers defining what kind of fold is being read or written. The values written as attributes must be chosen from those defined in appendix D.6

6 A parameter called **errornum** may at any time return an error code (see appendix E) to the caller. If any of the folded stream output procedures returns **errornum <> fi.ok** then the stream must be immediately closed or the program will deadlock trying to read the next command from the receiver.

Write folded stream

This group of procedures gives sequential write access to a folded stream for output. They are designed for use with a receiver process which obeys the folded stream protocol used by the user filer. This section should be read in conjunction with section 16.7.6 describing the user filer communications.

As supplied the procedures are restricted to creating a new fold at the end of an existing bundle and writing sequentially into this new fold. The sequential stream may include nested folds and filed folds created by calls of procedures which create the creases around the nested folds.

A procedure similar to **create.new.fold** may be written for opening an existing empty filed fold for writing in folded stream mode.

Similar procedures could be written for writing folds in data stream mode. As this mode is better supported by using screen protocol and the interface procedures **so.scrstream.to.fold** and **scrstream.to.file**, these variants are left as exercises for the reader.

A user filer channel pair may be used to access one or more file streams in sequence. Procedures applicable in file stream output modes must be called in sequence according to the kinds of items being output. These calls must be bracketed between calls of **create.new.fold** and **finish.folded.stream** which change the channel pair from user filer control mode to folded file stream output mode and vice versa, respectively.

The procedures which create new folds will return an error **fi.not.on.a.valid.fold** if called on a fold bundle whose type attribute is **ft.foldset**. If a program really wishes to write into such a bundle this may be done by converting it into a **ft.voidset** before attempting to write to it. This may be done by:

```
send.command (from.uf, to.uf,
              uf.unmake.fold.set, 0, result)
```

create.new.fold   **userio**

```
PROC create.new.fold (CHAN OF ANY from.ws, to.ws, INT fold.number,
                      VAL []BYTE comment, VAL []INT attributes,
                      VAL []BYTE fileid, INT errornum)
```

This procedure may only be called when the user filer channel pair **from.ws** and **to.ws** are in user filer control mode. Creates a new fold at the end of the bundle at position **fold.number**. (Counting from the first embedded fold as 1). Writes the comment and attributes provided. Makes the fold filed and opens a stream for writing. If successful the channel pair are then in file stream output mode. Any error is signalled in **errornum**.

write.record.item   **userio**

```
PROC write.record.item (CHAN OF ANY from.ws, to.ws,
                        VAL []BYTE record, INT errornum)
```

A record item is a line of text to be written to the output stream.

This procedure may only be called when the user filer channel pair **from.ws** and **to.ws** are in file stream output mode. Reads a command from the receiver and writes the record, if possible. If the record includes a trailing '***c**', this is removed.

write.number.item   **userio**

```
PROC write.number.item (CHAN OF ANY from.ws, to.ws,
                        VAL INT number, INT errornum)
```

This procedure may only be called when the user filer channel pair **from.ws** and **to.ws** are in folded file stream output mode. Reads command and writes the (non-negative) number, if possible. See appendix G for the representation of numbers in folded files.

write.top.crease   **userio**

```
PROC write.top.crease (CHAN OF ANY from.ws, to.ws,
                       VAL []BYTE comment, VAL []INT attributes,
                       VAL BYTE file.or.fold, VAL []BYTE fileid,
                       INT errornum)
```

For internal use by the following two procedures.

write.fold.top.crease   **userio**

```
PROC write.fold.top.crease (CHAN OF ANY from.ws, to.ws,
                            VAL []BYTE comment,
                            VAL []INT attributes, INT errornum)
```

This procedure may only be called when the user filer channel pair **from.ws** and **to.ws** are in folded file stream output mode (folded stream or data stream). Reads a command from the receiver and writes an unfiled top crease if possible. The comment string may contain a trailing '***c**' which will be removed.

`write.filed.top.crease`   userio

```
    PROC write.filed.top.crease (CHAN OF ANY from.ws, to.ws,
                                 VAL []BYTE comment,
                                 VAL []INT attributes,
                                 VAL []BYTE fileid, INT errornum)
```

This procedure may only be called when the user filer channel pair **from.ws** and **to.ws** are in folded file stream output mode. Reads a command from the receiver and writes a filed top crease if possible. The **comment** and **fileid** strings may contain a trailing '***c**' which will be removed.

`write.bottom.crease`   userio

```
    PROC write.bottom.crease (CHAN OF ANY from.ws, to.ws, INT errornum)
```

This procedure may only be called when the user filer channel pair **from.ws** and **to.ws** are in folded file stream output mode. Reads a command from the receiver and writes a bottom crease if possible.

Note that it is not necessary to distinguish between a filed and an unfiled fold as the receiver keeps track of this.

`finish.new.fold`   userio

```
    PROC finish.new.fold (CHAN OF ANY from.ws, to.ws,
                          VAL INT fold.number, VAL BOOL must.unfile,
                          INT errornum)
```

This procedure may only be called when the user filer channel pair **from.ws** and **to.ws** are in file stream output mode (folded or data stream). This procedure closes the newly written stream, unfiling it if **must.unfile** is **TRUE**. The channel pair will then be in user filer control mode.

Must be entered with **errornum** containing the result from the most recent write command, so correct action can be taken if the file stream has been prematurely closed.

Read folded stream

This group of procedures gives read access to a folded stream. They can only be used in conjunction with a source which obeys the folded stream protocol used by the user filer. They are thus principally applicable in programs which will run as executable procedures (**EXE**s) in the TDS, or which communicate with such a system at run time across INMOS links.

The procedures have been designed to give the programmer a view of a folded data stream corresponding to the displayed form of a fold structure presented by the editor.

A folded data stream may be opened, its contents may be read sequentially (with the option to skip folds of certain kinds), and it may be closed. It may be a complete fold pointed at by the editor cursor when the program is run, or it may be a fold immediately contained within such a fold.

In the simplest style of use, one procedure call corresponds to each line to be read into the program. These lines are text lines or crease lines and there is a procedure for each type of line. Each procedure reads the current line, and also determines the nature of the subsequent line so that the appropriate call may be made to obtain it. The nature of the next line is returned by a **BYTE** parameter **next.item** which takes one of a set of values conventionally associated with the names:

`fsd.record`	normal text line
`fsd.fold`	top crease of an unfiled fold
`fsd.filed`	top crease of a filed fold
`fsd.endfold`	bottom crease of an unfiled fold
`fsd.endfiled`	bottom crease of a filed fold
`fsd.endstream`	end of folded stream — must now be closed

Two additional values are also in general possible, **fsd.number** and **fsd.error**.

All these values are supplied as a set of constant declarations in the library **filerhdr** or **uservals**.

A user filer channel pair may be used to access one or more file streams in sequence. Procedures applicable in file stream input mode must be called in sequence according to the value of **next.item**. These calls must be bracketted between calls of **open.folded.stream** and **close.folded.stream** which change the channel pair from user filer control mode to file stream input mode and vice versa, respectively.

The first group of procedures supports reading the heading and attributes of a fold, opening a stream, and exhaustive sequential access to the stream. Such exhaustive access involves entering any embedded folds or filed folds and reading all of their contents also.

If a decision not to proceed, or not to enter a fold, is to be taken after reading part of the stream, then the second group of **input.** procedures should be used rather than the **read.** procedures.

read.fold.heading   userio

```
PROC read.fold.heading (CHAN OF ANY from.rs, to.rs,
                        VAL INT fold.number, INT len.comment,
                        []BYTE comment, []INT attributes,
                        INT errornum)
```

This procedure may only be called when the user filer channel pair **from.rs** and **to.rs** are in user filer control mode. Reads comment and attributes of the fold **fold.number** in the bundle pointed to by the cursor. The number 0 refers to the whole bundle, 1 to the first fold embedded within it, and so on. The **attributes** are an array of 3 integers defining the *fold.type*, *fold.contents* and relative *fold.indent* of this fold. Values of these attributes and their meanings are defined in appendix F. The effects of other attribute values are undefined. Any error is signalled in **errornum**. This will be zero (**fi.ok**) for success or a value from the list in appendix D.

read.file.name   userio

```
PROC read.file.name (CHAN OF ANY from.rs, to.rs,
                     VAL INT fold.number, INT len.file.id,
                     []BYTE file.id, INT errornum)
```

This procedure may only be called when the user filer channel pair **from.rs** and **to.rs** are in user filer control mode. According to which version of the TDS is in use, files may have names (as in a conventional operating system directory structure) or not. This procedure reads the file name of the indicated fold into **[file.id FROM 0 FOR len.file.id]**. If the development system does not have a named file store **len.file.id** will be zero. Any error is signalled in **errornum**.

open.folded.stream   userio

```
PROC open.folded.stream (CHAN OF ANY from.rs, to.rs,
                         VAL INT fold.number, BYTE first.item,
                         BOOL not.filed, INT errornum)
```

This procedure may only be called when the user filer channel pair **from.rs** and **to.rs** are in user filer control mode. The procedure opens a fold for folded reading, and if successful the channel pair are then in folded file stream input mode.

File stream modes require the fold being accessed to be filed and so if the fold identified by **fold.number** is found not to be filed the parameter **not.filed** is set TRUE and it is filed by the procedure. (Note that this attempt will fail if the fold is of a kind which cannot be filed). **errornum** is returned as 0 (**fi.ok**) if it is already filed, −1206 (**fi.not.filed**) if it was unfiled (and has not been successfully filed by this procedure); other negative values indicate other error conditions.

first.item is one of **fsd.record**, **fsd.number**, **fsd.fold** or **fsd.filed** according to the identity of the first item in the fold.

This value is used to choose the appropriate read procedure to call the first item in the folded stream. In a similar way each subsequent read operation also defines the type of the following item. Any failure to read an item will result in the value **fsd.error** being returned instead of a valid item tag. At the end of the stream the next item will have the tag **fsd.endstream**. No further read operations are allowed after **fsd.endstream** is returned. After **fsd.error** the procedure **read.error.item** should be called to obtain the error number.

A similar procedure **open.data.stream** exists in the library **ufiler** to open a fold for data stream reading. In this case there is only one kind of item — **fsd.record**. This mode of reading is used in the interface procedure **keystream.from.file**.

read.record.item   **userio**

```
PROC read.record.item (CHAN OF ANY from.rs, to.rs, INT len,
                       []BYTE record, BYTE next.item)
```

This procedure may only be called when the user filer channel pair **from.rs** and **to.rs** are in a file stream input mode (folded stream or data stream). Must be entered with **next.item = fsd.record**. Reads the record into **[record FROM 0 FOR len]** and indicates the type of the next item.

The record will contain the text of a line from the folded file. In folded stream mode the text will contain leading spaces only if the line is indented relative to the immediately enclosing fold. In data stream mode indentation spaces are provided relative to the indentation of the root fold.

If a previous call of a stream input procedure has indicated that the next item is a top crease, the fold may be skipped by explicitly changing the variable corresponding to **next.item** to **fsd.record** before calling this procedure. The record returned will then be the fold comment and **next.item** will correspond to the item after the fold.

read.number.item   **userio**

```
PROC read.number.item (CHAN OF ANY from.rs, to.rs,
                       INT number, BYTE next.item)
```

This procedure may only be called when the user filer channel pair **from.rs** and **to.rs** are in folded file stream input mode. Must be entered with **next.item = fsd.number**. This procedure is provided for completeness only as number items are rare in ordinary fold structures in the TDS.

Reads the number and indicates type of the next item.

read.error.item   **userio**

```
PROC read.error.item (CHAN OF ANY from.rs, to.rs,
                      INT status, BYTE next.item)
```

This procedure may only be called when the user filer channel pair **from.rs** and **to.rs** are in file stream input mode. Must be entered with **next.item = fsd.error**. This procedure is provided for completeness only as error items will not occur unless stream input procedures have been called in the wrong context.

Reads the error status and indicates type of the next item.

read.fold.top.crease   **userio**

```
PROC read.fold.top.crease (CHAN OF ANY from.rs, to.rs,
                           INT len.comment, []BYTE comment,
                           []INT attributes, BYTE next.item)
```

This procedure may only be called when the user filer channel pair **from.rs** and **to.rs** are in folded file stream input mode. Must be entered with **next.item = fsd.fold**. Reads comment and attributes from fold line and returns the type of the first item within the fold.

read.filed.top.crease   userio

        PROC read.filed.top.crease (CHAN OF ANY from.rs, to.rs,
                                    INT len.comment, []BYTE comment,
                                    []INT attributes, INT len.fileid,
                                    []BYTE fileid, BYTE next.item)

This procedure may only be called when the user filer channel pair **from.rs** and **to.rs** are in folded file stream input mode. Must be entered with **next.item = fsd.filed**. Reads comment and attributes from fold line and returns the type of the first item within the fold. On a named file store also reads the file name into **[fileid FROM 0 FOR len.fileid]**, otherwise sets **len.fileid** to zero.

read.bottom.crease   userio

        PROC read.bottom.crease (CHAN OF ANY from.rs, to.rs,
                                 []INT attributes, BYTE next.item)

This procedure may only be called when the user filer channel pair **from.rs** and **to.rs** are in folded file stream input mode. Must be entered with   **next.item = fsd.endfold** or **fsd.endfiled**. Reads attributes of the enclosing fold (the one whose end has been encountered) and returns the type of the next item within the fold outside that one.

The following procedures may be used instead of or in addition to the above procedures in programs where more control is required over the sequence of read operations on the file stream.
The procedures whose name begins **input.** start to do the same as the corresponding procedure with a name beginning **read.**, but do not ask what kind of item the next item will be.
The user then has the option to exit the current fold, repeat the current fold as well as to skip the current item, and (when the current item is a fold) to enter the item. This option is exercised by calling one of the procedures **exit.fold**, **repeat.fold**, **skip.item** or **enter.fold**. These all have a **BYTE next.item** parameter and one of these must be called in sequence before the next call of an **input.** procedure. The input stream may be closed by calling **close.folded.stream** after any of the **input.** procedures.

input.record.item   userio

        PROC input.record.item (CHAN OF ANY from.rs, INT len,
                                []BYTE record, VAL BYTE next.item)

This procedure may only be called when the user filer channel pair are in file stream input mode. It inputs a record item but does not advance to the next item, which should be done by a later call of **skip.item**. Must be entered with **next.item = fsd.record**. Reads the record into **[record FROM 0 FOR len]**.

This procedure does not prepare for the next input (it does not set **next.item**). It may therefore be used instead of **read.record.item** if it is necessary to inspect the record to determine whether any further lines in the current fold need to be read.

Note that this procedure has only one channel parameter **from.rs**.

input.number.item   userio

        PROC input.number.item (CHAN OF ANY from.rs,
                                INT number, VAL BYTE next.item)

This procedure may only be called when the user filer channel pair are in folded file stream input mode. It inputs the number but does not advance to the next item, which should be done be as a later call of **skip.item**. Must be entered with **next.item = fsd.number**. This procedure is provided for completeness only as number items are rare in ordinary fold structures in the TDS.

Note that this procedure has only one channel parameter **from.rs**.

`input.error.item`   `userio`

>   PROC input.error.item (CHAN OF ANY from.rs,
>   INT status, VAL BYTE next.item)

This procedure may only be called when the user filer channel pair **from.rs** and **to.rs** are in file stream input mode. Must be entered with **next.item** = **fsd.error**. This procedure is provided for completeness only as error items will not occur unless stream input procedures have been called in the wrong context. Reads the error status.

Note that this procedure has only one channel parameter **from.rs**.

`input.top.crease`   `userio`

>   PROC input.top.crease (CHAN OF ANY from.rs, to.rs,
>   INT len.comment, []BYTE comment,
>   []INT attributes, INT len.fileid,
>   []BYTE fileid, VAL BYTE next.item)

This procedure may only be called when the user filer channel pair **from.rs** and **to.rs** are in folded file stream input mode. Must be entered with **next.item** = **fsd.fold** or **fsd.filed**. Reads comment and attributes from fold line. An array of 3 integers is required for the attributes. On a named file store also reads the fileid.

This procedure does not prepare for the next input (it does not set **next.item**). It may therefore be used instead of **read.fold.top.crease** or **read.filed.top.crease** if it is necessary to inspect the comment and/or attributes to determine whether the fold should be entered or skipped.

`skip.item`   `userio`

>   PROC skip.item (CHAN OF ANY from.rs, to.rs, BYTE next.item)

This procedure may be called at any time immediately after one of the **input.** procedures. It will return the value of **next.item** corresponding to the following item. It may be used to skip any item including a fold.

`enter.fold`   `userio`

>   PROC enter.fold (CHAN OF ANY from.rs, to.rs, BYTE next.item)

This procedure may only be called immediately after a call of **input.top.crease**. On entry the value of **next.item** should be **fsd.fold** or **fsd.filed**. It will return the value of **next.item** corresponding to the first item within the fold.

`exit.fold`   `userio`

>   PROC exit.fold (CHAN OF ANY from.rs, to.rs, BYTE next.item)

This procedure may be called at any time immediately after one of the **input.** procedures. It will return the value of **next.item** corresponding to the item immediately after the bottom crease of the current fold, thereby causing the remainder of the contents of the fold not to be read.

`repeat.fold`   `userio`

>   PROC repeat.fold (CHAN OF ANY from.rs, to.rs, BYTE next.item)

This procedure may be called at any time immediately after one of the **input.** procedures. It will return the value of **next.item** corresponding to the item immediately after the top crease of the current fold, thereby causing the contents of the fold to be read again.

close.folded.stream   **userio**

> PROC close.folded.stream (CHAN OF ANY from.rs, to.rs,
>   VAL INT fold.number,
>   VAL BOOL must.unfile, INT errornum)

This procedure may only be called when the user filer channel pair **from.rs** and **to.rs** are in file stream input mode. This procedure should be called when **next.item = fsd.endstream**. If it is desired to close a folded input stream at any other time, the current item must first be consumed by calling the appropriate **input.** procedure, or if the current item is **fsd.error**, by reading an integer error result from the channel **from.rs**.

The procedure closes the folded stream, returning the channel pair to user filer control mode. If **must.unfile** is **TRUE** the fold is unfiled. Any error is returned in **errornum**.

For safe use of these procedures the procedure **number.of.folds** must be called first and must return a positive number.

get.stream.result   **ufiler**

> PROC get.stream.result(CHAN OF ANY fs, INT result)

Internal procedure called by fold stream procedures.

clean.string   **ufiler**

> PROC clean.string (INT len, []BYTE str)

Internal procedure called by fold stream procedures. Replaces any ASCII control characters in **str** by '`' and removes trailing ones.

truncate.file.id   **ufiler**

> PROC truncate.file.id (INT len, VAL[]BYTE id)

Internal procedure called by fold stream procedures. Truncates **id** at the first space or . character.

number.of.folds   **ufiler**

> PROC number.of.folds(CHAN OF ANY from.uf, to.uf, INT n, result)

If **result** is non-zero on entry does nothing. Otherwise examines the item at the current cursor position and returns **n** = -1, if not on a fold item, or returns **n** as the number of folds at the top level within the current fold item. The value of **result** is not changed. If **result** is not **fi.ok**. the value of **n** is undefined.

write.fold.string   **ufiler**

> PROC write.fold.string(CHAN OF ANY from.uf, to.uf, VAL INT seq.no,
>   VAL INT len, VAL []BYTE data, INT result)

If **result** is non-zero on entry does nothing. Otherwise writes [**str** FROM 0 FOR **len**] as the fold comment of the fold identified by **seq.no**. Any error is returned as **result**.

create.fold   **ufiler**

```
PROC create.fold(CHAN OF ANY from.uf, to.uf,
                 INT new.fold.number, VAL []INT attributes,
                 INT result)
```

If **result** is non-zero on entry does nothing. Otherwise creates a new fold, with the specified **attributes**, in the current bundle after all existing folds. The position of this new fold is returned as **new.fold.number**. Any error is returned as **result**.

send.command   **ufiler**

```
PROC send.command(CHAN OF ANY from.uf, to.uf,
                  VAL BYTE op, VAL INT seq.no, INT result)
```

If **result** is non-zero on entry does nothing. Otherwise sends a user filer command **op** to perform an operation on the member **seq.no** of the fold bundle which is the current item. Any error is returned as **result**.

make.filed   **ufiler**

```
PROC make.filed (CHAN OF ANY from.uf, to.uf, VAL INT seq.no,
                 VAL INT id.len, VAL []BYTE file.id, INT result)
```

If **result** is non-zero on entry does nothing. Otherwise assumes the current item is an unfiled fold and makes it filed with the name **[file.id FROM 0 FOR id.len]**. Any error is returned as **result**.

open.stream   **ufiler**

```
PROC open.stream(CHAN OF ANY fs, ts,
                 VAL BYTE op, VAL INT fold.no, INT result)
```

If **result** is non-zero on entry does nothing. Otherwise opens the indicated fold as a user filer input or output stream. The value of **op** should be a **uf.open** command as defined in the library **filerhdr**. Any error is returned as **result**.

read.fold.string   **ufiler**

```
PROC read.fold.string(CHAN OF ANY from.uf, to.uf,
                      VAL INT seq.no, INT len, []BYTE data,
                      INT result)
```

If **result** is non-zero on entry does nothing. Otherwise assumes the current item is a fold and returns the fold comment string of the fold **seq.no** in the bundle as **[data FROM 0 FOR len]**. Any error is returned as **result**.

read.fold.attr   **ufiler**

```
PROC read.fold.attr(CHAN OF ANY from.uf, to.uf,
                    VAL INT seq.no, []INT attributes, INT result)
```

If **result** is non-zero on entry does nothing. Otherwise assumes the current item is a fold and returns the **attributes** of the fold **seq.no** in the bundle. Any error is returned as **result**.

open.data.stream   `ufiler`

> PROC open.data.stream (CHAN OF ANY from.rs, to.rs,
>   VAL INT fold.number, BYTE first.item,
>   BOOL not.filed, INT errornum)

Opens the indicated fold as a user filer data input stream. The boolean **not.filed** is set according as the fold was an ordinary or a filed fold. This procedure files the fold if it was not already filed. **first.item** is returned as **fsd.record** if there is at least one text line in the fold, **fsd.endstream** otherwise. Any error is returned as **errornum**.

close.uf.stream   `ufiler`

> PROC close.uf.stream (CHAN OF ANY from.rs, to.rs,
>   VAL INT fold.number, VAL BOOL must.unfile,
>   INT errornum)

Closes a user filer data or folded input stream, unfiling it if **must.unfile** is TRUE. Any error Is returned as **errornum**.

read.data.record   `ufiler`

> PROC read.data.record (CHAN OF ANY from.rs, to.rs,
>   INT len, []BYTE record, BYTE next.item)

Must be entered with **next.item = fsd.record**. Reads the record and indicates the type of the next item.

14.13   Access to transputer board peripherals

The terminal driving hardware of IMS B002, IMS B006 or similar transputer boards may be accessed using the procedures of this library which use TDS **screenstream** and **keystream** channels.

B00x.term.p.driver   `t4board, t2board`

> PROC B00x.term.p.driver (CHAN OF SS from.user.scrn,
>   CHAN OF KS to.user.kbd,
>   VAL INT board.type, port,
>   baud.rate, screen.type)

N.B. an additional parameter has been added to this procedure, therefore existing programs must be modified before recompilation.

This interface procedure may be run in parallel with any application on an IMS B001 or IMS B002 evaluation board. It takes input in screen stream protocol on the channel **from.user.scrn** and sends it to an RS232 output, and sends the corresponding input in key stream protocol to the channel **to.user.kbd**.

board.type should be 1 for B001 or 2 for B002 or 6 for B006. The uart port is defined by passing 0(terminal) or 1(host) as the parameter **port**. **baud.rate** if non-zero causes the UART to be reset at startup, the value should be one of 38400, 19200, 9600, 7200, 4800, 2400, 2000, 1800, 1200, 1050, 600, 300, 200, 150, 134, 110, 75 or 50. If zero the reset is assumed to have been already performed (e.g. by code in the ROM).

screen.type may be 0 for ANSI terminals or 1 for TVI 920 terminals.

Other procedures that were in this library were directly dependent on the old TDS server and are not included in TDS3.

15 Tools

The tools described in this chapter are all TDS executable programs (**EXE**s). To run a tool place the cursor on the line in the **Tools** fold in the **Toolkit** fold to which the **CODE EXE** for the appropriate tool has been attached, and press GET CODE . Then return to the editing environment and apply the tool using RUN EXE to an appropriate fold as described in the specification of the tool below.

Any number of tools may be held in the TDS memory, and optionally included in the **AUTOLOAD** fold. The only restrictions arise when there is insufficient memory left after loading tools for running one of them or one of the utilities.

The tools provided with the TDS are:

1 **debugger** the post mortem occam debugger

2 **nettest** the network tester

3 **list** the selective lister (see section 4.9.1)

4 **unlist** the unlister (see section 4.9.1)

5 **linkcopy** the tool for transferring files across transputer links

6 **memint** the memory interface table generator

7 **epromhex** the tool for preparing code for EPROM

8 **hextoprg** the tool for writing to a PROM programmer

9 **promfile** the tool for writing EPROM code to a file

10 **addboot** the bootstrap adder for a standard hosted PROC

11 **wocctab** the tool to convert an SC into an occam table of bytes

Users are encouraged to add any tools they create for their own purposes to the same fold within the toolkit fold. If tools are to be offered to other TDS users they should, if possible, be compiled for the **TA** transputer class so they can run on TDSs running on any 32-bit transputer.

15.1   Debugger

Introduction

The Debugger can be used to debug any of the following:

TDS style programs running on a network of transputers attached via a link,

Programs written under the TDS, but executing outside the TDS environment,

TDS style programs running on the host transputer (**EXE**s and **UTIL**s),

occam SC modules.

15.1.1   Debugging a PROGRAM on a network which may include the host

The Debugger can be used to debug a TDS **PROGRAM** which has been loaded and run on a transputer network. The network may contain large numbers of transputers — the debugger has been tested on 1024 processors! If it is able to assert **Analyse** on the transputers in the network (e.g. by using a subsystem port on the host) it will do so, otherwise the transputers which constitute the network must be in an analysed state.

If the TDS has been rebooted the analyse may have been performed by the server before rebooting.

The debugger reads the state of the network described in the **PROGRAM** fold, retrieving state information and a copy of an area of memory, starting at the lowest memory address, from every processor in the network. It then loads a program into this area, which sets up a communication path through the network, and allows the debugger to retrieve the contents of memory from any transputer in the network.

Start up procedure for a PROGRAM

Place the cursor on the **PROGRAM** fold and press RUN EXE. The initial display is the title and version, and the prompt:

```
Debugging a PROGRAM

Transputer link, Dummy, Analyse, Host, Network dump
                              or Quit  (T,D,A,H,N,Q)  ?
```

T — **Transputer link**

This option is used for debugging transputer networks loaded from the TDS.

If you choose option 'T', you will then be prompted:

```
Link number, or Quit  (0,1,2,3,Q)  ?
```

Type in the link number through which the host is connected to the network. You do not need to press ENTER.

```
How to analyse network? Is host a B004 or a B002,
                  Ignore analyse, or Quit  (4,2,I,Q)  ?
```

Type   **4** if the host system is an IMS B004 or TRAM,
       2 if the host system is an IMS B002 (or equivalent),
       I if you do not wish to assert **Analyse**,
or   **Q** to quit.

This information is required because the subsystem ports are different; the debugger must know which is being used so that it can assert the network's **Analyse** signal.

While examining the network, the debugger displays:

```
Reading logical name table ...
Analysing network ...
```

D — **Dummy network**

The response '**D**' runs the debugger in parallel with a process which acts as a network of transputers. This process simulates an arbitrary network in the analyse state providing artificial data for display. The user may thus explore some of the options available without being connected to a target network. While initialising itself, the debugger displays:

```
Reading logical name table ...
Analysing network ...
```

H — **Network including Host**

This option is used to debug a standalone **PROGRAM** which has been developed in the TDS, but is executed outside the TDS, and therefore includes the host transputer (i.e. the transputer which is now running the TDS) in the network. The TDS should be restarted with an appropriate option to analyse the network. See section 15.1.8 for details of how to create a core dump file.

In reply to the first prompt, you will then be asked:
> **Read Core dump file, Ignore core dump, or Quit (C,I,Q) ?**

If you reply 'C', the debugger will then prompt for the name of the core dump file:

> **Core dump filename ("core.dmp", or "QUIT") ?**

>   Press ENTER to use the default filename
> or   enter a filename (any filename extension will be replaced by '.dmp')
> or   type 'QUIT' (uppercase) to exit the debugger

If you type 'I',the debugger will not read a core dump file, and therefore provides no useful information about the root processor.

You will be asked what type of board is being used, so that the debugger can assert **Analyse** on the network. If you have a modified IMS B004 or a TRAM/B008 which propagates the **Reset** and **Analyse** signals through to the subsystem port, you should type 'I' so that the subsystem is not analysed twice.

While reading the core dump file, the debugger will display:

> **Reading logical name table ...**
> **Reading Core dump file "filename.dmp" ...**

A — Assert analyse

This option allows you to assert the **Analyse** signal on the transputer network attached to the host's subsystem port. You will be asked what type of board is being used, or given the option to quit.

After successfully asserting the analyse signal, you will see the message:

> **Subsystem has been analysed**
> **Press any key to return to the TDS**

N — Network dump

This option allows you to read a previously created 'network dump' file. The debugger uses the information in this file for its target information. This allows the complete state of a network to be dumped to a file, and the debugging session to be suspended, so that debugging can be resumed at a later date, when the target network may no longer even be present. How to create a network dump file is covered in section 15.1.6.

After pressing 'N' you will be prompted:

> **Network dump filename ("network.dmp", or "QUIT") ?**

>   Press ENTER to use the default filename.
> or   enter the name of the network dump file (any filename extension is replaced by '.dmp')
> or   type 'QUIT' (uppercase) to exit the debugger

While reading the network dump file, the debugger will display:

> **Reading logical name table ...**
> **Reading Network dump file "filename.dmp" ...**

What the debugger does

After this initial interaction, the debugger uses the configuration description folds in the **PROGRAM** fold set to build a data base for the network. The complete network is then analysed and some data retrieved from every processor. If the 'H' or 'N' options were selected, the information will be read from the appropriate file.

The debugger will then determine which processor (if any) had its error flag set, and will continue with that processor selected as the current processor. If none is set, the 'root' processor will be selected. The debugger will then display the occam source in the vicinity of the error, or the last instruction executed, as explained later.

15.1.2   Debugging an EXE (or UTIL)

The Debugger can be used to debug **EXE** programs running on the host transputer. To do this it uses a 'core-dump' feature.

If the transputer error flag is set while executing an **EXE**, you should re-boot the TDS, and request the 'C' option for a normal core dump. This will save the memory of the transputer in a core dump file, whose name you may specify, rather than use the default '**core.dmp**'. If the debugger is subsequently executed while positioned over that **EXE**, it can read the core dump file, and you may then use all of the symbolic features of the debugger. See section 15.1.8 for more details about creating a core dump file.

Start up procedure for an EXE

Place the cursor on the **EXE** fold, and press RUN EXE. The initial display is the title and version, and the prompt:

> **Debugging an EXE**
> **Read Core dump file, or Quit (C,Q) ?**

Type either 'Q' to quit, or 'C' to confirm that you wish to continue to debug the **EXE**. If you type 'C', you will be asked for the core dump filename:

> **Core dump filename ("core.dmp", or "QUIT") ?**

>   Press ENTER to accept the default filename
> or   enter a filename (any filename extension will be replaced by '.dmp')
> or   type '**QUIT**' (uppercase) to exit the debugger.

If the core dump file does not exist, it will be treated as though you had typed 'I'.

The debugger will then display:

> either   **Reading Core dump file "filename.dmp" ...**
> or   **Analysing EXE ...**

15.1.3   Debugging an SC

The Debugger can also be used on **SC** modules, to find the occam source line corresponding to any particular offset within this **SC**. The **SC** may contain nested **SC**s. No other facilities are available.

Start up procedure for an SC

The cursor should be positioned on the **SC** fold, and the debugger executed by pressing RUN EXE. The initial display is the title and version, and the prompt:

> **Display any offset within an SC**
> **Display occam source, or Quit (O,Q) ?**

Type either 'Q' to quit, or 'O' to locate to an occam source line.

If you type 'O', you will be prompted for the offset of the instruction you wish to find. You should type this in in decimal, and the debugger will display the occam source line corresponding to that offset from the start of the SC. The SC must have been compiled with the compiler's debugging option enabled. Press MONITOR to return to this prompt, or EXIT FOLD and FINISH to leave the debugger.

15.1.4   Symbolic functions

Once you have started to debug a **PROGRAM**, or an **EXE**, the debugger will automatically display the occam source corresponding to the error. If the program was still executing correctly when 'analyse' was asserted, the debugger will display the last source line executed. However, if the transputer had stopped, rather than halted upon finding an error, or was deadlocked, there will be no 'last instruction', so you will be left at the main 'Monitor page' (see section 15.1.5).

While it is looking for the required source line, the debugger will display 'Locating ...' at the top of the screen.

If the location which is to be displayed is in a compacted library for which the source code is not present, but which was compiled with the debugging option enabled, the debugger will instead locate the line corresponding to the library call, and will repeat until it finds some source code to display. As this is done, the original message will be changed to 'Backtracing ...'. When the debugger has successfully located some source, it will display the name of the library which it first tried to display, which SC within that library (counting from one), and the offset within that SC.

If the ultimate location is in a library containing source, the occam in the vicinity of that location is displayed, and the cursor is left at the start of the correct source line. The debugger also displays the name of that library, the SC number, and the offset within that SC. You may scroll through the source, and the special debugging features are available via the utility tool keys.

If the location is in a section of normal occam source, its context is displayed, and the user is left within the TDS editor, providing read-only access to the source. You can use the BROWSE key to allow modification of the source. In addition there is a set of debugging features available via the utility keys.

Note that in certain situations the location displayed may not correspond to the expected location. In particular, if no valid branch of an **IF** or **CASE** has been found or all branches of an **ALT** have **FALSE** guards, the debugger will locate to the *following* statement. See section 15.1.7

Note also that only the SCs which are to be inspected via this debugger need to have been compiled with the debugging option enabled; the remainder need not. It should also be pointed out that compiling an SC with debugging enabled does not affect the code which is produced in any way; it merely controls whether the debug fold is produced, containing the information for the debugger to use. This means that no extra bugs will be introduced (or hidden) by re-compiling with a different debugging option.

Debugging utilities

The extra debugging utilities are available via the following utility keys (see appendix A):

INSPECT	Display the type and value of an occam symbol.
CHANNEL	Locate to the process waiting on a channel.
TOP	Locate back to the error, or last occam location.
RETRACE	Retrace the last BACKTRACE etc.
RELOCATE	Locate back to the last location line.
INFO	Display some extra information.
LINKS	Display the link connections.
MONITOR	Change to the 'Monitor page'.
BACKTRACE	Locate to the procedure or function call.
CODE INFORMATION	Display a summary of utility key uses.

N.B. a number of editor function keys are disabled while using the debugger.

INSPECT

This function allows you to find the type and contents of any occam symbol. You should use the cursor keys to position the cursor on the required symbol, then press INSPECT.

If the cursor is not positioned over an occam symbol or keyword when you press INSPECT, you will instead be prompted for the symbol name at the top of the screen. You may type ENTER here to abort the INSPECT operation, or type a name, followed by ENTER. The case of the letters of the name is significant, as are spaces. If the name is an array, it may also be followed by constant integer subscripts in square brackets ('[' and ']').

It is then checked that the symbol is in scope at the line last 'located to'. Note that this is not necessarily the same as the current cursor position, and this must be understood for this feature to be useful. If the symbol is not in scope at that location, or not found at all, one of the following messages will be displayed:

      Name 'symbol' not in dynamic scope
  or **Name 'symbol' not found**

Inspecting arrays

If the symbol is an array name, and you have not already supplied subscripts, you will be prompted for them. The debugger will display the size and type of the array, and ask for the subscripts. For example:

      [5][4]INT ARRAY 'a', Subscripts ?

      Press ENTER to obtain the address of the array
  or enter the required subscripts, which must be in the correct range

The subscripts should be typed either as decimal constant integer values within square brackets, or as integers separated by commas (e.g. '**[3][2]**', or '**3, 2**'). Spaces are ignored.

To simplify access to values such as '**a[i]**' you may type '**a[!]**'; the '**!**' is replaced by the value of the last integer displayed.

Instead of supplying subscripts for an array element, the debugger allows you to scroll through the elements of an array while in symbolic mode. It also allows you to see a short 'segment' of a BYTE array. You can move this segment up and down like a window into the array.

When asked for a subscript, you may add '**++**' onto the end (or even type '**++**' on its own; this assumes a subscript of zero). Then instead of just displaying that element of the array, the debugger also puts the following message onto the second line on the screen:

      Press [UP] or [DOWN] to scroll, any other to exit :

You may use the up and down arrow keys to scroll through the elements of that array. The debugger will not allow you to scroll past the beginning or end of the array. Pressing any other key puts you back into normal symbolic mode. You can then press REFRESH to re-draw the second screen line.

BYTE arrays have another feature. If you add a single '**+**' to the subscripts, the debugger displays a 'segment' of 16 bytes starting at those subscripts. You may again scroll through the array by using the up and down arrow cursor keys. Again you cannot scroll past the beginning or end of the array. If you use the single '**+**' on a non-BYTE array, it is treated exactly like '**++**'.

If the INSPECT key is used on the name of a channel or variable which has been placed, the message shows the PLACE address in occam terms and the absolute memory address, but not the contents of the location.

Information displayed

If the name is in scope, its type and value will be displayed, together with its address in memory. If it is an array, and subscripts were supplied, its type, value, and address will be displayed. If it is a short **BYTE** array, it is displayed in ASCII. If it is any other type of array, its dimensions are displayed. If it is a channel, and is not empty, the **Iptr** and **Wdesc** of the process waiting for communication, and its priority, are displayed. If it is a **PROC** or **FUNCTION** name, its entry address, and nested workspace and vectorspace requirements are displayed (no address is displayed for library names). Only the types of protocol names and tags, timers, and ports are displayed.

If there is too much information to be displayed on one line, it will be displayed in two parts. Firstly the symbol's name and type will be displayed, then, after a short pause, its value and address.

Inspecting memory

To inspect the contents of any location in memory, specify an address rather than a symbol name. Type the address as a decimal number, a hexadecimal number (preceded by '#'), or the special short form %h...h, which assumes the prefix #8000... . Any letters (A to F) in a hexadecimal number typed at this prompt must be in upper case. The debugger displays the contents of the word of memory at that address, in both decimal and hexadecimal. For more versatile displays of memory contents, use the options available from the 'Monitor page' (see section 15.1.5).

CHANNEL

Use this function as you would INSPECT, but specify the name of the channel. Instead of displaying the **Iptr** and **Wdesc** of the process waiting for communication on that channel, the debugger locates to the corresponding line of occam source, from where you can continue debugging that process. This function is invalid if the symbol specified is not a channel.

'Hard' channels

The CHANNEL key also allows you to 'jump' from one processor to another along hard channels (channels mapped onto transputer links). If a process waiting for communication from the processor at the other end of the specified channel, the debugger will change to that processor. It will then display the new processor number to inform you that it has changed processor. If there is no process waiting you are informed, and if the debugger is already located at the waiting process the following message is displayed:

```
Already located - No process is waiting at the other end
                  of this link
```

TOP

This function forces the debugger to locate back to the line containing the original error that crashed the program, or to the line located to by the Monitor page 'G', 'O' or 'X' commands (see section 15.1.6).

RETRACE

This function forces the debugger to retrace its steps. It will locate back to the previously displayed location. Repeated use of RETRACE will reverse the effect of successive BACKTRACE, CHANNEL, and TOP operations.

RELOCATE

This function relocates to the last location point. This allows you to return to the original source line after examining a section of the source code.

| INFO |

This function displays the **Iptr** of the last location, the corresponding **Wdesc**, in hex, the process priority and the current processor's number and type. For example:

```
Located to Iptr #80001564, and Wdesc #80000124,
            (Hi pri), Processor 2 (T800)

or  Located to Iptr #80001564, and Wdesc #80000124,
            (Hi pri), EXE (T800)
```

If a **Wdesc** has not been supplied, it will be given as 'invalid'.

If this key is pressed when the debugger has been invoked on an **SC**, you will see a message of the form:

```
Located to offset 450 of this SC
```

| LINKS |

This function provides a quick means of determining the connections on the currently displayed processor. It lists each link in turn, and the processor and link to which it is connected. For example:

```
Links: L0 to host. L1 to P3 L2. L2 ---. L3 to P45 L0.
```

| CODE INFORMATION |

This function displays a summary of the debugger function keys.

| MONITOR |

This function transfers the user to the debugger 'Monitor page'.

| FINISH |

To leave the debugger use the [EXIT FOLD] and/or [FINISH] keys. You may also leave by using the '**Q**' option from the Monitor page.

| BACKTRACE |

This function locates to the line corresponding to the call of the currently displayed procedure or function. If the current location is in the processor's top level procedure, the following message is displayed:

```
Error : Cannot backtrace from here
```

Debugging an SC

The operations [BACKTRACE], [INSPECT], [CHANNEL], [TOP], [LINKS], and [RETRACE] have no meaning when the debugger has been invoked on an **SC**. If you press any of these keys, you will be informed:

```
This key is invalid on an SC
```

Invalid Wdesc

If you are debugging an **EXE**, without a core dump file, the debugger cannot read the contents of memory at the time of the error. This means that it cannot read the contents of variables and channels, nor find the return addresses of procedures. It flags this by leaving the **Wdesc** as an invalid value — that of the transputer's most negative address. Also, if you do not supply a valid **Wdesc** when using the Monitor page 'o' (occam) option, you will not be able to access memory contents. However, you may still determine the values of scalar constants, and some other symbols.

Any attempt to inspect variables or channels, or to backtrace, will cause one of the following messages to be displayed:

```
      Wdesc is invalid - Cannot backtrace
   or Wdesc is invalid - Cannot inspect variables
```

Also, if the location to be displayed is in a compacted library, and the **Wdesc** is invalid, the debugger will not be able to find the call of that library function or procedure. You will then be informed:

```
   Wdesc is invalid so cannot backtrace out of compacted library
```

15.1.5 Monitor page

When you leave the symbolic mode to enter the low level mode, the debugger displays a 'Monitor page' containing information about the current processor. The information displayed lists:

Iptr	Contents of instruction pointer (address of the last instruction executed)
Wdesc	Contents of workspace descriptor
IptrIntSave	Contents of saved low priority instruction pointer
WdescIntSave	Contents of saved low priority workspace descriptor
Error	Whether the error flag was set
FPU Error	Whether the FPU error flag was set (if it exists)
Halt On Error	Whether the halt on error flag was set
Fptr1	Pointer to the front of the low priority active process queue
Bptr1	Pointer to the back of the low priority active process queue
Fptr0	Pointer to the front of the high priority active process queue
Bptr0	Pointer to the back of the high priority active process queue
TPtr1	Pointer to the low priority timer queue
TPtr0	Pointer to the high priority timer queue
Clock1	Value of the low priority clock
Clock0	Value of the high priority clock

It also displays the current processor number and type, the cause of any error, and last instruction executed, and the current transputer's memory map.

Iptr points to the last instruction executed. Low priority **Iptr** and **Wdesc** are only displayed if the processor was running in high priority mode when it was halted.

If **Wdesc** contains the most negative address value, it will be described as 'invalid'. This normally means that no process was executing on that processor when it was halted (e.g. it may have been deadlocked). Try using the 'L' option to find processes waiting for communication on the links. The **Wdesc** is also flagged as invalid when debugging an **EXE** with no core dump file.

If **Wdesc** contains the address of 'Memstart' it is displayed as such. This normally means that the network's analyse signal has been asserted more than once. This may be because your host transputer board (e.g. IMS B004) has been modified to assert its subsystem signal when it is itself analysed. If this happens try re-running your program, then when re-running the debugger type 'I' to ignore analyse, rather than '4' to indicate that you wish to assert the IMS B004's subsystem signal.

An asterisk displayed next to either **Iptr** or **Wdesc** indicates that they do not correspond to a valid code and data pointer for the program. It may be possible to find the source of the problem by using the 'M' option to display a memory map for each transputer. If debugging an **EXE** this is normally because the last instruction executed was part of the TDS itself; your **EXE** may have deadlocked. See section 15.1.7.

Summary of commands

Key		Description
A	ASCII	View a portion of memory in ASCII.
C	Compare	Compare the code on the network with the code that should be there, to ensure that the code has not become corrupted.
D	Disassemble	Display the transputer instructions at a specified area of memory.
E	Next Error	Switch the current display to data from the next processor in the network which has halted with its error flag set.
G	Go to process	Go to source level debugging at an address already visible on the screen.
H	Hex	View a portion of memory in hexadecimal.
I	Inspect	View a portion of memory in any occam type (e.g. **REAL32**).
L	Links	Display the instruction pointers and workspace descriptors for the processes currently waiting for input or output on a transputer link, or for a signal on the **Event** pin.
M	Memory map	Display the memory map of that transputer.
N	Network dump	Copy the entire state of the transputer network into a 'network dump' file, so that you can continue debugging later.
O	occam	Resume the occam source level at an address to be typed in.
P	Processor	Switch the current display to data from a different processor.
Q	Quit	Leave the debugger, and return to the TDS.
R	Run queue	Display the instruction pointers and workspace descriptors of the processes on either the high or low priority active process queue.
T	Timer queue	Display the instruction pointers, the workspace descriptors and the wake-up times of the processes on either the high or low priority timer queue.
X RETRACE RELOCATE	Exit	Return to source level at the previous place.
CURSOR UP CURSOR DOWN	Scroll display	Scroll the currently displayed memory, disassembly, or queue
LINE UP LINE DOWN	Scroll display	Scroll the currently displayed memory, disassembly, or queue (restricted to single line)
CURSOR LEFT CURSOR RIGHT	Change processor	Scroll the currently displayed processor.
? CODE INFO	Help	Display a help screen.
REFRESH	Refresh	Re-draw the screen.
TOP		Locate to the last instruction executed on the current processor.

15.1.6   Monitor page commands

A full description of the Monitor page commands follows, with the options listed in alphabetical order. These options are not available when the debugger has been invoked on an **SC**.

A — ASCII

The ASCII command gives the following prompt:

> **Start address (#hhhhhhhh) ?**

> Press ENTER to accept the default address

or   enter the desired address (a decimal number, a hexadecimal number preceeded by '**#**',
or the short form '**%h...h**', which assumes a prefix of **#8000...**).

The memory is displayed as sixteen rows of 32 ASCII bytes. The bytes are displayed in order, with a '.' replacing any unprintable characters.

The address at the start of each line is an absolute address displayed as a hexadecimal number. The byte containing the specified start address is the top leftmost byte of the display. CURSOR UP and CURSOR DOWN keys may be used to scroll the display.

C — Compare memory

Selecting the Compare memory command allows you to check whether the code on the network agrees with the code which was loaded, in case memory has been corrupted. It can also be used for an **EXE**. You will be offered the following options:

```
        Compare memory
Number of processors in network    . 'n'

A - Check whole network for discrepancies
B - Check this processor for discrepancies
C - Compare memory on screen
D - Find first error on this processor
Q - Quit
```

or

```
Checking an EXE

A or B - Check this EXE for discrepancies
C - Compare memory on screen
D - Find first error on this processor
Q - Quit

Compare memory option (A,B,C,D,Q) ?
```

You should type one of the options **A**, **B**, **C**, **D**, or **Q**. Option '**Q**' will return you back to the Monitor page.

Checking the whole network – option A

Option '**A**' checks the whole network to ensure that the code in the network is the same as the code which was originally loaded onto each processor. All the top level SCs in the occam **PROGRAM** must have been extracted to **CODE SC** folds (this is performed automatically by the configurer).

As it is checking, the debugger will display the following messages for your information:

```
No of processors checked so far : 'n'
Checking processor : 'p' ...
Bytes to test       : nnn
Checking memory     : #hhhhhhhh to #hhhhhhhh ...
Checking address    : #hhhhhhhh ...
Checked processor   : 'p' OK
Checked processor   : 'p', 'e' errors
```

When it has finished checking, it will display either

```
    Checked whole network OK
or  'n' Errors, first at #hhhhhhhh on processor 'p'
```

Checking a single processor – option B

Option '**B**' checks just the current processor. In all other respects it is similar to option '**A**'.

Compare memory on screen – option C

Option '**C**' allows you to display both the correct contents, and the actual contents, on screen side by side. It displays each block of memory as sixteen lines of 8 bytes, with the contents of the network on the left, and the correct code on the right. Any discrepancies are marked with an asterisk ('*****'). At the end of each 128 byte block, type either '**Q**' to quit, or CURSOR DOWN to read the next block. The display will look like:

```
                Network Code        Correct Code
#800001234 :  0011223344556677   7766554433221100 *
#80000123C :  0011223344556677   0011223344556677
#800001244 :  0011223344556677   7766554433221100 *
   ...            ...                 ...
#8000012AC :  AABBCCDDEEFF0011   AABBCCDDEEFF0011

Press [DOWN] to scroll memory, [SPACE] for next error,
                                  or Q to quit :
```

Find first error – option D

Option '**D**' allows you to let the debugger look for any discrepancy itself, and then display it on the screen. It will search this processor, as in option '**B**', until it finds a discrepancy. If it does, it switches into option '**C**', and allows you to continue displaying the memory on screen.

[D] — Disassemble

The Disassemble command gives the following prompt:

```
Start address (#hhhhhhhh) ?
```

Press ENTER to accept the default address

or   enter the desired address (a decimal number, a hexadecimal number preceeded by '**#**', or the short form '**%h...h**', which assumes a prefix of #8000...).

The memory is displayed in batches of sixteen transputer instructions, starting with the instruction at the specified address. If the specified address is within an instruction, the disassembler begins at the start of that instruction. Note that this may not work correctly if data precedes that instruction, rather than other transputer instructions. This is because the data may end with a byte corresponding to a transputer 'pfix' or 'nfix' instruction, and therefore is indistinguishable from a real instruction.

Each instruction is displayed on a line preceeded by the address corresponding to the first byte of that instruction. The disassembly is a direct translation of memory contents into instructions, it does not insert labels, nor provide symbolic operands. CURSOR UP and CURSOR DOWN keys may be used to scroll the display 16 bytes at a time.

E — Next Error

Next Error searches forward through the network for the next processor which has both its error and halt-on-error flags set. Processors are searched in the order in which the processors are stored in the debugger's internal data base, not in the order of processor number. If a processor is found with both flags set the display is updated to the new processor as if the '**P**' option had been used. Press TOP to display the occam source line which caused the error.

If you press this key when debugging an **EXE**, or if there is only one processor in the network, you are informed:

> **This is an EXE - There are no other processors**
> or **There is only one processor in the network**

G — Go to process

This command locates to the source code for any process which is currently shown on the screen. The cursor is positioned next to the **Iptr**, and permitted responses are listed on the screen as follows:

> **[CURSOR] then [RETURN], or 0 to F, (I)ptr, (L)o, or (Q)uit**

> Select the desired process with the cursor keys or '0' to 'F' keys then press ENTER to select the process
> or   enter '**I**' to select the saved **Iptr**
> or   enter '**L**' to select the interrupted low priority process when currently in high priority
> or   type '**Q**' to abort this choice.

H — Hex

The Hex command gives the following prompt:

> **Start address (#hhhhhhhh) ?**

> Press ENTER to accept the default address
> or   enter the desired address (a decimal number, a hexadecimal number preceeded by '**#**',
>   or the short form '**%h...h**', which assumes a prefix of **#8000...**).

The memory is displayed as rows of words in hexadecimal format. Each row contains four or eight words, depending on the transputer word length. Words are displayed in hexadecimal (four or eight hexadecimal digits depending on word length), most significant byte first. For a four byte per word processor the sequence of bytes in a single row would be:

> **: 3 2 1 0   7 6 5 4   11 10 9 8   15 14 13 12**

For a processor with two bytes per word the sequence would be:

> **: 1 0   3 2   5 4   7 6   9 8   11 10   13 12   15 14**

The address at the start of each line is an absolute address displayed as a hexadecimal number. The word containing the specified start address is the top leftmost word of the display. The address will be aligned to the start of that word. CURSOR UP and CURSOR DOWN keys may be used to scroll the display.

I — **Inspect**

This command allows you to inspect the contents of an entire occam array. The Inspect command gives the following prompt:

> **Start address (#hhhhhhhh) ?**

> Press ⎵ENTER⎵ to accept the default address

or enter the desired address (a decimal number, a hexadecimal number preceeded by '**#**', or the short form '**%h...h**', which assumes a prefix of **#8000...**).

The start address of an array can be found in the symbolic mode by pressing ⎵INSPECT⎵ while the cursor is positioned over the array name, then simply pressing ⎵ENTER⎵ when asked for a subscript.

When a start address has been given the following prompt is displayed:

```
        Typed memory dump
0  -  ASCII
1  -  INT
2  -  BYTE
3  -  BOOL
4  -  INT16
5  -  INT32
6  -  INT64
7  -  REAL32
8  -  REAL64
9  -  CHAN

Which occam type (1 - INT) ?
```

Type the number corresponding to the occam type you wish to display, or press ⎵ENTER⎵ to accept the default type.

The memory is displayed as sixteen rows of data. ASCII arrays are displayed in the format used by the Monitor page ASCII command. Other occam types are displayed both in their normal representation and in hexadecimal format.

The address at the start of each line is an absolute address displayed as a hexadecimal number. The value containing the specified start address is on the top row of the display. It will be aligned to the nearest valid boundary: **BYTE** and **BOOL** to the nearest byte; **INT16** to the nearest even byte; **INT**, **INT32**, **INT64**, **REAL32**, **REAL64**, and **CHAN** to the nearest word. ⎵CURSOR UP⎵ and ⎵CURSOR DOWN⎵ keys may be used to scroll the display.

When displayed as **CHAN** words containing the special empty channel value or a valid **Wdesc** are displayed. If a **Wdesc** is valid the value of the **Iptr** saved in that workspace is also shown.

L — **Links**

Selecting the Links command displays the instruction pointer, workspace descriptor, and priority, of the processes waiting for communication on the links, or for a signal on the **Event** pin. If no process is waiting, it is described as 'Empty'.

The link connections are also displayed; each link is described as 'unconnected', 'connected to host', or 'connected to processor ..., link ...'. This information is taken from the software description and does not necessarily reflect the actual hardware connections.

Finally the link by which that processor was booted is also displayed. The display will look something like
this:

```
Link 0 out Empty
Link 1 out Empty
Link 2 out Iptr: #80000256 Wdesc: #80000091 (Lo)
Link 3 out Empty
Link 0 in  Empty
Link 1 in  Empty
Link 2 in  Iptr: #80000321 Wdesc: #80000125 (Lo)
Link 3 in  Iptr: #80000554 Wdesc: #80000170 (Hi)
Event in   Empty

Link 0 connected to Host
Link 1 not connected
Link 2 connected to Processor 88, Link 1
Link 3 connected to Processor 23, Link 3

Booted from link 0
```

M — Memory map

Selecting the Memory map command displays the memory map of the current processor. This is the same
as that provided by the utility COMPILATION INFORMATION, when applied to a PROGRAM.

It lists the start and finish addresses of the program's code, libraries, and real arithmetic library. It also
includes the configuration code, and the program's workspace and vectorspace. If any of these components
are not used, they will not be listed. The size of each component is then listed, in bytes, or rounded up to
the next K. The debugger also displays the total memory usage on this processor.

It also lists the size and address range of that processor's on-chip RAM, and 'MemStart', the first free location
after the RAM reserved for the processor's own use.

It then lists the maximum size network which can be accommodated by the debugger's buffer space. This
will depend on the memory size of the host system, and on the other code in memory at the same time.

The complete display looks like this:

```
        Memory map
Workspace          : #80000064 - #800000F3  (  144 )
Configuration code : #800000F4 - #80000117  (   36 )
Program body       : #80000234 - #80012373  (   73K)
Libraries          : #80012374 - #80012773  ( 1024 )
Vectorspace        : #80012774 - #80024643  (   72K)

Total memory usage : 149060 bytes   (146K)

On-chip memory (2K) : #80000000 - #800007FF
Mem Start           : #80000048

Debugger has enough memory for 1271 processors
```

The value which is displayed for MemStart is the value actually found on the transputer in the network. If this
does not correspond to that expected by the configuration description, for example because a T414 and a
T800 have been mixed up, you will be informed:

```
MemStart should be : #80000070 (T800) !!!!!
```

[N] — Network dump

This command allows you to save the state of the transputer network, so that you can continue debugging later. If you leave the debugger without creating a network dump file, you will not be able to continue debugging from the same point without re-running the application program. This is because the debugger itself corrupts parts of the memory on each transputer in the network.

Once you have created a network dump file, you may continue debugging from the file instead of from the target network. The debugger will take all relevant information from the network dump file, and from the program's source code and descriptors, and it does not even need to be still connected to the target network.

You will be informed how much space a network dump file would take up, and asked whether to continue. The space required depends on how much memory is actually used on each processor in the network.

```
                     Create network dump file
        Number of processors : 10
        File size will be     : 89673 bytes

        Continue with network dump (Y,N) ?
```

If you type 'N', no file will be created, and the operation is aborted. Otherwise you will be asked:

```
        Filename ("network.dmp", or "QUIT") ?
```

       Press ENTER to use the default filename
or   enter a filename (any filename extension will be replaced by '.dmp')
or   type 'QUIT' (uppercase) to exit the debugger.

If the file already exists, you will be warned:

```
        File "network.dmp" already exists
        Overwrite it (Y,N) ?
```

If you type 'N', you will be prompted for the filename again.

While dumping the state and memory contents of each processor in the network, it will display:

```
        Dumping network to file "network.dmp" ...
        Processor 99 (T800)
        Memory to dump : 10456 bytes ...
```

This command can not be used while debugging an **EXE**; this is because you can use a TDS core dump file instead.

[O] — occam

Selecting the occam command allows you to resume symbolic debugging, either at the same occam line, or at another location. You will be prompted:

```
        Iptr (#hhhhhhhh) ?
```

The default suggested is the last occam line located to on this processor, or the address shown as the last instruction executed.

       Press ENTER to accept the default address
or   enter the desired address (a decimal number, a hexadecimal number preceeded by '#',
                   or the short form '%h...h', which assumes a prefix of #8000...).

Useful values are displayed by the 'R', 'T', and 'L' commands from the Monitor page, or the value of the saved low priority **Iptr**.

If the supplied **Iptr** is not within the program body, one of the following errors is displayed:

> **Error : Cannot locate to configuration level code**
> **Error : Location is not in program or a library**

After pressing any key you are returned to the Monitor page.

Otherwise, you are then prompted:

> **Wdesc (#hhhhhhhh) ?**

If you used any **Iptr** which was shown on screen at that time, its corresponding **Wdesc** will be offered as a default. Otherwise you must supply it yourself, in the same format as the **Iptr**.

If no symbolic features other than a single 'locate' are required, the **Wdesc** is not needed, and any value may be given, so you should accept the default by typing [ENTER] on its own. Note that if an invalid **Wdesc** is given, most of the symbolic features will not work, or will give incorrect answers.

Once the **Iptr** and **Wdesc** have been supplied, the debugger will display the occam source at the required location, and the full range of symbolic features are then available.

[P] — Processor

This command is used to change the Monitor page to show details for a different processor in the network. Specify the processor number after the prompt:

> **New processor number ?**

Type a processor number (the number used to identify the processor in the configuration description of the program). This is checked against the data base to make certain the processor exists. If that processor is found, the display is changed to provide the same information for the new processor. If memory is being displayed, but the new processor's word length is different to that most recently displayed, the start address will be reset to the bottom of memory. If the processor is not in the configuration, the following message is displayed:

> **Error : That processor number does not exist**

If there is only one processor in the network, you are informed:

> **There is only one processor in the network**

If you press this key when debugging an **EXE** you are informed:

> **This is an EXE - There are no other processors**

[Q] — Quit

This command leaves the debugger and returns to the TDS. Once quit, the debugger cannot be used to debug the same program without reloading the program unless a 'network dump' file has been created. This is because using the debugger overwrites much of the contents of the network.

[R] — Run queue

This command allows you to see a list of the processes waiting on the processor's active process queues. If both high and low priority front process queue pointers are empty, the following message is displayed:

> **Both process queues are empty**

If neither are empty, you will be prompted:

High or low priority process queue ? (H, L)

You should then type '**H**' or '**L**' as required. Otherwise the debugger will assume the non-empty queue.

The instruction pointers and workspace descriptors of the first page full of processes on the queue will be displayed. If there are more processes than can fit on the screen, the following message(s) will be displayed:

<<< Scroll up for more >>>
and/or **<<< Scroll down for more >>>**

The CURSOR UP and CURSOR DOWN keys can be used to see the extra processes.

T — **Timer queue**

This command allows you to see a list of the processes waiting on the processor's timer queues. If both high and low priority Front Timer queue pointers are empty, the following message will be displayed:

Both timer queues are empty

If neither are empty, you will be prompted:

High or low priority timer queue ? (H, L)

You should then type '**H**' or '**L**' as required. Otherwise the debugger will assume the non-empty queue.

The instruction pointers, workspace descriptors, and wake-up times of the first page full of processes on the queue will be displayed. If there are more processes than can fit on the screen, the following message(s) will be displayed:

<<< Scroll up for more >>>
and/or **<<< Scroll down for more >>>**

The CURSOR UP and CURSOR DOWN keys can be used to see the extra processes.

X
RETRACE
RELOCATE — **Exit**

These commands return to the debugger's symbolic mode. They can not be used if you have changed processor while in the Monitor page.

CURSOR UP
CURSOR DOWN — **Scroll display**

Typing CURSOR UP or CURSOR DOWN scrolls the display of either the ASCII or hex memory dump, disassembly, occam typed memory, or queue, whichever was last displayed. The memory dump is scrolled by eight lines (256 bytes of ASCII data, 128 bytes of hex data) up or down, or sixteen lines for the typed memory dump. The disassembly is scrolled by sixteen bytes, then aligned to the start of that instruction. The memory display wraps round when the highest memory address is reached. The process and timer queues are scrolled by fourteen lines.

LINE UP
LINE DOWN — **Scroll display**

As CURSOR UP and CURSOR DOWN but with the scroll increment restricted to a single line.

CURSOR LEFT
CURSOR RIGHT — Change processor

Typing CURSOR LEFT sets the current processor to the preceding processor in the data base and displays the same information for the new processor. Typing CURSOR RIGHT sets the current processor to the succeeding processor in the data base and displays the same information for the new processor.

This next processor may not correspond to the next processor number given in the configuration details, but depends on the internal database in the debugger. The display shows the processor number, as given in the configuration details.

If you press these keys when debugging an **EXE**, or if there is only one processor in the network, you will be informed:

This is an EXE - There are no other processors
or **There is only one processor in the network**

TOP — Display last instruction

The TOP key can be used from the Monitor page to display the occam source corresponding to the last instruction to be executed on the current processor. Its use is as if you typed 'O', then gave the **Iptr** and **Wdesc** as displayed on the Monitor page.

?
CODE INFORMATION — Help

These commands display a page of help information, which lists the commands available at the monitor page.

REFRESH — Refresh

This command redisplays the screen.

15.1.7   Hints

Invalid pointers

Any time an instruction pointer and workspace descriptor are displayed, they are checked to be within correct code and data limits, as defined by the memory map command. Any invalid **Iptr** and **Wdesc** pair is flagged by an asterisk ('*').

This can occur when displaying: the **Iptr** and **Wdesc**; the saved low priority **Iptr** and **Wdesc**; the processes waiting for communication on any of the links; processes waiting on any of the queues; a typed memory dump as **CHAN**s; or when you use INSPECT on a channel.

Failure to communicate

The debugger uses the extraordinary link handling library routines for communication with the target network. This means that the debugger can recover if communication fails for any reason. This will normally be because the debugger has failed to reset the network, or because it has been executed on the wrong **PROGRAM** fold. This will be reported by the message:

Error : Cannot read processor 'n' (T414)

Default addresses

The debugger's 'Monitor page' maintains two default addresses. These are the address of the last disassembly, and the address of the last other memory display. This means that you can disassemble a portion of memory, then look at its workspace as a hex dump, for example, then simply by typing 'D' again, you will still have the correct address to disassemble from.

IF, CASE and guarded ALT

The semantics of occam 2 state that an **IF** construct with no **TRUE** guards behaves like **STOP**. Similarly a **CASE** construct with no matching selection also behaves like **STOP**. In both cases it is not necessary to insert an explicit default case which simply **STOP**s. However, it can be a very good idea to do so, to aid debugging. The way in which the debugging information is generated means that if either of these defaults are taken, and there is no explicit default, the debugger can only locate to the line *following* the **IF** or **CASE** construct.

However, if the default is explicitly stated, the debugger will locate to the **STOP** statement, which provides a more immediate indication of the cause of the error. There is no object code size penalty in providing the explicit case.

An **ALT** with all guards **FALSE** also behaves like **STOP**, and the debugger can only locate to the line following the **ALT** construct.

ALT

Due to the way that **ALT** constructs are implemented on a transputer, all channels and timers waiting in a single **ALT** will wait at the same location. The debugger will indicate this by locating to the *first* alternative of the **ALT**, no matter which channel or timer is requested.

CASE input

In some circumstances a **CASE** input will stop due to an incorrect protocol tag being received, yet the sender will appear to be sending a valid tag. Consider the following example executing on a single transputer:

```
PROTOCOL protocol
  CASE
    tag1
    tag2 ; INT
:
CHAN OF protocol c :
PAR
  INT x :
  c ? CASE tag2 ; x

  SEQ
    c ! tag1
    c ! tag2 ; 42
```

This will **STOP** on the **CASE** input, since 'tag1' has been sent. Suppose the inputting branch is executed before the other branch of the **PAR** (Note that occam does not define which branch will be executed first). It will then deschedule, waiting for communication on channel 'c'. The other branch of the **PAR** will then proceed. It will communicate 'tag1', and return the waiting process to the active process queue. It will also proceed to the next communication, where it will deschedule since there is now no process waiting to input.

The first process will then resume execution. It will test the value of the tag it received, namely 'tag1', and hence **STOP**. The debugger can then be used to locate to this **CASE** input. However, if you use the debugger to look at the channel 'c' to determine which process was outputting on that channel, the debugger will indicate that the second output ('tag2') is waiting.

This problem can only occur with communications involving variant protocols, when a tag with no data is communicated.

Deadlocks

There is a simple method which can be used to help find the cause of a deadlock. Since the debugger can inspect the transputer's links, it can be used to detect deadlocks which occur across more than one transputer (use the Monitor page 'L' command). A problem only arises when a single processor has deadlocked. Then there will be no active process from which the programmer can inspect channels, and hence jump to the waiting process.

In practice, it is almost always known (or guessed) which channel or channels are causing deadlock. This means that we can add a simple routine to help keep track of those channels. Consider the following simple procedure:

```
PROC p ()
  CHAN OF INT c :
  PAR
    SEQ
      c ! 99
      c ! 101
    INT x :
    c ? x
:
```

This procedure will deadlock, and the debugger will not be able to find out where the channel is stored in memory.

The procedure can be transformed to:

```
PROC p ()
  CHAN OF INT c :
  CHAN OF INT stopper :
  PAR
    VAL one.second   IS 15625 :  -- Low priority
    VAL secs.per.day IS (60 * 60) * 24 :
    VAL one.day      IS one.second * secs.per.day :
    TIMER time :
    INT now :
    SEQ
      time ? now
      ALT
        time ? AFTER now PLUS one.day  -- will locate to here
          SKIP
        stopper ? now
          SKIP
    SEQ
      PAR
        SEQ
          c ! 99
          c ! 101                      -- will jump to here

        INT x :
        c ? x
      stopper ! 0
  :
```

When this modified procedure is executed, it will appear to deadlock, as before. However, there is now a 'way in' to the program. The debugger can be used to inspect the transputer's timer queue (using the Monitor page 'T' command), which will have a process waiting on it. You can then use the Monitor page 'O' command, and give it the **Iptr** and **Wdesc** of that waiting process. The debugger will then locate to the **ALT** statement. You can then use the symbolic INSPECT key to inspect the channel 'c', which will be found to have a process waiting inside it. Use the CHANNEL key to jump to the process waiting for communication, where the program has deadlocked.

Note that the compiler does not insert these modifications automatically for many reasons. Firstly, one philosophy behind this debugger is that the code being debugged is identical to that which is eventually used. Note that the extra code portion inserted in the above example can be safely inserted permanently; on a transputer, a process waiting on a timer consumes no CPU time. Secondly, in a typical program, there would be many channels, and it would significantly increase the channel and **PAR** execution overhead, not to mention the code size. Thirdly, if every channel had this type of extra debugging provided, there would be so many processes waiting on the timer queues that it would be difficult to detect which was actually required. Deadlocked channels can also be found using INSPECT from the Monitor page.

15.1.8   Creating a core dump file

The debugger can read a file to find the contents of a transputer's memory. This is useful either when debugging an **EXE**, or when debugging a **PROGRAM** which has used the host processor as part of the network.

Debugging EXEs

The debugger can be used to debug **EXE**s which have been written within the TDS. When an **EXE** fails, the server will detect that the transputer's error flag has been set, and allow you to re-boot the TDS. The TDS loader will then allow you to create a core dump file. It will prompt you:

```
Options :
    c : normal core dump
    f : normal core dump + freespace
    a : standalone core dump - all of memory
    s : standalone core dump - part of memory
    <RETURN> to skip
```

You should then select an option — when debugging an **EXE** you should use either 'C' or 'F'. If you type 'C', the TDS will save all the relevant memory contents of a normal **EXE**. The length of the file will be the size of the **EXE**'s code, plus its workspace and vectorspace, plus about 25K bytes of extra information. If your **EXE** uses the extra **freespace** parameter, as a dynamic buffer, you should type 'F', and the buffer will be saved to the file too.

The TDS will then ask for the name of the core dump file:

```
Core dump file name ("core.dmp") ?
```

    Press ENTER to accept the default name
or enter another filename (any filename extension will be replaced by '.dmp')

While writing the file, the TDS will display:

```
Writing core dump file "core.dmp" ...
```

with further notes identifying the blocks of data written to the file.

Debugging standalone programs

The debugger can be used to debug **PROGRAM**s which have been written within the TDS, but are booted directly by a server. These programs will use the host transputer (i.e. the transputer which runs the TDS) as part of their network. This means that simply re-booting the TDS will corrupt the contents of the first transputer in the network. Instead, the debugger has the ability to read the first transputer's state from a file held on the host filing system, and read the rest of the network directly as normal.

The TDS can be directed to save the state of the transputer as it starts up. This tells the server to *analyse* the transputer, rather than *reset* it. The TDS will then allow you to create a 'standalone core dump' file in the same way as for an **EXE**. You should use either the 'A' or 'S' commands from the core dump menu.

Use option '**A**' if your program uses all of the memory on the host transputer. If you type option '**S**', you will be asked:

> **Memory size in kilobytes :**

You should type in the amount of memory which your program uses on the host transputer board. This must include the code size, workspace, and vectorspace.

Next you will be asked for a filename, in exactly the same way as when creating an **EXE**'s core dump file.

The TDS will then save that amount of memory in the core dump file, starting at the bottom of memory. The file length will simply be the amount of memory saved, plus about 500 bytes for register contents, etc.

15.1.9   occam run time errors

This section lists the possible causes of run time errors. All the errors will have their effect defined by the compilation error mode:

- In **HALT** mode, they will halt the transputer.

- In **STOP** mode, they will stop that process, allowing other processes executing on the same transputer to continue.

- In **REDUCED** mode most of these errors will not be detected.

The compiler will perform as many of these checks as possible when compiling. For instance, if an array is subscripted by a constant value, the range check is performed by the compiler and no extra code is inserted to check at run time.

STOP   The **STOP** process is implemented to behave as though an error has occurred. So too are occam constructs defined to behave like **STOP**:

> **IF**   An **IF** construct with no *true* guard will **STOP**.
>
> **CASE**   A **CASE** construct with no **ELSE** option will **STOP** if no option is matched.
>
> **ALT**   An **ALT** construct with none of the boolean guards of its alternatives *true* also
>   behaves like **STOP**.

Arithmetic errors   Arithmetic overflow, divide by zero, etc., cause an error.

> Also any Floating-point calculations will cause an error if any of their inputs are either infinity, or 'Not-a-Number'. This can be avoided by explicit use of the **IEEEOP** library routines. See the occam 2 reference manual for details.

Shifts   Shifting an integer by more than the number of bits in its representation will cause an error.

Type conversions   When converting a value from one type to another, the value must be able to be represented in the target range, or an error will be caused (e.g. a **BYTE** must lie in the range 0–255).

Replicators   Any replicated construct (**SEQ**, **PAR**, **IF**, or **ALT**) with a *negative* replicator count will cause an error. A zero replicator is permitted.

Array accesses   Any accesses to elements outside the range of an array will cause an error. This also applies to segments of arrays.

> If a segment of an array is assigned to another segment of the same array, the two segments must not overlap.

The sizes of an array must correspond when an array is passed as a parameter to a procedure or function, or when an array is assigned or abbreviated. Zero length segments are allowed.

The **range.checking** compiler option can be used to disable these forms of error checking.

Abbreviations If the same element of an array is abbreviated twice in the same scope, an error will be caused. The **alias.checking** compiler option can be used to disable this form of error checking.

Communications Attempting to communicate a zero length array on a channel of type **CHAN OF ANY** will cause an error. However, you may use a zero length *counted array* communication.

A **CASE** input process, where the communicated tag does not match any of those supplied, will cause an error.

RETYPES Any **RETYPES** expression must be aligned to the correct word or byte boundary (e.g. you may not **RETYPE** bytes 5, 6, 7 and 8 of a **BYTE** array as an **INT32**, since **INT32**s must be aligned on a word boundary).

PRI PARs If a **PRI PAR** is executed from within a high priority process, an error will be caused.

15.2   Transputer network tester – `nettest`

Introduction

A number of 'worm' programs have been developed, for use in exploring, testing and debugging various transputer hardware systems. Some of these have been put into a single program, **nettest**, for use as a general purpose diagnostic tool. An algorithm for relating the physical network found to the one which the user specifies is also included.

This section describes the transputer network test program, gives interpretations of the error messages, and describes some of the more common problems encountered in running multiple transputer networks.

15.2.1   What the network tester does

This section describes the use of the transputer network test program, both for checking the configuration in which a number of transputers have been connected, and for pinpointing any hardware problems. The program is run as an **EXE** under the transputer development system. Any of the available links may be connected into a network of transputers, with a reset cable from the subsystem socket of the master transputer controlling the rest of the network. One of the links connects the master transputer to the host computer, and this link should not be tested (If it is the program will crash). In the rest of this section, the term 'master transputer' is used to describe the processor on which the network test program is run.

The network test program executes by sending a worm into a network of transputers. The worm explores the network, reaching every single transputer that is connected, no matter what configuration, and reports back the configuration which it finds. No initial assumption is made about the network. This should be contrasted with the loading of a network of transputers with a program of fixed configuration, which is the normal approach to developing programs using the transputer development system.

Connections made via a C004 link switch are invisible to the worm, but a link connected to the control link of a C004 can be identified as such.

The worm algorithm is described in INMOS technical note 24, 'Exploring multiple transputer arrays'. It is important to realise that the worm numbers transputers in the order in which it finds them, which may bear no relation to any conceptual order which the user has in mind.

For ease of use, however, the network test program can compare the network it has found against a user's **PROGRAM** specification, and give results in terms of the user's numbering, together with a statement on whether or not the two networks match.

The worm proceeds in two phases. Initially, each transputer in the network is loaded with a copy of the worm program. As this happens, information about each new transputer found — the *loading data* — is relayed back to the master and displayed. At the same time, the error flag is briefly set high on the newly found transputer (halt on error has been set to false!) which may light up an error light, and is detected by the master in order to determine that error signals are being propagated back correctly. The error flag (and, in the case of transputers with a floating point unit, the FPU error flag) is then left cleared.

Once the entire network has been explored, any further tests are performed on all transputers in the network in parallel. The 'network test data' thus found, including a complete list of link connections, is reported.

Having completed testing, the program starts again by resetting or analysing the system and sending in a fresh worm.

15.2.2   Using the network test program

The program is loaded and run as an **EXE**. If you want the results to be filed, then run the program while the cursor is pointing at a fold bundle.

If you want to match results against your own network configuration, run the program with the cursor pointing at your **PROGRAM** fold. If the matched results are to be filed, then pick the **descriptor** fold from your **PROGRAM** fold, and put it inside a fold bundle.

You are then prompted for an option. These are listed in the next section. Different options are appropriate to different circumstances, but for a quick check, try option 'C'.

After selecting an option, you will be prompted for a link from which to send the worm. When the master transputer controls a module motherboard, this will usually be link 2. The link connected back to the host computer must not be selected; if it is the program will crash (this is usually link 0).

Some options allow both links 2 and 3 of the master to be tried. Usually only one connection is made from the master into the rest of the system, but it is often useful to be able to explore a network from two different directions, in order to pin-point an error.

Finally, you will be asked whether you want results displayed in brief or in full. Full mode presents all results available, and is described in the following sections. Brief mode simply presents a summary of results — whether a hardware error has been found, and whether the network found will match the one specified by the user (if, indeed, one was specified). This may be useful for long test runs, for example for a 24 hour burn-in.

After a 3 second delay, testing is repeated, until a key is pressed. If results are being filed, a new file will appear for each run.

A note on matching

The problem of matching two networks is not trivial.

The worm uses its own numbering as it explores the network of transputers, and matches what it has found against the description of the user's **PROGRAM** configuration (if given). So long as the two networks match, the worm gives results using the user's numbering scheme. If, however, no match is found (which is always the case if no **PROGRAM** configuration is given), the worm's own number will be given, suffixed by a *****.

The master transputer is never included in a **PROGRAM** description, but is reported by the worm as **MT***.

The matching algorithm is as follows. While the worm loads the network, each time it finds a new transputer it consults the configuration specified in the descriptor fold to see whether it matches. However, if the physical network contains more transputers, or more links, than the network specified, the matching may be incomplete; consider for example the situation when an extra connection is present in the physical network, and the worm loads through it onto a transputer which does have a counterpart in the **PROGRAM** descriptor. The matching algorithm cannot know immediately that the worm has found a transputer which is described in the descriptor.

As the worm returns the network test data, together with a complete map of the link connections, results are reported using the (possibly incomplete) numbering equivalence discovered above.

However, once all the results are returned, and a complete list of link connections established, the link map is used to discover any new equivalences between the network found and the network specified. If new equivalences are indeed found, then the network test data and link map is again displayed, using the user's numbering scheme as far as possible.

While all transputer types share the same basic instruction set, there are some differences (wordlength, on-chip FPU, 2d block move, etc.) which affect compatibility. For the sake of matching the type of a transputer found in the network against a **PROGRAM** descriptor, the following compatibility sets are distinguished:

{T212, T222, M212}, {T414}, {T800, T801}, {T225}, {T425}, {T805}.

Limitations of use

Different pre-programmed options are available to suit different classes of configurations. Systems with memory mapped peripherals, IMS M212 disk controller chips, IMS T414 revision A transputers, and the control links of IMS C004 link switches require careful selection of the right option in order that the state is not corrupted.

The term T2 is used below to describe the IMS T212, IMS T222 and IMS T225 range of chips, similarly for T4 and T8. Reference to the C004 means connections to the control link of the C004 — connections made via a C004 are of course transparent to all options.

15.2.3   Options available

The following options are available:

A — Check T4/T8/C004 network using internal RAM

The network is explored, but no testing is performed, by a program which only requires less than 2Kbytes of memory on each transputer (except, of course, the master). The network may consist of any combination of 32-bit transputers and C004's. This is the only option which can explore networks containing IMS T414's of revision A and is used both to locate such devices (which should ideally be upgraded) and to check configurations in which one or more transputer has no external memory.

B — Check T2/C004 network using internal RAM

As option A, but for T2's and C004's. This option again requires only 2Kbytes of memory on each transputer — it fits inside internal RAM and makes no access to external RAM.

C — Check IMS M212/T2/T4/T8 networks

A network of mixed wordlength transputers is explored, but no testing is performed. The size and speed of external memory is found. This is the only option which will explore networks of mixed wordlength (16 and 32-bit transputers), and it is also safe to use with IMS M212 disk controller chips.

4 — Test T4/T8 networks

A network of 32-bit transputers is explored, and all devices found are tested. The parameters used for testing can be varied, or certain suitable defaults used. These are outined below. Section 15.2.7 describes the testing in more detail. The worm requires 8Kbytes of memory to operate. The following sections outline the different modes of testing T4/T8 networks.

2 — Test T2 networks

A network of T2 transputers is explored, and all devices found are tested. The worm requires 8Kbytes of memory to operate. The following sections outline the different modes of testing T2 networks.

$\boxed{\text{D}}$ — Development mode

This mode allows you to explicitly set certain testing values. These are described in section 15.2.7.

$\boxed{\text{E}}$ — Error light testing

This mode proceeds slowly, flashing the error light on and off on each transputer as it is found. All modes toggle error, but this one holds the light on for long enough for you to see it clearly. This is useful both for testing the error LED and for following the progress of the worm.

$\boxed{\text{F}}$ — Full testing

This mode tests links and all the memory it finds on each transputer, with pauses to test for data retention. This means that it loads up the network very slowly, for it is testing the bottom 8Kbytes of memory (the area to be occupied by the worm) thoroughly before loading each transputer.

An algorithm is used to determine the size of memory. The untested remainder of the memory, and the links, are thoroughly tested once the whole network of transputers have been loaded. Since memory is not tested on the master transputer, there may be a pause after results come back from the master transputer, before data from the rest of the network is returned.

$\boxed{\text{L}}$ — Link test

This mode loads up the network, and then tests all the links in the network in parallel. Because the network is loaded quickly, and a lot of power is drawn due to all links and the processor working flat out in parallel, this mode is useful to 'warm up' a network if a temperature-dependent problem is suspected.

$\boxed{\text{M}}$ — Memory test

As in mode F, 8Kbytes of memory is tested on each transputer before loading, and the remainder is tested at the end. However, there is no pause to test for data retention. This makes this mode much faster but less thorough than mode F. Links are not tested.

$\boxed{\text{H}}$ — Help

A comprehensive set of help windows are available to help the user.

$\boxed{\text{Q}}$ — Quit

The program terminates without exploring the network. However, it does reset all input links of the master transputer, and also resets the subsystem.

15.2.4   Interpretation of loading data

This section covers the table of data which appears as the network is loaded. Each new entry in the table corresponds to a new transputer which has been found. The first entry is from the host transputer.

Typically, the table looks somtething like this:

Id	Boot Link	Booted by Id	Link	Type	Speed	Analysed or Reset	Error Line	Mem. Speed	
--	----	--	----	----	-----	--------	-----	-----	
MT*	--	--	--	T414b	-15	Reset	ok	5 cycle	..
0	1	MT*	2	T800d	-20	ok	ok	3 cycle	..
1	0	0	3	T801a	-22.5	ok	ok	2 cycle	..

A classic problem is that a network is found on the first run of the worm, but not on subsequent runs. This indicates that the reset cable is not connected to the subsystem socket of the master transputer correctly. The network of transputers always powers up in a reset state, ready to run a program, but if the reset cable is not connected correctly, then the network cannot be reset for another run of the worm.

Id

As far as possible, the numbering scheme is as specified in the user's **PROGRAM**. However, if no **PROGRAM** descriptor fold is found, or the networks don't match exactly (see the note on matching, above), then the worm's own numbering will be given, suffixed by a *. In certain cases, such as a module motherboard or IMS B003, the worm's numbering will conveniently match the standard numbering of transputers. **MT*** is the master transputer.

Link

Links are numbered 0, 1, 2, 3. The boot link is the link on which the transputer was booted by the worm. (Note that this is not necessarily the same as the way a **PROGRAM** would be booted.)

Booted by id, link

These entries indicate which parent loaded the transputer. The master was loaded by no-one. So, in the example given above, link 2 of the master transputer booted transputer 0 on link 1, and link 3 of transputer 0 booted transputer 1 on link 0.

Type

The type and revision of the transputer is given. The main reason for giving the revision number (a, b, etc) is to identify early sample devices which the user may still be using. Note that, at this stage, the worm does not distinguish an IMS M212 from an IMS T212 (see section 15.2.5).

Speed

The speed of the part (in MHz) is reported.

Processor analysed

Every transputer has an internal flag which indicates whether the transputer was most recently reset or analysed. The worm reads this flag, and sends the results back. The purpose of this is to check that both the reset and analyse control signals are correctly propagated through the system.

On alternate runs, the master will reset, then analyse the system. If the flag matches what was expected, then the message **ok** is given. Otherwise, a message **Reset not Analysed** or **Analysed not Reset** is given, and a fault in the reset/analyse chain should be suspected.

The flag on the master transputer, however, is either **Reset** or **Analysed**, but should not change during repeated testing.

Error Line

The error line is tested each time the worm finds a new transputer. If it is working, the message **ok** is reported. If the line is broken, the message **Not set** will appear.

On the master transputer, however, it is expected that the error line is clear. If this is not so, then the message **Not clear** will appear. The same message may also appear the first time that the worm is run — this is quite normal, and is due to the fact that error may have been set on one of the transputers (transputers power up with the error flag in a random state).

Since the worm leaves the error flag low, option C is useful when a user wants to clear error (and, in the case of transputers with floating point units, the FPU error flag) on all transputers in a network.

Memory Speed

The speed of external memory is given. Since it is possible to buy transputer modules with different banks of memory operating at different speeds, it should be noted that it is the first word above 4kbytes that is used for this test.

15.2.5   Description of network

Having completed loading the network, further tests may be performed, according to the option selected. Sometimes, the program may appear to pause while returning results. This is because it is still testing some transputers. Results are then returned, together with a complete list of link connections.

Here is an example of some results returned (using option F), from a particularly bad network:

```
                     Memory   Link:   0          1          2          3
 Id   Type     Found  Tested    Id Link    Id Link    Id Link    Id Link
 --   ----     -----  ------    -------    -------    -------    -------
 MT*  T414b     --      --      ----       ----        0  1      ----
  0   T800d    256 k    ok      oooo       MT  2      oooo        1  0
  1   T801a     64 k  #80008014  0  3     Link Err 17  oooo     #80000800
```

Memory is tested in two phases. Firstly, as the network is loaded, a transputer tests a section of its neighbour's memory before loading the program. Then, once all the transputers have been loaded, each transputer tests the remainder of its own memory. If a memory error is found at the first stage, it is reported under the link which was doing the testing. At the second stage, the error is reported under **Memory Error**.

Type

IMS M212 is now distinguished from the IMS T212, otherwise the type of transputer is repeated.

Memory found

All options except A, B (which assume that no external memory is present) will find out how much memory is available. In the example above, 256 Kbytes of memory have been found for transputer 0, 64 Kbytes for transputer 1. The memory of the master transputer is never investigated.

Memory tested (applies to options D, F, M)

If an option is selected which tests memory, it will first find out how much memory is present, and then test that amount of memory. A given memory size can be explicitly selected by using option **D** (see section 15.2.7).

Testing memory can take some time, (up to 30 seconds per megabyte), and the program will pause while this is happening.

ok indicates that memory has been tested successfully, while -- indicates that memory was not tested. If an error is found, it will be reported as a hexadecimal address. This address should give the actual byte which is at fault. In the example given, there is a problem in the memory of transputer 1 at address #80008014.

Network connections

Each transputer has four entries, corresponding to its four links. Each entry may be

   ---- indicating that the link has not been tried (applies to master transputer);

   oooo indicating that the link is unattached;

   x y indicating that the link is attached to link y of transputer x;

or an error message. Such a message indicates either a problem on the link, or on a transputer attached to that link. It does not necessarily imply a problem on the transputer being reported. Read section 15.2.7 for a background to the error messages.

15.2.6   Error messages

The error messages are listed below. Some of the messages may refer to a particular 'stage', which is the stage of testing at which the error occurred. Certain errors tend to be revealed at particular stages, but for completeness the stages are listed in section 15.2.8.

#8000$abcd$ — Options D, F, M

Before a neighbouring transputer is loaded with code, part of its memory is tested using the peek and poke facility of the transputer. This error indicates that a neighbouring transputer has indeed been found, but that its memory is faulty (or does not exist) at the address given. If the value is in the range #80000000 to #800007FF (IMS T414) or #80000FFF (otherwise), then the problem lies in internal RAM. Otherwise, up to a highest tested address of #80001FFF (8K), the problem lies in external RAM. Don't forget that a transputer must have 8Kbytes of RAM for this test to succeed.

In the example, the entry #80000800 indicates that the first byte of external RAM is faulty on the transputer which is connected to transputer 1 link 3. The most likely explaination in this example is that it has no external memory!

Token error x — all options

When waiting for a reply on a link under test, an unknown token has been returned, at stage x. This may indicate a problem on the links (e.g. they are communicating at different speeds, or noise) — this usually appears as **Token Err 1**.

If an option is selected which doesn't check memory, a Token Error may indicate a memory problem on the adjacent transputer; when a transputer is first booted, it returns a copy of the program code for confirmation, which may have been corrupted. This usually occurs as **Token Err 9**.

Time out error x — all options

When waiting for a reply on a link under test, no reply has been received within a reasonable time, at stage x. For some reason, the neighbour failed before it was properly loaded.

If an option is selected which doesn't check memory, **Time Out 9** may indicate a memory problem on the adjacent transputer; the transputer has been loaded, but does not run. This might happen, for example, if option C is used on a network which includes a transputer with no external memory.

Time Out 18 indicates that a link, which was expecting to pass back results from further down the chain, has not received anything. This error will usually be part of a line of results which is otherwise a repeat of a previous line — the transputer has reported results, but now wishes to revise its report to indicate an error. It means that, although the worm was successfully loaded, it has subsequently failed somewhere down the chain from the link indicated.

Alt error y — all options

While waiting for a reply on link y, an unrecognised token was input on this link. This frequently occurs when two links are communicating at different speeds. It can also occur if a link is unconnected and floating (i.e. is not pulled down using an appropriate resistor). Check link y on the same transputer, as well as the link which reported this error.

Link error x — options D, F, L

When testing the links, corrupted data was transmitted at stage x (probably stage 17, which is when the links are tested exhaustively). This may be due to noise on the line, because of insufficient decoupling, or strong electical interference, for example. Or it might indicate a problem with the transputer link at either end, though this is rare.

More often, this error indicates a fault in the section of memory where test data is prepared. This occurs when the links but not the memory are being tested (i.e. option L). Try again with option M or F.

In the example, the entry **Link Err 17** indicates a problem on link 1 of transputer 2. Since option F was used in the example, this implies that data was being corrupted during transmission between transputer 2 link 1 and the attached neighbour.

Output error x — all options except A, B

The worm has failed to output data on a link, despite the fact that it has already discovered the link to be attached to another transputer. This implies that the neighbouring transputer has died, for some reason.

? #z

An unknown error message has been returned to this link. The hexadecimal error value, z, may or may not be useful. If an option has been used which performs no testing, then try again with, for example, option F.

15.2.7   Testing specifications

To understand how a worm works, it is essential that Technical Note 24 is read. The program explores the network, using one of five worms, corresponding to options A, B, C, 2, and 4. These have been called the 'Skinny' (T2, T4), 'Mixed Network', and 'Fat' (T2, T4) worms. All five will exercise the reset, analyse and error lines, but the first three do not have any means of testing memory or links.

The fat worm, on the other hand, performs tests on memory and links using parameters supplied to it. The various options E, F, M, L, set defaults which have found to be suitable, while option D allows the user to alter the defaults. The sections below which describe the testing of memory and links refer to the fat worm.

Order of loading

The worm explores the transputer links in the order 2, 3, 0, 1 (i.e. from a particular transputer, it first tries to explore any network off link 2, then off link 3, etc.). This contrasts with the order given in technical note 24. The order of exploration means that the numbering scheme which the worm uses matches the actual numbering of transputers when module mother boards and IMS B003s are explored.

Size of system

The limit on the size of system (number of transputers) which can be explored is displayed by the program.

Speed of part

The speed of the processor is found by performing $1 << x$ for a suitable value of x. This is known to take x cycles, and by using the transputer's internal clock, which ticks at a constant rate of once every 64 microseconds at low priority, the number of cycles performed per second can be deduced.

Memory size and speed

The speed of memory is found by timing repeated executions of Block Move of data onto itself. This is performed at 4Kbytes up from the base of memory.

The memory size algorithm sweeps through memory, writing a marker, until either it encounters the top of memory or finds its own marker, indicating a wrap around. Hence this algorithm violates parity. If parity checking is in use, then use option D to specify the size of memory to be tested.

Type

Recent transputer versions now support the **LDDEVID** instruction which returns a value giving the type and revision of the transputer. Early transputers do not have this, and for them the type is worked out from the wordlength and location of **MemStart**. A 16-bit transputer is easily distinguished from a 32-bit transputer, for example, by using the transputer **BCNT** instruction (see 'Transputer instruction set — a compiler writer's guide').

Reset and analyse

On alternate runs, the master will either reset or analyse the subsystem. This makes no difference to the function of this program, but a register exists on the transputer which is read to tell whether the transputer was reset or analysed, and hence confirm that these signals have been propagated correctly.

Error line

As each transputer (except the master) is loaded, its error flag is set, the master reads the subsystem error line, and the error flag is then cleared before the worm proceeds to the next transputer. Thus, the Error Line should be TRUE for every transputer, except for the master, when it should be FALSE.

Note that, on the first run, the error line may be TRUE for the master transputer — error flags may have been set by a previous program, and not yet cleared. Indeed, when transputers power up, the state of the error flag is not initialised. If, however, the value is TRUE for the master transputer on subsequent runs, or FALSE for any other transputer, then a fault on the error line should be suspected.

Memory

Before loading a transputer with code, the parent tests the lowest 8Kbytes of memory using peek and poke. This allows it to verify that the space which will be occupied by the worm, when it is loaded onto that transputer, is indeed safe. An error which occurs at this stage is reported in the network test data as an entry under the link which was performing peek and poke. It is possible to increase this value using option D.

Once loaded, the program uses an algorithm to determine how much memory it has. This algorithm has an 8Kbytes resolution, and may violate parity. In development mode (D), the user may specify the amount of memory to be tested on each transputer, in Kbytes, with no risk of violating parity.

The remainder of memory, up to the largest memory address found, is tested on all transputers in parallel, once all the transputers in the network have been loaded. Any error will be reported in the network test data under the memory error column.

In both cases, the memory is tested as follows:

1 The address is written, as a word, to each word in the block to be tested.

2 After a pause of 1000 milliseconds, each word is read back, and checked, in turn, being replaced by the BITNOT conjugate word.

3 After a further pause, each word is checked, and replaced by the value #55555555 (#5555 for T2's).

4 After a further pause, the words are checked and replaced by #AAAAAAAA (#AAAA for T2's).

5 Finally, after a pause, the words are read and checked.

Option M does not use any pause. A different pause may be specified using option D.
The memory of the master transputer is not tested. The program does not perform detailed tests on the memory (e.g. march tests, etc.) except as described above.

Links

Each link is tested for the existence of a neighbour by attempting to output a probe sequence, and waiting 250 milliseconds for a reply. This default is more than ample. Communication takes place using a byte protocol, and if at any stage incorrect data is returned, the link is assumed to be bad, and no further communication takes place on that link. If a communication with incorrect protocol takes place on the link which is being probed, the error is reported as a Token Error, the entry being made against the link which was doing the probing. If some unrecognised data appears on a different link, it is reported as an Alt error.

When a transputer is loaded, it immediately returns the program for checking. If the program has been corrupted in transmission, this will show up as Token Error 9.

After all transputers in the network have been loaded, on receipt of a synchronise token, all links in the whole network are tested in both directions in parallel. A test block of data, which is 128 bytes long and consists of a section of the orginal program, is transmitted in both directions on each link. The input is checked, and the exchange is repeated 2000 times on each link, independently. As far as possible, the constructs **OutputOrFail.t** and **InputOrFail.t** are used, so that the program can recover from, and report the communication of bad data. An error appears as the entry Link Error 17 against the link which discovered that its input data was corrupted.

Any links of the master transputer which are found to be connected into the rest of the network are tested in the same way as the other links in the network

15.2.8   Stages of loading

In technical note 24, the worm algorithm is described with reference to a number of different stages. These stages are also useful in telling when an error was detected. The following list of stages refers to the fat worm. Other worms may use a subset of these stages. The meaning of the tokens is described in technical note 24.

1   Send a probe sequence from a link, to determine whether there is another transputer connected.

2   Set the bottom 8K of memory of the neighbour to word addresses.
Pass back a **GreenLight.t** token to the master. Pause for one second.

3 – 6   Read back and check data, writing a new word as we go.
After each stage, pass back a **GreenLight.t** token, and pause for one second.

9   Having determined that there is an unbooted neighbour with at least 8K of good memory, boot that transputer with a copy of the worm program. The neighbour will return the program for checking.

11   Send down a set of initalisation data to the newly booted transputer. The booted transputer will return a set of **loadingData**. Pass this back to parent, and synchronise with the master.

12   The neighbour, or someone further down the chain, is now testing its links.
Pass back **GreenLight.t** tokens. Do not timeout the link at this stage.

12   Also be prepared to pass back **loadingData**, and forward a **Synchronise.t** token.

12   When the neighbour sends back **ReturnControl.t** and the number of transputers found so far, it is assumed that the branch off that link has been completly explored. Try another link.

14   Once all links have been explored, return control to parent.

15   Synchronise the whole network, prior to final testing.

17   Test all links and memory in parallel.

18   Send results of testing, **networkData**, back to parent. Forward **networkData** from each link in turn, reading from a link until **NoMoreData.t** is encountered. When all links have been read, return **NoMoreData.t** to parent.

The dots which appear while the worm is loading indicate the return of the token **GreenLight.t**.

15.3   Memory interface program – `memint`

Introduction

The External Memory Interface Program allows the designer of a system to get the best out of the configurable external memory interface on the IMS T414, IMS T425 and IMS T800 transputers.

The program allows the system designer to modify the values of simulated timing parameters of the memory interface and see what effect the changes have. The program's user interface has a number of pages, some of which have inputs on them, some of which have outputs and some of which have both. It is possible to switch between pages at the touch of a button, and therefore to be able to see the effect of the input values very quickly and change them easily.

It is possible to store both the current input parameters and the contents of the pages in folds. The input parameters may then be read in by the program allowing continued development of a system. The pages, stored in folds, can for instance be sent to a printer and the configuration table can be used as input to the EPROM programmer program for placing in ROM.

The configuration table can also be used to generate PAL equations.

15.3.1   Capabilities

Input to the program can be divided into three broad categories:

1 Values of the various parameters of the memory interface itself, such as the periods of the various Tstates.

2 Parameters of the system, such as the processor type and speed.

3 Parameters of the program itself, such as the labels and names of signals and the name to be used if the parameters and pages are written to folds.

Given the above input, the program displays on its various pages:

1 General timings useful for any memory.

2 Times specific for DRAMS.

3 The waveforms of the address/data pins, programmable strobes and other timings.

4 The table of bits that makes up the memory configuration data and the addresses that those bits occupy in memory.

15.3.2   Using the program

This section explains how to get started with the program and lists the various commands that it will obey.

Getting started

The program is in the form of a **CODE EXE** fold. First it is necessary to get the program into memory by pressing the GET CODE key. Place the cursor on an empty fold and press the RUN EXE key.

The program will initialise itself, then the title page will be displayed.

Commands

Key	Description
0..5	Move to and display the corresponding page.
CURSOR UP CURSOR DOWN	Use to select the current input; the cursor moves to the next or previous input on the page. If there is only one input on a page then nothing happens. If there are no inputs on the page then the help window is displayed.
CURSOR LEFT CURSOR RIGHT	Used to scroll the waveforms page (page 4) horizontally. If used on any other page or an attempt is made to scroll off the edge of the screen then the help window is displayed.
C	Change the value of a parameter. It's action depends upon the type of the parameter (see section 15.3.3). If there are no input parameters on the current page then the help menu is displayed.
R	Reset all the parameters to their default values; these are the values that the program uses immediately after starting unless the parameters were read from a fold.
F	Store the current values of the parameters and the contents of all the pages, other than the title page, in folds (see section 15.3.5).
Q	Quit from the program. The program will ask for confirmation, press **Q** again to confirm the command.
HELP	Displays the help window. This key may be used at any time to get the help window (e.g. half way through typing in a new label for a strobe).
other	Pressing any other key will cause the help window to be displayed.

The commands may be typed in upper or lower case.

15.3.3   Input

There are three main types of input parameter, which can be changed using **C**:

> **Cycle** — **C** is used to obtain the next value in the cycle, wrapping round at the end of the cycle.

> **Number** — A number is expected, terminated by any character other than **0** to **9** and DELETE. DELETE may be used to delete the last digit entered. The program prevents too many digits being typed in. The value is checked and if it is outside the range for the parameter an error is produced and the number must then be re-entered.

> **String** — A string is expected, terminated by ENTER. DELETE may be used to delete the last character entered. The string is displayed between a pair of ". Entry of strings too long to fit between the "'s is prevented, excess characters being ignored.

The durations of the Tstates, the strobes and the wait period are measured in periods Tm. One period Tm is half of the processor cycle time.

Memory interface parameters

1 The length of each of the Tstates, T1 to T6, is entered as a number of Tm periods between 1 and 4.

2 The time periods of each of the programmable strobes, S1 to S4, is entered as a number between 0 and 31. Note that 0 is a special case, if the period of S1 is set to 0 then **notMemS1** stays high throughout the memory cycle and if the period of S2, S3 or S4 is set to 0 then the corresponding signal will stay low throughout the memory cycle.

3 The refresh period cycles between 0, 18, 36, 54 and 72 clockin periods. If the value is 0 then refresh is disabled.

4 The write mode cycles between early and late.

5 The configuration cycles between 0 to 11 and 31 for the IMS T414 and 0 to 15 and 31 for the IMS T800 and IMS T425. This indicates whether the current parameters match one of the preset memory configurations of the transputer. These configurations may be chosen by cycling through the values of this parameter. When modifying other parameters, it is possible for the resulting parameters to match one of the preset configurations, if so the value will indicate which preset configuration this is, if not then '–' is displayed instead of a number.

System parameters

1 The type and speed cycles between IMS T414-15, T414-17, T414-20, T800-17, T800-20, T800-22, T800-25, T800-30, T800-35, T425-17, T425-20, T425-22, T425-25, T425-30, and T425-35. For the IMS T400 use IMS T425-20.

2 The clock frequency should not be changed; it should be left at 5000kHz.

3 The wait parameter may be set either to a number greater than or equal to 0, or to one of the programmable strobes, S2 to S4, simply by either typing in the number or **s** followed by a number 2 to 4. Note that connection to S1 is meaningless and therefore not allowed. Connection to a number means that number of Tm periods of delay have been inserted by external hardware.

Program parameters

1 Each strobe has two labels each of which is a string. One is 9 characters long and is used on the waveforms page to label each waveform. The other is only 1 character long and is used extensively in the timing pages.

2 The file name is a 20 character string used as a label on the folds produced by **F**.

15.3.4   Output

There are two types of output:

- Numeric output
- Waveform output

Numeric output

Three pages consist entirely of numeric output; for basic times, DRAM times and configuration table.

Basic times

The basic times page contains general times useful for every type of memory:

T0L0L	Cycle time (in both nanoseconds and processor cycles)
TAVQV	Address access time
T0LQV	Access time from notMemS0
TrLQV	Access time from notMemRd
TAV0L	Address setup time
T0LAX	Address hold time
TrHQX	Read data hold time
TrHQZ	Read data turn off
T0L0H	notMemS0 pulse width low
T0H0L	notMemS0 pulse width high
TrLrH	notMemRd pulse width low
TrL0H	Effective notMemRd width
T0LwL	notMemS0 to notMemWrB delay
TDVwL	Write data setup time
TwLDX	Write data hold time 1
TwHDX	Write data hold time 2
TwLwH	Write pulse width
TwL0H	Effective notMemWrB width

DRAM times

The DRAM times page contains information useful when using drams:

T1L1H	notMemS1 pulse width
T1H1L	notMemS1 precharge time
T3H3H	notMemS3 pulse width
T3H3L	notMemS3 precharge time
T1L2L	notMemS1 to notMemS2 delay
T2L3L	notMemS2 to notMemS3 delay
T1L3L	notMemS1 to notMemS3 delay
T1LQV	Access time from notMemS1
T2LQV	Access time from notMemS2
T3LQV	Access time from notMemS3
T3L1H	notMemS1 hold (from notMemS3)
T1L3H	notMemS3 hold (from notMemS1)
TwL3H	notMemWrB to notMemS3 lead time
TwL1H	notMemWrB to notMemS1 lead time
T1LwH	notMemWRB hold (from notMemS1)
T1LDX	Write data hold from notMemS1
T3HQZ	Read data turn off
TRFSH	Time for 256 refresh cycles (in microseconds)

Configuration table

The configuration table page contains a list of bits that make up the memory interface configuration respresented by the input parameters together with the addresses that those bits occupy when placed in ROM.

The basic and DRAM times pages each have a list of parameters. Each of these parameters consists of a JEDEC symbol, a description of the parameter, and the minimum and/or maximum times for that parameter.

The number of wait states is displayed on the parameters page.

Waveform output

The waveforms page displays a diagram of the waveforms of each of the external memory interface pins.

At the top of the page is displayed the processor clock and the Tstates, a number indicating the Tstate, 'W' indicating a wait state, and 'E' indicating a state that is inserted to ensure that T1 starts on a rising edge of the processor clock.

Below this are displayed the waveforms of the programmable strobes, the read and write strobes and address/data pins. Each of these is labelled with the corresponding label parameter.

The point at which the read data is latched is indicated by a '^' beneath the read cycle address/data waveform.

The **MemWait** waveform shows the input to the **MemWait** pin. If the wait input is a number then it goes low n Tm periods after the end of T1 and high again at the end of T6, if the wait input is connected to a strobe it goes low and then high when that strobe does so.

15.3.5   Storing and retrieving parameters and pages

The **F** command causes the program to write out two folds. These two folds are inserted as the last items in the fold bundle on which the cursor was placed when the program was run. Repeated use of **f** results in a pair of folds each time.

The first of the folds contains the current values of the input parameters. If the program is run with the cursor pointing to a fold bundle with one of these folds as the first item in the bundle then the parameters will be read from the fold and a message displayed. This enables continued development of a memory configuration. It is strongly recommended that no changes be made to the contents of this fold directly as this may cause problems should it later be used as input to the program.

The second of the folds contains a fold for each page apart from the title page. These folds contain the text of the pages making, for example, printing the waveforms very easy. The fold containing the configuration table can be used as part of the input to the EPROM hex program to make placing the configuration in ROM easy. It is strongly recommended that no changes be made to the contents of the configuration fold directly as this may cause problems should it later be used as input to the EPROM hex program.

15.3.6   Examples

This is some sample output, taken directly from the folds produced by pressing **F**:

```
   Page 1                        EMI   Configuration   Parameters
                                 ===================================


                              Device selection - T414-20
                    External Memory Interface clock period (Tm) = 25 ns
                           Input clock frequency = 5000khz
                               Wait States =  0
            Address setup time                        T1 =  1   periods Tm
            Address hold time                         T2 =  1   periods Tm
            Read cycle tristate / Write data setup    T3 =  1   periods Tm
            Extended for wait                         T4 =  1   periods Tm
            Read or write data                        T5 =  1   periods Tm
            End tristate / Data hold                  T6 =  1   periods Tm
            Programmable strobe   "notMemS1 " "1"     S1 = 30   periods Tm
            Programmable strobe   "notMemS2 " "2"     S2 =  1   periods Tm
            Programmable strobe   "notMemS3 " "3"     S3 =  3   periods Tm
            Programmable strobe   "notMemS4 " "4"     S4 =  5   periods Tm


            Refresh period  72   clockin periods            Wait   0
            Write mode      Late                       Configuration   0


               Non-Programmable strobe  (S0)  "notMemS0 " "0"
               Read  cycle strobe             "notMemRd " "r"
               Write cycle strobe             "notMemWrB" "w"
```

Page 2   Basic Times
                                              ============

Symbol	Parameter	min(ns)	max(ns)	notes
T0L0L	Cycle time	150	–	= 3 processor cycles
TAVQV	Address access time	–	125	
T0LQV	Access time from 0	–	100	
TrLQV	Access time from r	–	50	
TAV0L	Address setup time	25	–	
T0LAX	Address hold time	25	–	
TrHQX	Read data hold time	0	–	
TrHQZ	Read data turn off	–	25	
T0L0H	0 pulse width low	100	–	
T0H0L	0 pulse width high	50	–	
TrLrH	r pulse width low	50	–	
TrL0H	Effective r width	50	–	
T0LwL	0 to w delay	50	–	
TDVwL	Write data setup time	25	–	
TwLDX	Write data hold time 1	75	–	
TwHDX	Write data hold time 2	25	–	
TwLwH	Write pulse width	50	–	
TwL0H	Effective w width	50	–	

Page 3   Dram Times
                                              ============

Symbol	Parameter	min(ns)	max(ns)	notes
T1L1H	1 pulse width	125	–	
T1H1L	1 precharge time	25	–	
T3L3H	3 pulse width	25	–	
T3H3L	3 precharge time	125	–	
T1L2L	1 to 2 delay	25	–	
T2L3L	2 to 3 delay	50	–	
T1L3L	1 to 3 delay	75	75	
T1LQV	Access time from 1	–	100	
T2LQV	Access time from 2	–	75	
T3LQV	Access time from 3	–	25	
T3L1H	1 hold (from 3)	50	–	
T1L3H	3 hold (from 1)	100	–	
TwL3H	w to 3 lead time	50	–	
TwL1H	w to 1 lead time	75	–	
T1LwH	w hold (from 1)	100	–	
T1LDX	Wr data hold from 1	125	–	
T3HQZ	Read data turn off	–	25	
TRFSH	256 refresh cycles	–	3650	Time is in microseconds

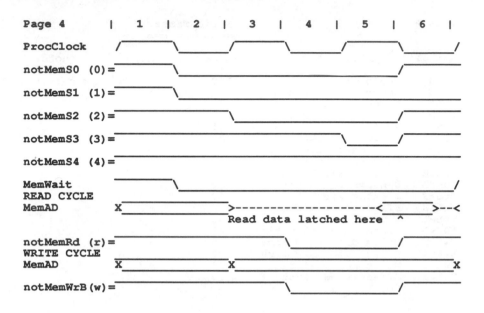

Page 4 | 1 | 2 | 3 | 4 | 5 | 6 |

ProcClock

notMemS0 (0)=

notMemS1 (1)=

notMemS2 (2)=

notMemS3 (3)=

notMemS4 (4)=

MemWait
READ CYCLE
MemAD

Read data latched here

notMemRd (r)=
WRITE CYCLE
MemAD

notMemWrB(w)=

Page 5   Configuration Table
                                           ====================

#7FFFFF6C	–	0		#7FFFFFB4	–	0
#7FFFFF70	–	0		#7FFFFFB8	–	0
#7FFFFF74	–	0		#7FFFFFBC	–	0
#7FFFFF78	–	0		#7FFFFFC0	–	0
#7FFFFF7C	–	0		#7FFFFFC4	–	1
#7FFFFF80	–	0		#7FFFFFC8	–	1
#7FFFFF84	–	0		#7FFFFFCC	–	0
#7FFFFF88	–	0		#7FFFFFD0	–	0
#7FFFFF8C	–	0		#7FFFFFD4	–	0
#7FFFFF90	–	0		#7FFFFFD8	–	1
#7FFFFF94	–	0		#7FFFFFDC	–	0
#7FFFFF98	–	0		#7FFFFFE0	–	1
#7FFFFF9C	–	0		#7FFFFFE4	–	0
#7FFFFFA0	–	1		#7FFFFFE8	–	0
#7FFFFFA4	–	1		#7FFFFFEC	–	1
#7FFFFFA8	–	1		#7FFFFFF0	–	1
#7FFFFFAC	–	1		#7FFFFFF4	–	1
#7FFFFFB0	–	1		#7FFFFFF8	–	1

15.3.7   Caveats

Please note that the values supplied by the program are subject to alteration when IMS T414, IMS T425 and
IMS T800 characterisation data is added to the program.

15.3.8   Error and warning messages

The following is a list of error and warning messages the program can produce:

Wait race

> If one of the programmable strobes is used to extend the memory cycle then the strobe must be taken low an even number of periods Tm after the start of the memory interface cycle. If the strobe is taken low an odd number of periods after the start then a wait race warning will appear. Should this warning appear, it will remain on display on all pages apart from the title and waveforms page until the race condition is removed. See the IMS T414 and/or IMS T800 data sheet for more details.

Input out of range

> If the value entered for a numeric parameter is outside the range valid for that parameter, an input out of range warning is displayed, the value cleared from the screen and the program waits for a new value.

S1 to MemWait

> If an attempt is made to connect S1 to the **MemWait** input an error is displayed because it is a meaningless operation.

Unable to access fold

> This can occur on startup and indicates that the program is unable to read and write folds, usually because the program has not been run on a fold bundle. The program waits for a key press and then terminates.

Filer unusable

> This indicates that the program is unable to create the folds due to a previous error.

Filer error

> This indicates that when trying to store the parameters and pages in folds an error occured. The filer will be unusable for the rest of the program's execution.

15.4   EPROM hex program – epromhex

15.4.1   Using the program

The program must be run on a fold bundle. The fold bundle may contain up to three folds.

   1 A **CODE SC** fold.

   2 A **CODE PROGRAM** fold.

   3 A memory configuration fold.

A **CODE SC** fold and a **CODE PROGRAM** fold are produced by the EXTRACT utility of the compiler, when applied to an occam **SC** or occam **PROGRAM** fold set respectively. The memory configuration may be produced as output from the 'Memory interface program' or may be generated by hand.

The only item which must be present in the bundle is the **CODE SC** fold. If no memory configuration fold is present, no memory configuration will be loaded into the EPROM. Similarly, if no **CODE PROGRAM** fold is present, no **CODE PROGRAM** fold will be loaded into the EPROM.

Two occam procedures are provided in source form as example loaders. One of these interfaces to a host and loads a network from information received via a serial line and the other loads a network from information obtained by scanning through a **CODE PROGRAM** held in the EPROM. These programs should be modified for the environment in which they are to run, if the user wishes to create an EPROM which is to be used as a loader.

15.4.2   What the EPROM hex program does

The EPROM hex program builds a buffer containing the future contents of the EPROM in the order:

1 Contents of the **CODE PROGRAM** fold (if present)

2 Contents of the **CODE SC** fold (the main procedure)

3 Transputer initialisation code

4 Memory configuration (if present)

5 Entry jump to the inititalisation code

These items, padded to word boundaries, are placed adjacent in the buffer. A transputer booting from ROM starts executing with the instruction pointer set to location **MOSTPOS INT − 1**. The two bytes from this address are loaded with a jump to the entry point of the initialisation code, or to a longer jump if the full jump cannot be achieved using a two byte instruction. The position of this jump, therefore, governs the actual address for all the other components of the EPROM. The size and start address of the contents of the EPROM and how large an EPROM is required can then be calculated.

The buffer is built in the following way:

Stage 1
The fold bundle is checked to see if a **CODE PROGRAM** fold is present. If it is, the contents of this fold are read directly into the buffer, which is then padded to a word boundary.

Stage 2
The contents of the **CODE SC** fold are added to the buffer. At the same time the workspace requirement and the entry point of the **SC** are noted. The buffer is again padded to a word boundary.

Stage 3
The bundle is checked for a memory configuration fold. Note that this fold must be filed. A memory configuration is a list of pairs of numbers in the format:

configuration.address configuration.value

A configuration address may be in hex (preceded by '#') or in decimal. A configuration value must be either 0 or 1. A memory configuration fold must contain all of the configuration address value pairs, a fold which contains some but not all of the values is treated as an error. A 'Configuration Table' page produced by the memory interface program, which has been written to a fold, may be used without modification.

Stage 4
The code which initialises the transputer is added to the buffer directly after the **CODE SC**. The initialisation code does the following:

1 Copy 600 bytes from internal RAM to top of RAM if required.

2 Read and save the current state of the transputer if required.

3 Set up local workspace.

4 Initialise process queues.

5 Clear *Error* and *HaltOnError*.

6 Clear *FP.Error* on a T800.

7 Initialise all link process words, the event process word and the high and low priority timer queues to **Empty**.

8 Start the processor clock

9 Initialise the work space and parameters for the procedure in the SC.

10 Call the procedure.

The local workspace is initialised for the main procedure as though it had just been called from an outer level process. The workspace is initialised for the following parameter list

```
PROC EPROM.SC (INT            entry.point,
          [60]BYTE            buffer,
          VAL[600]BYTE        memory.copy,
          VAL[]BYTE           program.buffer)
```

The workspace initialisation depends on the presence of a **CODE PROGRAM** fold in the bundle and on the response to prompts given by the EPROM interface program as it is constructing its internal buffer. All of the parameters must be present to enable the **EPROM Hex Program** to supply a pointer to a separate vector space if the SC has been compiled with a separate vector space. The use of the parameters in relation to the actions of the initialisation code is described in more detail below.

entry.point is used when the main procedure is going to be used as a loader. This variable must then be set by the procedure to the offset (from **MOSTNEG INT**) of the initial workspace pointer AND entry point of the loaded code. This value is obtained from the loading information that the EPROM loader interprets from the host, or from the included **CODE PROGRAM**. INMOS technical note 34, 'Loading transputer networks', describes the transputer development system loading protocol in detail and shows how the entry point is sent to each processor in a network. On exit from the loader procedure, the transputer's instruction pointer and workspace pointer will be set to the value contained in **entry.point**.

buffer is used when the main procedure will load and analyse the transputer network. It has two uses. On entry to the procedure, **buffer** contains the processor state information, retrieved by the initialisation code, which enables the EPROM to emulate the actions of a processor which is analysing from link. After any analyse function required has been completed, this buffer may then also be used as an intermediate buffer for passing on code packets to processors later in the network, and passing back data packets from processors later in the network.

memory.copy is used if the procedure analyses the transputer network. It contains a copy of the 600 bytes of memory starting at **MOSTNEG INT**. This copy is the first action performed by the initialisation code if the question in stage 5 '**Insert copy for analyse (y/n) ?**' is answered yes. The copy is performed, because the area of RAM copied contains information vital when analysing a transputer, and the EPROM uses the area as local workspace and thus destroys its contents.

program.buffer is used if the procedure loads a network from the contents of the EPROM. It is a reference to the location of the contents of the **CODE PROGRAM** fold if there was one in the fold bundle. If there was not a **CODE PROGRAM** fold in the fold bundle, then **(SIZE program.buffer) = 0**.

When the required components have been input and inserted into the buffer, the following information is displayed:

- The total ROM requirement for code and data, in bytes.

- The workspace requirement of the main procedure, in words and bytes.

- The total RAM requirement of the code in the EPROM, in bytes.

The total RAM requirement of the EPROM is made up from:

- Transputer reserved locations from MOSTNEG INT.

- The (small) workspace of the initialisation code.

- The 60 bytes of **buffer**.

- The workspace of the main procedure.

- The separate vector space of the main procedure (if any).

If the EPROM is intended for use as a loader, the total RAM requirement must not exceed 560 bytes on a T4 or T8 and 464 bytes on a T2. The program produces a warning message if these values are exceeded. If, however, the main procedure is not a loader the warning can be ignored.

Stage 5
The program will write the contents of the buffer into a new last fold of the bundle, labelled **EPROM hex**. The first line of the new fold contains the start address for the code in the EPROM followed by the transputer type. A typical example is:

```
.7FFFF22C    --    T4
```

The rest of the fold consists of lines containing the EPROM contents as bytes written out in hexadecimal:

```
73 41 F7 72 30 AC 71 73 72 30 F7 72 30 71 F2 D1
 .
 .
00 00 66 02
```

Messages

The initial display is the title and version followed by the prompt:

```
Create Hex Table For EPROM Program

Insert copy for analyse (y/n) ?
```

If **n** is typed then the initialisation code will not copy the bottom 600 bytes of RAM to the top of RAM and the **memory.copy** parameter will not contain information for use by analyse.

If **y** is typed then the initialisation code will copy the bottom 600 bytes of RAM up to the top of RAM; this is the copy that is passed into the **SC** as the parameter **memory.copy** and contains information which is needed when analysing the board. The program then prompts the question:

```
RAM size (in k-bytes) ?
```

The size may be entered in hex (preceded by '#') or in decimal and is terminated by carriage return. The initialisation code will copy to the 600 bytes immediately below the top of the RAM, calculated from this value of the RAM size.

The program will then display the message:

> **Building table in buffer...**

and then either

> **Configuration table read ok**
> or **No configuration table in bundle**

indicating whether or not a configuration table has been read and inserted into the buffer. This is followed by the message

> **EPROM hex created OK**

indicating that the EPROM code has been built. Information about the EPROM code is then displayed.

> **Total ROM space requirement =** *number* **bytes**
> **SC's work space requirement =** *number* **words,** *number* **bytes**
> **Total RAM space requirement =** *number* **bytes**

If the total RAM requirement is too large for the code to act as a loader, the program produces the following warning message:

> **WARNING: total RAM space requirement exceeds maximum allowed for a loader (limit =** *number* **bytes)**

If the main procedure is not a loader the warning can be ignored. The next message displayed is:

> **Writing hex to fold**

This is followed by the termination message:

> **Press a key to exit**

and waits for the user to press a key before terminating.

Error messages

Error messages produced by the program are:

> **Cannot open fold**
>
> **Unreadable fold in bundle**
>
> **Cannot create file for write**
>
> **Cannot open file for write**
>
> **Code buffer overflow**
>
> **Incorrect configuration value**
>
> **All config values have not been filled**
>
> **Fold bundle does not contain code file**
>
> **Unknown transputer target**

15.5   Hex to programmer program – `hextoprg`

The Hex to Programmer program inputs a fold in the format output by the EPROM hex program, and outputs it to the serial port (device name **COM1:**) in Intel Hex format. The program has been tested using the GP XP640 EPROM programmer.

The program is provided in source form to illustrate how a fold prepared for EPROM may be transmitted to a device or file. Many users will find that the program can be used without modification but if, for example the EPROM programmer being used does not support Intex Hex format, then the program will need to be modified. The program has been designed to make such modifications easy to make. The format specific parts of the program are in **PROC send.buffer** and are also listed at the end of this section.

15.5.1   Using the program

The Hex to Programmer program should be applied to a fold containing data in the format output by the EPROM hex program.

The first action of the program is to read in the data and check that the start address is correct for the amount of data in the fold. The start address is checked by making certain that the code length read in, added to the start address, places the last two bytes at the correct boot from ROM entry point for the specific transputer type.

The startup title displayed by the program is:

> **Hex Table To EPROM Programmer Program**

This is followed by version and copyright messages, and then the message:

> **Reading table into buffer...**

When this is completed, information about the input data is displayed:

> **Transputer target is T2/T4/T8**
> **Start address** = *hex.number*
> **Code size** = *hex.number* **bytes**

If the transputer target is **T2**, this will be followed by the question:

> **Number of EPROMs on T2 board (1 (byte access) or 2 (word access)) ?**

This question is to determine whether consecutive bytes are to be written to a single EPROM (byte access), or alternate bytes are to be written to two EPROMs (word access). Only word access is allowed for **T4** and **T8** and so four EPROMs are assumed.

Depending on the response to the previous question, one of the following messages is displayed:

> **In word access mode**
> or   **In byte access mode**

This is followed by the prompt:

> **EPROM size (in K bytes): (2, 4, 8, 16, 32, 64) (0 quits) ?**

The EPROM size is used by the program to determine the start offset within the EPROM which corresponds with the start address of the code read in. Only EPROM sizes large enough to hold the data are offered to the user.

The next message displayed is the help message for communicating with the EPROM programmer. The contents of this message will depend on the interface provided to the programmer.

```
XP640 EPROM programmer
Connect XP640 to IBM's serial port (COM1)
Set XP640 to :
   9600 baud
   Intel Hex
   8 data bits, 1 stop bit, no parity, handshake on

Commands are :
   ?           : type this information
   D           : display DEFine area (part of EPROM to be written to)
   W           : send full words      (only in byte access mode)
   {byte.no}   : send selected byte   (only in word access mode)
   Q           : terminate session
   S{byte.no}  : display selected byte on screen
```

The program then prompts for a command:

Ready for command:

Before use the serial port (**COM1:**) must be set up using the DOS command:

mode com1:9600,n,8,1

For other baud rates etc. the parameters to the **mode** command will need to be changed, see the DOS manual for more information.

This is a more detailed version of the help information:

? displays the help information that has been listed above.

D displays the range of address in the EPROM that are to be programmed. This makes it easy to only program the area of the EPROM that needs to be programmed thus saving time.

W sends every byte to be programmed out to the programmer. It is only usable in byte access mode as it sends out both bytes of the word (remember this is only possible on a T2) one after the other.

byte.no sends only the selected byte from each word to the programmer (0 or 1 for a T2, 0 to 3 for a T4 or T8) to enable each one of the EPROMs to be programmed individually with a single byte from each word.

Q terminate the programming session (i.e. exit from the program).

Sbyte.no displays the selected byte from each word on the screen.

15.6   Write EPROM file program – promfile

The Write EPROM File program inputs a fold in the format output by the EPROM hex program, and creates up to four files containing the Intel hex of the EPROMS to be created. These files are compatible with the input requirements of many EPROM programmers.

The files created are **ROMBYTE0.HEX** and **ROMBYTE1.HEX** for a **T2** in word access mode, **FULLROM.HEX** for a **T2** in byte access mode and **ROMBYTE0.HEX**, **ROMBYTE1.HEX**, **ROMBYTE2.HEX** and **ROMBYTE3.HEX** for a **T4** or **T8**. Note that word/byte access mode refers to the way the ROMs are accessed on the target board and thus in word access mode the bytes in the word are placed into separate ROMs and in byte access mode the bytes are placed contiguously in a single ROM.

15.6.1 Using the program

The Write EPROM File program should be applied to a fold containing data in the format output by the EPROM hex program.

The first action of the program is to read in the data and check that the start address is correct for the amount of data in the fold. The start address is checked by making certain that the code length read in, added to the start address, places the last two bytes at the correct boot from ROM entry point for the specific transputer type.

The startup title displayed by the program is:

WRITE EPROM FILE PROGRAM

This is followed by version and copyright messages, and then the message:

Reading eprom contents fold

When this is completed, information about the input data is displayed:

> **Transputer target is T2/T4/T8**
> **Start address** = *hex.number*
> **Code size** = *hex.number* **bytes**

If the transputer target is **T2**, this will be followed by the question:

Number of EPROMS on T2 board (1 (byte access) or 2 (word access)) ?

This question is to determine whether consecutive bytes are to be written to a single EPROM (byte access), or alternate bytes are to be written to two EPROMs (word access). Word access is assumed for **T4** and **T8** and so four EPROMs are assumed.

Depending on the response to the previous question, one of the following messages is displayed:

> **In word access mode**
> or **In byte access mode**

This is followed by the prompt:

EPROM size (in K bytes): (2, 4, 8, 16, 32, 64) (0 quits) ?

The EPROM size is used by the program to determine the start offset within the EPROM which corresponds with the start address of the code read in. Only EPROM sizes large enough to hold the data are offered to the user.

The next message displayed is the help page for examining the potential EPROM contents or writing the files:

```
Commands are :
   ?          : type this information
   D          : display DEFine area (part of EPROM to be written to)
   B          : write buffer as bpw files (only in word access mode)
   W          : write buffer as single file
                                        (only in byte access mode)
   Q          : terminate session
   S{byte.no} : display selected byte on screen
```

The program then prompts for a command:

Ready for command:

This is a more detailed version of the help information:

?   displays the help information that has been listed above.

D   displays the range of address in the EPROM that are to be programmed. This makes it easy to only program the area of the EPROM that needs to be programmed thus saving time.

B   writes the bytes to be programmed as either two or four files, depending on the word length of the target processor. The files are named ROMBYTE0.HEX, ROMBYTE1.HEX (and ROM-BYTE2.HEX and ROMBYTE3.HEX for a 32-bit processor). The files are written using INTEL HEX format and so can be used as input directly by many EPROM programmers. The bytes included in a file are only the bytes at a particular offset in each word in the EPROM. For example ROMBYTE1.HEX contains all the Byte 1's for the whole EPROM. This is the way the EPROMS must be written when they will be accessed as the individual bytes in word-wide memory.

W   writes the complete program to a single file named FULLROM.HEX. It is only usable when the EPROM will be accessed as a single bank of byte-wide memory by a 16-bit processor. This option is not available when the program is intended for a 32-bit processor. The file is written using INTEL HEX format.

Q   terminate the programming session (i.e. exit from the program).

Sbyte.no   displays the selected byte from each word on the screen

15.7   Preparing a bootstrap and adding it to a program – addboot, wocctab

A pair of tools is provided to support the conversion of a compiled separate compilation unit (a **CODE SC**) into a bootable program file for a hosted single processor application. An example bootstrap loader is also supplied for use with those tools, which enables a program coded as an **SC** with a particular formal parameter list, to be bootstrapped on a transputer attached by any link to a host running **iserver** and supporting a mechanism for determining the memory size at run time.

The tools and example loader are supplied as source and may be used as the basis for developing alternative bootstrapping schemes.

15.7.1   The code to occam table converter wocctab

This tool may be applied to any **SC** foldset, in which there is a **CODE SC** fold created by means of the EXTRACT utility. A new text fold will be added to the bundle, containing the contents of the **CODE SC** expressed as an occam **VAL []BYTE** declaration, with the name **Table**.

The principal purpose of this tool is to enable the code of a procedure to be treated as constant data in a program which intends to call it by means of the predefined procedure **KERNEL.RUN** (see section 11.3) An example of the use of this tool is given in section 8.4

The user will be asked the following questions:

        Include CODE SC header?   (Y,N)

        Y causes the header (see appendix G.4) to be included in the table.
        Any other answer will cause the table to contain only the code itself.

        Table of BYTE or String or Quit?   (B,S,Q)

        B causes the table to be expressed as hexadecimal bytes.
        S causes the table to be expressed as an occam string, where characters are hexadecimal escape sequences.
        Any other response abandons the tool.

15.7.2   The bootstrap adder `addboot`

This tool may be applied to a **CODE SC** fold, or to an **SC** foldset containing a **CODE SC** fold called the application program. It creates a bootable program in a host file, outside the TDS folded file store, by prepending to the code a bootstrap composed of two parts. The first part is the primary bootstrap which initialises the transputer timer, queue pointers, etc and calls the secondary bootstrap which it expects to read on the same channel down which it was itself booted. The secondary bootstrap is read from another host file which was created by applying WRITE HOST to a **CODE SC** fold. The application program must have the form of a standard hosted **PROC** (see section 8.1).

The user is given the option of making the primary bootstrap set or clear the HaltOnError flag. Otherwise the primary bootstrap is fixed. The user is then asked for the name of the secondary bootstrap. Two precompiled example secondary bootstrap loaders are provided, compiled for T2 and for TA transputers respectively. They are used by pressing RETURN.

It is possible to create a customised secondary bootstrap from a **CODESC** using WRITE HOST. The formal parameter list of the secondary bootstrap must have a particular form:

```
PROC procname   (CHAN OF ANY from.link,
                 CHAN OF ANY to.link,
                 VAL INT bytes.per.word,
                 VAL INT word.mem.start,
                 VAL INT word.free.mem.offset)
```

where: **from.link** and **to.link** are the input and output channels respectively of the transputer link down which the transputer was booted.

bytes.per.word is the number of bytes per word on which the process is to run (4 for the T414, T425 and T800 and 2 for the T212, T222 and M212).

word.mem.start is the word offset of **MemStart** from **MOSTNEG INT** (18 for the T414, T212 and T222, 28 for the T425 and T800).

word.free.mem.offset is the word offset from **MOSTNEG INT** of the start of *free* memory, that is, memory unused by the primary and secondary bootstrap code and workspaces.

The third and fourth parameters are determined by the transputer type.

15.7.3   The example two-stage loader

The secondary bootstrap supplied as an example for use with **addboot** works as follows:

It reads the header of the **CODE SC** which is to be made into a bootable file, to determine its workspace requirements etc. This header must have the form described in appendix G.4 and must correspond to a standard hosted procedure whose formal parameter list has one of the forms:

For occam code:

```
PROC program (CHAN OF ANY from.link, to.link,
              []INT user.buffer)
```

For non-occam code:

```
PROC program (CHAN OF ANY from.link, to.link,
              []INT user.buffer, stack.buffer)
```

where: **from.link** and **to.link** are the input and output channels respectively of the transputer link down which the transputer will have been booted.

user.buffer is the free memory buffer.

stack.buffer is the stack buffer for non-occam code. This is not relevant for TDS occam programs.

The parameter **user.buffer** is a vector that represents the amount of *free* memory that is still available on the board for use by the program, that is, memory not already used by the program for its code and workspace.

To calculate the actual memory available, the loader first reads the total memory size from the host environment variable **IBOARDSIZE**. This communication with the host is performed after the program has been loaded onto the transputer and before the program is started. The size of the free memory vector passed to the program is given by **IBOARDSIZE** minus the combined program code and workspace allocation.

It then copies a relocatable loader from a table within itself (made by means of wocctab) to a place in memory which will not be occupied by code of the application program. The memory is allocated according to the following scheme and the application program is read from the boot channel and entered.

15.7.4   Memory allocation

The default bootstrap loader attempts to optimise placement of the programs, and its own code and workspace. The rules it uses are as follows:

1   Application program code is placed as low as possible in memory, taking into account the scalar work space requirement of the program and the requirements of the non-occam stack if present.

2   Application program code is placed above the memory required for the program's scalar work space and the C, FORTRAN and Pascal stack if specified. The memory reserved by the bootstrap for the program's code and scalar workspace will overlap the bootstrap's own work and code space.

3   If the program uses a separate vector workspace, the bootstrap reserves a portion of the program's memory as vector workspace. From the size of this workspace, the size of the program code, and the size of its scalar workspace, the bootstrap determines the offset, from the start of memory of *free* (unused) memory. This offset is used in conjunction with the environment variable **IBOARDSIZE** to determine the amount of memory available to the application program, which is then passed as a vector parameter for the program to use.

4   If the program uses a separate stack for C, FORTRAN or Pascal code, the stack is placed at the base of memory, beginning at **MemStart**.

Figure 15.1 shows the memory map of the loaded code as created by the default bootstrap.

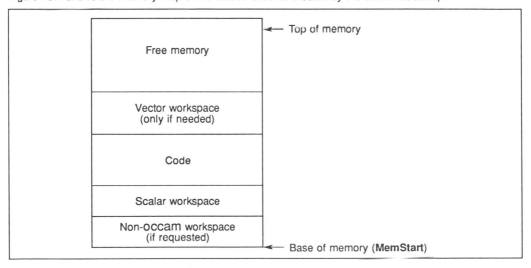

Figure 15.1 Memory map

16 System interfaces

This chapter describes a number of system interfaces associated with components of the transputer development system. These are provided for reference; it is expected that users will normally use the command files and I/O procedures provided to carry out sequences of operations on these interfaces.

The interfaces described are:

1 Host environment variables.

2 The TDS loader.

3 The ITERM configuration file used by the TDS editor and debugger.

4 The host file server interface available across the link to the host when a standalone program is booted up by the host file server.

5 The terminal interfaces provided over the channels **keyboard** and **screen** available to an **EXE** running within the TDS.

6 The user filer interface provided over the channel array pairs **to.user.filer** and **from.user.filer** available to an **EXE** running within the TDS.

7 Other interfaces between an **EXE** and the TDS.

16.1 Use of host environment variables

The TDS assumes five environment variables on the host system. These are listed below.

Variable	Description
COMSPEC	DOS command shell spawner.
IBOARDSIZE	The memory size of the transputer board.
ITERM	The file containing terminal keyboard and screen codes.
TDSSEARCH	The list of directories on which the TDS will search for certain files if the full pathname is not specified.
TRANSPUTER	The address at which the transputer board is connected to the host if not at the default address. Only used by the server.

On the IBM PC running DOS these are defined as environment variables using the **set** command. Host commands defining these variables are included in the TDS3 command file.

16.2 The TDS loader and TDS start up process

The TDS loader is loaded by the **iserver** and in turn loads the code of the TDS itself. It is a bootable program created by adding a specially crafted bootstrap to a **CODE SC**. The source of the loader and the tool for adding the bootstrap are included in the software for information only. It is not expected that users will need to modify these.

The loader can operate in reset or analyse mode. Analyse mode is required when a program running in the TDS has stopped by setting the error flag, or when execution has been interrupted by user action at the keyboard. When the loader is started in analyse mode it gives the user the opportunity to save the contents of the root transputer in a core dump file for possible subsequent use by the debugger.

The DOS command file for calling the TDS contains commands such as:

```
iserver /se /sb \tds3\system\tds3load.b4
    -f \tds3\system\tds3.xsc %1 %2 %3 %4
:analyse
if ERRORLEVEL 0 goto exit
iserver /sa /sp 16 /se/si/ss/sc \tds3\system\tds3load.b4
    -f \tds3\system\tds3.xsc
goto analyse
:exit
```

The first call of **iserver** resets the root transputer, whose address is determined by the environment variable **TRANSPUTER** or an additional server parameter **/sl** *number* which may be added.

The second call of **iserver** analyses the root transputer and peeks 16k bytes of its memory into its own workspace. This memory may then be used to load the loader which in turn gives the user the option of saving this and other aspects of the state of the root transputer in a core dump file. An alternative command file including only the second type of call is provided for use after a standalone program has failed and needs debugging.

The loader must be told how much memory is available on the root transputer if this differs from the default (2 megabytes). This may be done by setting the host variable IBOARDSIZE.

The TDS loader loads the TDS code at the top of available memory, see the table in section 6.4. It then allocates an array for use as fold manager buffer immediately below the TDS workspace. By default this buffer is allocated (for historical reasons) as 10/57 of the total free space available. However a larger or smaller size can be allocated by adding a command line parameter pair **-m** *number*, where *number* is the buffer size in bytes. The size of the fold manager buffer may be seen by using the CODE INFO key.

The command line parameter pair **-f** *filename* defines the file where the TDS code itself is found. This is a CODE SC file which has been written out as a host file by WRITE HOST.

The TDS loader detects what transputer type it has been loaded on, and displays this and other useful data on the screen during its starting sequence. By testing the analyse state of the root transputer it decides whether to give the user the option of creating a core dump file. This option is fully described in the description of the debugger in section 15.1.

When the TDS starts up it searches the current directory for all files whose names end with **.top** and presents these to the user as closed folds. If there are no such files in the directory a new empty file **toplevel.top** is created. Any existing TDS text file may be used as a top level file by changing its name to end **.top**. The environment variable **ITERM** is used to locate the terminal configuration file that will be used for translation of keyboard and screen codes. A keyboard help file is also located either in a file whose name is derived from the ITERM file by forcing the suffix to **.hlp** or in a file found by using the environment variable **TDSSEARCH** as a search path and a filename **toplevel.hlp**. (for details of the path searching mechanism see the description of the procedure **so.popen.read** in section 14.11) The search path defined by **TDSSEARCH** is also used to locate the toolkit fold **toplevel.tkt**. If this is not found a new empty toolkit fold is created in the current directory.

16.3   The ITERM terminal configuration file

This section describes the format of ITERM files; it is included for people who need to write their own ITERM because they wish to change keyboard mappings or are using terminals that are not supported by the standard ITERM file supplied with the TDS.

ITERMs are ASCII text files that describe the control sequences required to drive terminals. Screen oriented applications that use ITERM files are terminal independent.

ITERM files are similar in function to the UNIX *termcap* database and describe input from, as well as output to, the terminal. They allow applications that use function keys to be terminal independent and configurable.

Within the TDS, the ITERM file is used by the editor, by the debugger, for the **keyboard** and **screen** channels of an **EXE**, and by the procedures **so.pollkey** and **so.getkey**.

16.3.1   The structure of an ITERM file

An ITERM file consists of three sections. These are the *host*, *screen* and *keyboard* sections. Sections are introduced by a line beginning with the section letters 'H', 'S' or 'K'. Case is unimportant and the rest of the line is ignored. Sections consist of a number of lines beginning with a digit. A section is terminated by a line beginning with the letter 'E'. The *host* section must appear first; other sections may appear in any order in the file. Sections must be separated by at least one blank line.

The syntax of the lines that make up the body of a section is best described in an example:

>   3:34,56,23,7.   comments

Each line starts with the index number followed by a colon and a list of numbers separated by commas. Each line is terminated by a full stop ('.') and anything following it is treated as a comment. Spaces are not allowed in the data string and an entry cannot be split across more than one line.

Comment lines, beginning with the character '#', may be placed anywhere in an ITERM file. Extra blank lines in the file are ignored.

The index numbers in each section correspond to an agreed meaning for the data. In the following sections the meaning of the data in each of the three sections is described in detail.

16.3.2   The host definitions

ITERM version

This item identifies an ITERM file by version. It provides some protection against incompatible future upgrades.

>   e.g.   1:2.

Screen size

This item allows applications to find out the size of the terminal at startup time. The data items are the number of columns and rows, in that order, available on the current terminal.

>   e.g.   2:80,25.

Screen locations should be numbered from 0, 0 by the application. Terminals which use addressing from 1, 1 can be compensated for in the definition of goto X, Y.

16.3.3   The screen definitions

The lists of values in the screen section represent control codes that perform certain operations; the data values are ASCII codes to send to the display device.

ITERM version 2 defines the indices given in table 16.1. These definitions are used in the example ITERM file; for a complete listing of the file see section 16.3.6.

For example, an entry like: '8:27,91,75.' indicates that an application should output the ASCII sequence 'ESC [K' to the terminal output stream to clear to end of line.

Index	Screen operation	Index	Screen operation
1	cursor up	10	insert line
2	cursor down	11	delete line
3	cursor left	12	ring bell
4	cursor right	13	home and clear screen
5	goto x y	20	enhance on (not used by TDS)
8	clear to end of line	21	enhance off (not used by TDS)
9	clear to end of screen		

Table 16.1 ITERM screen operations

Goto X Y processing

The entry for 5, 'goto X Y', requires further interpretation by the application.
A typical entry for 'goto X Y' might be:

        5:27,-11,32,-21,32

The negative numbers relate to the arguments required for X and Y.

        ..., -ab, nn, ...

where: a is the argument number (i.e. 1 for X, 2 for Y).

        b controls the data output format.
        If b=1 output is an ASCII byte (e.g. 33 is output as !).
        If b=2 output is an ASCII number (e.g. 33 is output as **3 3**).

        nn is added to the argument before output.

As a complete example, consider the following ITERM entry in the screen section:

        5:27,91,-22,1,59,-12,1,72. ansi cursor control

This would instruct an application wishing to move the terminal cursor to X=14, Y=8 (relative to 0,0) to output the following bytes to the screen:

        Bytes in decimal: 27   91 57 59 49 53 72
        Bytes in ASCII: ESC [9   ;   1   5   H

16.3.4   The keyboard definitions

Each index represents a single 'cooked' keystroke. The values are in the same sequence as the **ft.** codes tabulated in appendix D.5 The data specified after each index defines the sequence of key values associated with that keystroke. Multiple entries for the same index indicate alternative keystrokes for the operation.

The meanings of the keystrokes are defined in the specification of the editing interface and the debugger.

The layout of typical physical keyboards is shown in appendix A. Sequences of key values which do not correspond to any of these cooked keystrokes are passed to programs as individual values.

In TDS3 the [SET ABORT FLAG] key is always [CTRL] [A] and is not affected by the ITERM file.

16.3.5   Setting up the ITERM environment variable

To use an ITERM the application has to find and read the file. An environment variable called 'ITERM' should be set up with the pathname and filename of the file as its value. For example, under MS-DOS the command would be:

```
set ITERM=C:\TDS3\SYSTEM\TDS3.ITM
```

For more details see the Delivery Manual.

16.3.6   An example ITERM

This is the TDS3 ITERM file for the IBM PC using an enhanced ANSI screen driver, which supports character and line insert and delete.

```
#
#    IBM PC (ANSI) ITERM data file
#    Assumes full support of insert/delete char/line and clear.to.eos
#
host
1:2.
2:80,25.
end host

#  screen control characters

screen
1:27,91,49,65.              up
2:27,91,49,66.              down
3:27,91,49,68.              left
4:27,91,49,67.              right
5:27,91,-22,1,59,-12,1,72.  goto x y
6:27,91,64.                 insert char
7:27,91,80.                 delete char
8:27,91,75.                 clear to end of line
9:27,91,74.                 clear to end of screen
10:27,91,76.                insert line
11:27,91,77.                delete line
12:7.                       beep
13:27,91,50,74.             clear screen
end of screen stuff

keyboard
6:0,72.        up
7:0,80.        down
8:0,75.        left
9:0,77.        right
10:8.          del.chl
11:0,83.       del.chr
12:0,110.      del.line     ALT F7
12:25.         del.line     CTRL Y    -- this works too
13:0,111.      undel.line   ALT F8
14:0,65.       sol          F7
15:0,66.       eol          F8
16:0,61.       move         F3
17:0,62.       copy         F4
18:0,67.       line.up      F9
19:0,68.       line.down    F10
20:0,112.      page.up      ALT F9
21:0,113.      page.down    ALT F10
22:0,82.       create.fold  NUMERIC 0
23:0,132.      remove.fold  CTRL NUMERIC 9
24:0,79.       open.fold    NUMERIC 1
```

```
25:0,81.        close.fold    NUMERIC 3
26:0,71.        enter.fold    NUMERIC 7
27:0,73.        exit.fold     NUMERIC 9
28:27.          refresh       ESC
29:0,85.        file.fold     SHIFT F2
#30             unfile
31:0,117.       finish        CTRL NUMERIC 1
31:24.          finish        CTRL X
32:0,104.       edit.params   ALT F1
33:0,60.        fold.info     F2
34:0,59.        help          F1
#35             locate
36:0,63.        get.code      F5
37:0,97.        save.macro    CTRL F4
38:0,96.        get.macro     CTRL F3
39:0,64.        run           F6
40:0,129.       tool 0        ALT 0
41:0,120.       tool 1        ALT 1
42:0,121.       tool 2        ALT 2
43:0,122.       tool 3        ALT 3
44:0,123.       tool 4        ALT 4
45:0,124.       tool 5        ALT 5
46:0,125.       tool 6        ALT 6
47:0,126.       tool 7        ALT 7
48:0,127.       tool 8        ALT 8
49:0,128.       tool 9        ALT 9
50:0,90.        word.left     SHIFT F7
51:0,91.        word.right    SHIFT F8
52:0,100.       del.wordl     CTRL F7
53:0,101.       del.wordr     CTRL F8
54:0,110.       delto.eol     CTRL NUMERIC 7
55:0,92.        topof.fold    SHIFT F9
56:0,93.        endof.fold    SHIFT F10
57:9.           select.param  TAB
58:0,105.       code.info     ALT F2
59:0,106.       pick          ALT F3
60:0,107.       copy.pick     ALT F4
61:0,86.        put           SHIFT F3
62:0,108.       next.util     ALT F5
63:0,98.        clear.util    CTRL F5
64:0,88.        autoload      SHIFT F5
65:0,109.       next.exe      ALT F6
66:0,99.        clear.exe     CTRL F6
67:0,89.        clear.all     SHIFT F6
68:0,84.        browse        SHIFT F1
69:0,118.       suspend.tds   CTRL 3
69:26.                        CTRL Z
70:0,102.       define.macro  CTRL F9
71:0,103.       call.macro    CTRL F10
72:0,130.       make comment  ALT -
#73             bad
```

end of keyboard stuff

 # eof

The actual file **TDS3.ITM** in the software includes some additional alternative keyboard code sequences.

A similar file for the NEC PC is also supplied.

The screen section above assumes that the host screen driver accepts the ANSI escape sequences shown. The standard Microsoft ANSI.SYS in DOS does not implement line/character insert/delete sequences. Either a special screen driver must be installed which does implement these codes or an appropriately modified ISERVER must be used. See delivery manual for further information.

16.4   The INMOS file server – `iserver` – command line interface

The host file server **iserver** loads programs onto transputers and transputer networks and provides a run-time environment through which programs may communicate with the host.

iserver is designed to be easily portable to all host computer types to which transputer boards may be attached.

The host file server **iserver** provides two functions:

- Loading programs and controlling transputer networks
- Runtime access to host services for application programs.

At the application program level, all communications with the host file server are through the libraries in the **hostio** group. These are described in chapter 14 and the underlying protocol is described in section 16.5.

16.4.1   **iserver** command line syntax

The syntax for the command line of **iserver** is as follows:

      iserver *[command.line] [program.parameter]*

Where *command.line* is defined as follows:

command.line	=	*option*
	|	*program.parameter*
	|	*option command.line*
	|	*program.parameter*
	|	*program.parameter.command.line*
program.parameter	=	any argument that is not an option
option	=	*−options*
	|	*/options*
options	=	**SA** Analyses the root transputer and peeks 8K of its memory.
	|	**SB** *filename* Boots the program contained in the named file.
	|	**SC** *filename* Copies the named file to the root transputer link.
	|	**SE** Terminates the server if the transputer error flag is set.
	|	**SI** Displays progress information as the program is loaded.
	|	**SL** *link.address* Specifies link address or device name.
	|	**SR** Resets the root transputer and subsystem on the link.
	|	**SP** *number* Changes the quantity of bytes peeked to *number* kilobytes.
	|	**SS** Serves the link, that is, provides host system support to programs communicating on the host link.
filename	=	*standard host file name*
link.address	=	*number*

program.parameter is supplied to resident programs on request by the program issuing a *command.line* command.

If **iserver** is invoked with no options, brief help information is displayed.

Loading programs

Before a program can be loaded onto a transputer network it must be compiled, linked and made bootable (see chapter 8).

The name of the file containing the program to be loaded is specified using the 'SB' option. If the file cannot be found an error is reported. This resets the board prior to loading the program. When the program has been loaded the server then provides host services to the program.

Note: Using the 'SB' option is equivalent to using the SR, SS, SI and SC options together.

To load a program onto a board without resetting the root transputer, use the 'SC' option. This should only be done if the transputer has already been reset, or has a resident program that can interpret the file. To reset the transputer subsystem use the 'SR' option.

To terminate the server immediately after loading the program use the 'SR' and 'SC' options together. This combination of options resets the transputer, loads the program onto the board, and terminates.

To load a board in analyse mode, for example when you wish to use the debugger to examine the program's state after execution, use the 'SA' option to dump the transputer's memory (starting from **MOSTNEG INT**). The data is stored in an internal buffer which is read by the TDS loader when programs that use the root transputer are to be debugged. The size of the memory read from the transputer may be changed from the default 8K by using an **SP** option with the number of kilobytes as a parameter.

Terminating the server

To terminate the server press the host system break key. When the key is pressed the following prompt is displayed:

> **(x)exit, (s)hell, or (c)ontinue?**

To terminate the server type 'x' or press $\boxed{\text{RETURN}}$.

To suspend the server and resume the program later, type 's'. On DOS-based systems this option may require a host environment variable. For further information see the Delivery Manual that accompanies the release.

To abort the interrupt and continue running the program, type 'c'.

Server termination codes

The server will return one of five status codes to the host operating system.

These codes distinguish:

    i User break

    ii Transfer error flag observed

    iii Failure to perform actions requested

    iv Termination at request of TDS (or other program on transputer)

           a with **sps.success** exit code

           b with **sps.failure** exit code

           c with other exit code

The values returned to DOS are listed in the delivery manual.

Specifying a link address – option SL

The server contains a default address or device name for communicating with boot from link boards. The address or name can be changed by specifying the '**SL**' option followed by the new value. Addresses can be given in decimal format, or in hexadecimal format by prefixing the number with '#'.

The default address (#150 on DOS machines) is overridden by the value of host environment variable **TRANSPUTER**, if this variable has been set. This variable is itself overridden by the address or name specified by the '**SL**' option.

Terminating on error – option OE

When debugging standalone programs it is useful to force the server to terminate when the subsystem's error flag is set. To do this use the '**SE**' option. This option should only be used for programs written entirely in occam and compiled in HALT system mode. If the program is not written entirely in occam then the error flag may be set even though no error has occurred.

16.4.2   Server functions

This section describes the basic set of server functions. All versions of the **iserver** will support these functions, enabling standalone programs to be used with any host computer on which **iserver** is available.

These functions are not intended for direct use by application programmers who should use **hostio** library procedures. They are briefly described here for those who wish to implement a server on a new host, or to add new facilities to the existing server.

The functions are divided into three groups:

    1 File system commands

    2 Host environment commands

    3 Server control commands

Commands in each group are summarised below. Formal definitions can be found in section 16.5.

File system commands

Command	Description
Fopen	Opens a file, and returns a stream identifier.
Fclose	Closes a file.
Fread	Reads a data block, in bytes.
Fwrite	Writes a data block, in bytes.
Fgets	Reads a line from an open stream.
Fputs	Writes a line to an open stream.
Fflush	Flushes an open stream to the destination device.
Fseek	Resets the file position.
Ftell	Returns the current file position.
Feof	Tests for end-of-file.
Ferror	Returns error status of a given stream.
Remove	Deletes a file.
Rename	Renames a file.

Host environment commands

Command	Description
Getkey	Reads a character from the keyboard.
Pollkey	Polls the keyboard.
Getenv	Retrieves a host environment variable.
Time	Returns local and universal time.
System	Runs a command on the host system.

Server control commands

Command	Description
Exit	Terminates the server.
CommandLine	Retrieves the server invocation command line.
Core	Retrieves the contents of a peeked transputer's memory.
Version	Retrieves revision data about the server.

16.4.3   `iserver` error messages

A list of error messages which **iserver** may produce follows. In many cases, the messages listed may be followed by an extra message which gives additional information. This information is host specific.

Aborted by user

> The user interrupted the server, by pressing Ctrl C or Ctrl Break.

Bad link specification

> The link name specified is not valid.

Boot filename is too long, maximum size is *number* **characters**

> The specified filename was too long. *number* is the maximum size for filenames.

Cannot find boot file *filename*

> The server cannot open the specified file.

Command line too long (at *string*)

> The maximum permissible command line length has been exceeded. The overflow occurred at *string*.

Copy filename is too long, maximum size is *number* **characters**

> The specified filename was too long. *number* is the maximum size for filenames.

Error flag raised by transputer

> The program has set the error flag. Debug the program.

Expected a filename after -SB option

> The 'SB' option requires the name of a file to load.

Expected a filename after -SC option

> The 'SC' option requires the name of a file to load.

Expected a name after -SL option

> The 'SL' option requires a link name or address.

Expected a number after -SP option

> The '**SP**' option must specify the number of Kbytes to peek.

Failed to allocate CoreDump buffer

> The '**SP**' option was used and the server was unable to allocate enough memory to allow the transputer's memory to be copied.

Failed to analyse root transputer

> The link driver could not analyse the transputer.

Failed to reset root transputer

> The link driver could not reset the transputer.

Link name is too long, maximum size is *number* **characters**

> The specified name was too long. *number* is the maximum length.

Protocol error, *message*

> Incorrect protocol on the link. This can happen if there is a hardware fault, or if an incorrect version of the server is used.

> *message* can be any of the following:
>
> > **got** *number* **bytes at start of a transaction**
> > **packet size is too large**
> > **read nonsense from the link**
> > **timed out getting a further** *dataname*
> > **timed out sending reply message**

> For more information about server protocols see section16.5.

Reset and analyse are incompatible

> Reset and analyse options cannot be used together.

Timed out peeking word *number*

> The server was unable to analyse the transputer.

Transputer error flag has been set

> The program has set the error flag. Debug the program.

Unable to access a transputer

> The server was unable to gain access to a link. This occurs when the link address or device name, specified either with the SL option or the **TRANSPUTER** environment variable, is incorrect.

Unable to free transputer link

> The server was unable to free the link resource because of a host error. The reason for the error will be host dependent.

Unable to get request from link

> The server failed to get a packet from the transputer because of some general failure.

Unable to write byte *number* **to the boot link**

> The transputer did not accept the file for loading. This can occur if the transputer was not reset or because the file was corrupted or in incorrect format.

16.5   The INMOS file server – `iserver` – program interface

The host file server **iserver** is implemented in C using ANSI standard run-time libraries to facilitate porting to other machines. This provides an easy method of porting hosted transputer software.

The source of the server and of the libraries used to communicate with the server is supplied with the toolset.

16.5.1   The server protocol

Every communication to and from the server is a packet consisting of a counted array of bytes. The count gives the length of the message and is sent in the first two bytes of the packet as a signed 16 bit number. The structure of a server packet is illustrated in figure 16.1.

Figure 16.1 **SP** protocol packet

This protocol has been given the name **SP**, and is defined in occam as follows:

```
PROTOCOL SP IS INT16::[]BYTE :
```

Packet size

There is a maximum packet size of 512 bytes and a minimum packet size of 8 bytes in the to-server direction (i.e. a minimum message length of 6 bytes). The server may take advantage of this knowledge.

The packet size must always be an even number of bytes. If the number of bytes is odd a dummy byte is added to the end of the packet and the packet byte count rounded up by one.

The hostio library contains routines that ensure that the size restrictions are met when sending a packet to the server (see section 16.5.2).

Protocol operation

Every request sent to the server receives a reply of the same protocol, in strict sequence, and no further requests are accepted until the reply has been sent.

All integer types used by the protocol are signed and are little endian. Numbers are transmitted as sequences of bytes (2 bytes for 16 bit numbers, 4 bytes for 32 bit numbers) with the least significant byte first. Negative integers are represented in 2s complement. Strings and other variable length blocks are introduced by a 16 bit signed count.

All server calls return a result byte as the first item in the return packet. If the operation succeeds the result byte is zero and if the operation fails the result byte is non-zero. The result is one (1) in the special case where the operation fails because the function is not implemented[1]. If the result is non-zero, some or all of the return values may not be present, resulting in a smaller return packet than if the call was successful.

16.5.2   The server libraries

The libraries **splib**, **solib**, **sklib** and **spinterf** contain all the routines provided in INMOS toolset products for communicating with the server. They also contain a set of basic routines, hidden from the user, from which the more complex user visible routines are built.

---

[1] Result values between 2 and 127 are defined to have particular meanings by occam server libraries. Result values of 128 or above are specific to the implementation of a server.

A naming convention has been adopted for the server libraries. The basic library routines use the server protocol directly and map directly to server functions. These have the prefix '**sp.**'. Routines which use the basic routines and are visible to the user have the prefix '**so.**'. The '**so.**' routines documented in this manual use underlying '**sp.**' routines, and in some cases the mapping is one to one.

The source of the hostio library is provided and serves as an example of how to use the **SP** protocol.

If you add your own libraries for server functions you are recommended to keep to the naming convention.

There are two '**sp.**' library routines included in **splib** to help you extend the set of available routines. These are **sp.send.packet** and **sp.receive.packet**. These are described below.

sp.send.packet

```
PROC sp.send.packet (CHAN OF SP ts,
                     VAL []BYTE packet,
                     BOOL error)
```

This procedure sends a packet on the channel **ts**, provided that it meets the requirements for a **SP** protocol packet. If the requirements are not met then the packet is not sent and **error** is set to **FALSE**.

sp.receive.packet

```
PROC sp.receive.packet (CHAN OF SP fs,
                        INT16 length,
                        []BYTE packet,
                        BOOL error)
```

This procedure receives a packet on the channel **fs**. The value **error** is set to **FALSE** if the packet exceeds the maximum packet size.

16.5.3   Porting the server

In order to port the **iserver** to a new machine you must have a C compiler for that machine with ANSI standard libraries.

The hostio library expects all the functions described below to be provided by **iserver**.

16.5.4   Defined protocol

The functions provided by the **iserver** are split into three groups:

>   1 File commands, for interacting with files

>   2 Host commands, for interacting with the host

>   3 Server commands, for interacting with the server.

In the descriptions that follow, the arguments and results of server calls are listed in the order that they appear in the data part of the packet. The size of a packet is the aggregated size of all the items in the packet, rounded up to an even number of bytes. occam types are used to define data items within the packet.

Input to the server consists of a named tag (see appendix D.3) to define the function required followed by data as given in the individual protocol definitions.

Reserved values

INMOS reserves the following values for its own use:

- Function tags in the range 0 to 127 inclusive.
- Result values in the range 0 to 127 inclusive.
- Stream identifiers 0, 1 and 2.

Some commands may return particular values, which may be reserved. The range of reserved values is given with each command as appropriate. The actual values of tags, etc. are held in the constant library **sphdr**, see appendix D.3

File commands

Open files are identified with 32 bit descriptors called *stream.id*s. There are three predefined open files:

    0 –   standard input
    1 –   standard output
    2 –   standard error

If one of these is closed then it may not be reopened.

Fopen

    Open a file.

    Protocol

    Fopen   =   *! to.Fopen ? from.Fopen*
    to.Fopen   =   **sp.open.tag** *filename type mode*
    from.Fopen =   *result stream.id*

    Description

    Open a named file and, if successful, return a stream identifier. This stream identifier will be used in all subsequent operations on the file, until it is closed.

    The number of streams that may be open at one time is host-specified, but will not be less than eight (including the three predefines).

    Parameters

    To server: **INT16::[]BYTE** *filename*
                   BYTE   *type*
                   BYTE   *mode*

    From server: **BYTE**   *result*
                   INT32   *stream.id*

type can take one of two possible values:

spt.binary   The file will contain raw binary bytes.
spt.text   The file will be stored as text records. Text files are host-specified.

mode can have 6 possible values:

spm.input	Open an existing file for input.
spm.output	Create a new file, or truncate an existing one, for output.
spm.append	Create a new file, or append to an existing one, for output.
spm.existing.update	Open an existing file for update (both reading and writing), starting at the beginning of the file.
spm.new.update	Create a new file, or truncate an existing one, for update.
spm.append.update	Create a new file, of append to an oxisting one, for update.

When a file is opened for update (one of the last three modes above) then the resulting stream may be used for input or output. There are restrictions, however. An output operation may not follow an input operation without an intervening Fseek, Ftell or Fflush operation.

The result returned can take any of the following values:

spr.ok	Open was successful.
spr.not.implemented	Operation not implemented by this version of file server.
spr.operation.failed	The open failed.
spr.bad.name	Null name supplied.
spr.bad.type	Invalid file type specified.
spr.bad.mode	Invalid open mode specified.
spr.bad.packet.size	File name was too big.

Fclose

Close a file.

Protocol

Fclose	=	*! to.Fclose ? from.Fclose*
to.Fclose	=	**sp.close.tag** *stream.id*
from.Fclose	=	*result*

Description

Fclose closes a stream *stream.id* which should be open for input or output. Fclose flushes any unwritten data and discards any unread buffered input before closing the stream.

Parameters

To server: **INT32**   *stream.id*

From server: **BYTE**   *result*

Fread

Read a block of data.

Protocol

Fread	=	*! to.Fread ? from.Fread*
to.Fread	=	**sp.read.tag** *stream.id count*
from.Fread	=	*result data*

Description

Fread reads *count* bytes of binary data from the specified stream. Input stops when the specified number of bytes are read, or the end of file is reached, or an error occurs. If *count* is less than one then no input is performed. The stream is left positioned immediately after the data read. If an error occurs the stream position is undefined.

Parameters

| To server: | `INT32` | *stream.id* |
| | `INT16` | *count* |

| From server: | `BYTE` | *result* |
| | `INT16::[]BYTE` | *data* |

result is always zero. The actual number of bytes returned may be less than requested and Feof and Ferror should be used to check for status.

Fwrite

Write a block of data.

Protocol

Fwrite	=	*! to.Fwrite ? from.Fwrite*
to.Fwrite	=	`sp.write.tag` *stream.id data*
from.Fwrite	=	*result written*

Description

Fwrite writes a given number of bytes of binary data to the specified stream, which should be open for output. If the length of *data* is less than zero then no output is performed. The position of the stream is advanced by the number of bytes actually written. If an error occurs then the resulting position is undefined.

Parameters

| To server: | `INT32` | *stream.id* |
| | `INT16::[]BYTE` | *data* |

| From server: | `BYTE` | *result* |
| | `INT16` | *written* |

Fwrite returns the number of bytes actually output in *written*. *result* is always zero. The actual number of bytes returned may be less than requested and Feof and Ferror should be used to check for status.

If *stream.id* is **1** (standard output) or **2** (standard error) then the write is automatically flushed.

Fgets

Read a line.

Protocol

Fgets	=	*! to.Fgets ? from.Fgets*
to.Fgets	=	`sp.gets.tag` *stream.id count*
from.fgets	=	*result data*

Description

Fgets reads a line from a stream which must be open for input. Characters are read until end of file is reached, a newline character is seen or the number of characters read is not less than *count*.

Parameters

To server: **INT32**   *stream.id*
                  INT16   *count*

From server: **BYTE**   *eesult*
                  INT16::[]BYTE   *data*

If the input is terminated because a newline is seen then the newline sequence is *not* included In the returned array.

If end of file is encountered and nothing has been read from the stream then Fgets fails.

Fputs

Write a line.

Protocol

Fputs   =   *! to.Fputs ? from.Fputs*
to.Fputs   =   **sp.puts.tag** *stream.id string*
from.Fputs   =   *result*

Description

Fputs writes a line of text to a stream which must be open for output. The host specified convention for newline will be appended to the line and output to the file. The maximum line length is host-specified.

Parameters

To server: **INT32**   *stream.id*
                  INT16::[]BYTE   *string*

From server: **BYTE**   *result*

Fflush

Flush a stream.

Protocol

Fflush   =   *! to.Fflush ? from.Fflush*
to.Fflush   =   **sp.flush.tag** *stream.id*
from.Fflush   =   *result*

Description

Fflush flushes the specified stream, which should be open for output. Any internally buffered data is written to the destination device. The stream remains open.

Parameters

To server: **INT32**   *stream.id*

From server: **BYTE**   *result*

Fseek

Set position in a file.

Protocol

Fseek   =   *! to.Fseek ? from.Fseek*
to.Fseek   =   **sp.seek.tag** *stream.id offset origin*
from.Fseek   =   *result*

Description

Fseek sets the file position for the specified stream. A subsequent read or write will access data at the new position.

Parameters

To server: **INT32**   *stream.id*
               INT32   *offset*
               INT32   *origin*

From server: **BYTE**   *result*

For a binary file the new position will be *offset* characters from *origin* which may take one of three values:

        spo.start, the beginning of the file

        spo.current, the current position in the file

        spo.end, the end of the file.

For a text stream, *offset* must be zero or a value returned by Ftell. If the latter is used then *origin* must be set to 1.

Ftell

Find out position in a file.

Protocol

Ftell   =   *! to.Ftell ? from.Ftell*
to.Ftell   =   **sp.tell.tag** *stream.id*
from.Ftell   =   *result position*

Description

Ftell returns the current file position for *stream.id*.

Parameters

To server: **INT32**   *stream.id*

From server: **BYTE**   *result*
               INT32   *position*

Feof

Test for end of file.

Protocol

Feof	=	_! to.Feof ? from.Feof_
to.Feof	=	**sp.eof.tag** _stream.id_
from.Feof	=	_result_

Description

Feof succeeds if the end of file indicator for _stream.id_ is set.

Parameters

To server: **INT32**   _stream.id_

From server: **BYTE**   _result_

Ferror

Get file error status.

Protocol

Ferror	=	_! to.Ferror ? from.Ferror_
to.Ferror	=	**sp.ferror.tag** _stream.id_
from.Ferror	=	_result error.no message_

Description

Ferror succeeds if the error indicator for _stream.id_ is set. If it is, Ferror returns a host-defined error number and a (possibly null) message corresponding to the last file error on the specified stream.

Parameters

To server: **INT32**   _stream.id_

From server: **BYTE**   _result_
                   INT32   _error.no_
                   INT16::[]BYTE   _message_

Remove

Delete a file.

Protocol

Remove	=	_! to.Remove ? from.Remove_
to.Remove	=	**sp.remove.tag** _name_
from.remove	=	_result_

Description

Remove deletes the named file.

Parameters

To server: **INT16::[]BYTE**   _name_

From server: **BYTE**   _result_

Rename

Rename a file

Protocol

Rename   =   *! to.Rename ? from.Rename*
to.Rename   =   **sp.rename.tag** *oldname newname*
from.Rename =   *result*

Description

Rename changes the name of an existing file *oldname* to *newname*.

Parameters

To server: **INT16::[]NAME**   *oldname*
               INT16::[]NAME   *newname*

From server: **BYTE**   *result*

16.5.5   Host commands

Getkey

Get a keystroke.

Protocol

GetKey   =   *! to.GetKey ? from.GetKey*
to.GetKey   =   **sp.getkey.tag**
from.GetKey =   *result key*

Description

GetKey gets a single character from the keyboard. The keystroke is waited on indefinitely and will not be echoed. The effect on any buffered data in the standard input stream is host-defined.

Parameters

To server:

From server: **BYTE**   *result*
              BYTE   *key*

Pollkey

Test for a key.

Protocol

PollKey   =   *! to.PollKey ? from PollKey*
to.PollKey   =   **sp.pollkey.tag**
from.PollKey =   *result key*

Description

PollKey gets a single character from the keyboard. If a keystroke is not available then PollKey returns immediately with a non-zero result. If a keystroke is available it will not be echoed. The effect on any buffered data in the standard input stream is host-defined.

Parameters

To server:

From server: **BYTE** *result*
             BYTE *key*

Getenv

Get environment variable.

Protocol

Getenv   =   *! to.Getenv ? from.Getenv*
to.Getenv   =   **sp.getenv.tag** *name*
from.Getenv   =   *result value*

Description

Getenv returns a host-defined environment string for *name*.

Parameters

To server: **INT16::[]BYTE**   *name*

From server: **BYTE**   *result*
                INT16::[]BYTE   *value*

If *name* is undefined then *result* will be non-zero.

Time

Get the time of day.

Protocol

Time   =   *! to.Time ? from.Time*
to.Time   =   **sp.time.tag**
from.time   =   *result LocalTime UTCTime*

Description

Time returns the local time and Coordinated Universal Time if it is available. Both times are expressed as the number of seconds that have elapsed since midnight on 1st January, 1970. If UTC time is unavailable then it will have a value of zero.

Parameters

To server:

From server: **BYTE**   *result*
             INT32   *LocalTime*
             INT32   *UTCTime*

System

Run a command

Protocol

System   =   *! to.System ? from.System*
to.System   =   **sp.system.tag** *command*
from.System   =   *result status*

Description

System passes the string *command* to the host command processor for execution.

Parameters

To server: `INT16::[]BYTE`   *command*

From server: `BYTE`   *result*
                `INT32`   *status*

If *command* is zero length then System will succeed if there is a command processor. If *command* is not null then *status* is the return value of the command, which is host-defined.

16.5.6   Server commands

Exit

Terminate the server.

Protocol

Exit   = *! to.Exit ? from.Exit*
to.Exit   = `sp.exit.tag` *status*
from.Exit = *result*

Description

Exit terminates the server, which exits returning *status* to its caller.

Parameters

To server: `INT32`   *status*

From server: `BYTE`   *result*

If *status* has the special value `sps.success` then the server will terminate with a host-specific 'success' result.

If *status* has the special value `sps.failure` then the server will terminate with a host-specific 'failure' result.

CommandLine

Retrieve the server command line.

Protocol

CommandLine   = *! to.CommandLine ? from.CommandLine*
to.CommandLine   = `sp.commandline.tag` *all*
from.CommandLine = *result string*

Description

CommandLine returns the command line passed to the server on invocation.

Parameters

To server: **BYTE**   *all*

From server: **BYTE**   *result*
               INT16::[]BYTE   *string*

If *all* is zero the returned string is the command line, with arguments that the server recognised at startup removed.

If *all* is non-zero then the string returned is the entire command vector as passed to the server on startup, including the name of the server command itself.

Core

Read peeked memory.

Protocol

```
Core      =   ! to.Core ? from.Core
to.Core   =   sp.core.tag offset length
from.Core =   result data
```

Description

Core returns the contents of the root transputer's memory, as peeked from the transputer when the server was invoked with the analyse option.

Parameters

To server: **INT32**   *offset*
               INT16   *length*

From server: **BYTE**   *result*
               INT16::[]BYTE   *data*

Core fails if *offset* is larger than the amount of memory peeked from the transputer or if the transputer was not analysed.

If *offset + length* is larger than the total amount of memory that was peeked then as many bytes as are available from the given offset are returned.

Version

Find out about the server.

Protocol

```
Version      =   ! to.Version ? from.Version
to.Version   =   sp.version.tag
from.Version =   result version host OS board
```

Description

Version returns four bytes containing identification information about the server and the host it is running on.

Parameters

To server:

From server: **BYTE**   *result*
                BYTE   *version*
                BYTE   *host*
                BYTE   *OS*
                BYTE   *board*

If any of the bytes has the value 0 then that information is not available.

version identifies the server version. The byte value should be divided by ten to yield the version number.

host identifies the host machine and can be any of the following:

    sph.PC
    sph.NECPC
    sph.VAX
    sph.SUN3
    sph.SUN4

OS identifies the host environment and can be any of the following:

    spo.DOS
    spo.HELIOS
    spo.VMS
    spo.SUNOS

board identifies the interface board and can be any of the following:

    spb.B004
    spb.B008
    spb.B010
    spb.B011
    spb.B014
    spb.DRX-11
    spb.QT0

Values of *host*, *OS* and *board* from 0 to 127, inclusive, are reserved for use by INMOS.

16.5.7   Extensions to `iserver` protocol supported within the TDS only

Getkey, Pollkey

Unless key cooking has been switched off by sending **tt.key.raw** on the screen channel (see section 16.6.2) the key values returned by **GetKey** in the TDS are cooked values. As these values can exceed 255 the special result value **ft.tag** (200) may be returned and this value should then be added to the key value returned.

Exit

This command is trapped by the TDS and not passed to the **iserver**. A message is sent to the screen and the TDS waits for any key to continue.

CommandLine

This is trapped by the TDS and returns the command line stored by the most recent **TDS/commandline** command.

TDS/commandline

Remember a command line.

Protocol

TDS	=	*! to.TDS ? from.TDS*
to.TDS	=	**sp.TDS.tag sp.commandline.tag** *string*
from.TDS	=	*result*

Description

Stores a string to be remembered by the TDS for subsequent **CommandLine** commands.

Parameters

To server:	**BYTE**	**sp.commandline.tag**
	INT16::[]BYTE	*string*

From server:	**BYTE**	*result*

16.6   The TDS screen and keyboard channels

The channels **keyboard** and **screen**, available within an executable procedure (**EXE**), communicate with the terminal used by the host. They are not connected directly to the devices of the terminal but to a terminal driver process called **term.p**. Various commands may be sent to **term.p** which implements a virtual screen and keyboard interface so that it is possible to write terminal-type independent code. The protocol also allows the user to drive the terminal directly and exploit features of a particular terminal which are not accessible using the simple virtual terminal interface.

The process **term.p** is actually two processes, running in parallel, one driving the screen and the other the keyboard. There are three occasions when these two processes have to communicate with each other; these are Initialisation, release and termination. These operations cause software generated values to be passed to the program on the **keyboard** channel. Apart from this the two processes are completely independent and asynchronous, (i.e. it is possible to output to **screen** in parallel with waiting for input from **keyboard** without either process being aware of the behaviour of the other). These interfaces are used by the procedures in the library **userio**.

16.6.1   Input from the keyboard channel

The **keyboard** channel returns integers. These are one of the following:

- ASCII values for simple keys

- positive values greater than ASCII values for special 'function keys' required by the folding editor

- negative values (sometimes followed by positive parameters) as responses to initialisation commands, error codes, etc.

- the special negative value **ft.terminated** generated at end of file by interface procedures producing a simulated keyboard stream from a file

The simple keys returned as ASCII values are the visible ASCII range from ' ' (32) to '~' (126), plus '*c' (13) and 'delete' (127). Programs reading from the keyboard should treat 127 as equivalent to the code **ft.del.chl**.

Special function keys result from the keyboard process recognising either multiple key strokes or terminal-specific keys that generate multiple character sequences.

The coded values for function keys are given in the library **userhdr** and tabulated in appendix D.5. They are all given names beginning **ft..** By sending the command *key.raw* to **screen** the recognition of multiple keystrokes and terminal-specific keys is disabled. This will be acknowledged by a value **ft.raw** on **keyboard**, preceded by any cooked keys that may be in a hardware or software type-ahead buffer. The character values that make up function key values are now passed direct to the user.

The recognition of function keys is re-enabled by sending *key.cooked* to **screen**. This will be acknowledged by a value **ft.cooked** on **keyboard**, preceded by any raw keys that may be in a hardware or software type-ahead buffer.

Negative values are used in responses from the software and also to convey errors in the hardware.

16.6.2   Screen stream and SS protocols

The protocol used by the **screen** output channel is an explicitly tagged protocol. A communication using such a protocol always starts with one of a limited set of constant values, coded as **BYTE**s. The rest of the communication depends on the particular value of the tag and may consist of zero or more further values of particular types. Arrays of up to 256 bytes may be communicated, if they are preceded by a count defining their size.

Note that an explicitly tagged protocol is not identical to the implicitly tagged protocols supported by the language. Hence for language purposes this protocol is coded as **CHAN OF ANY**.

The protocol of the parameter channel **screen** allows single characters (as bytes), strings (as arrays of bytes), cursor movement, absolute cursor addressing, insert and delete operations, etc., to be coded.

The channels called **sink** in all the simple output procedures use this protocol. It is defined below, but the meanings of all the tag values are not explained in detail until a later section. In this definition the values of the tags are represented by the names used in programs. These are given in the library **userhdr** and are listed in appendix D. Subsequent values are represented in the style of a variable declaration using a type and a name. Each of the lines of the table represents a command to the process at the other end of the channel. The use of the channel consists of a sequence of such commands.

The commands on the screen channel are as follows:

```
tt.reset
tt.up
tt.down
tt.left
tt.right
tt.goto; INT x; INT y
tt.ins.char; BYTE ch
tt.del.char
tt.out.byte; BYTE ch
tt.out.int; INT n
tt.out.string; INT::[]BYTE string
tt.clear.eol
tt.clear.eos
tt.ins.line
tt.del.line
tt.beep
tt.key.raw
tt.key.cooked
tt.initialise
tt.endstream
tt.set.poll; INT poll.gap
```

A corresponding occam protocol SS is declared in the library **strmhdr**. Library procedures are provided in the library **streamio** which specify screen stream channels as **CHAN OF SS**. The TDS implementation of occam allows a **CHAN OF ANY** to be passed as an actual parameter to a call of such a procedure. The protocol **SS**, listed in appendix D.4 includes some additional tags which are for internal use within the TDS only.

Screen stream protocol may be used for communicating text, in a form suitable for subsequent display, between arbitrary user processes. It is used as the input protocol in interface procedures which send text to the filing system. It is conventional to separate lines of text in screen stream protocol by the pair of characters **"*c*n"**.

In particular it allows text to be communicated as a mixture of single characters and arrays of characters communicated by block transfer. When the ultimate destination is a real terminal, screen commands for cursor movement, character and line deletion, etc., may be included. File interface procedures, however, while accepting the full protocol, are not able to perform all the control operations.

This section describes the behaviour of the TDS when the various commands available are sent to it via the channel **screen** of an **EXE**.

Any process written by the user to receive inputs in the same protocol should at least accept all possible commands, but may choose to perform modified or null actions where appropriate.

Each output to the screen channel must match one of the alternatives of the screen protocol. A command and its data is represented in this section by a command name in *italics*. For example *out.byte* represents a command that would be coded as **screen ! tt.out.byte; ch** where **ch** is a **BYTE** value.

Commands may have one or more of the following kinds of effect:

  1 text is displayed at the current cursor position,

  2 the cursor is moved relatively or absolutely to a new position,

  3 characters or lines are deleted

  4 blank characters or lines are created

  5 miscellaneous other actions.

Sending any **screen** commands that would result in moving the cursor or characters beyond the bounds of the screen has unspecified consequences. Subsequent commands also have unspecified consequences until a *goto* command or *reset* command is sent.

Some of the commands cause characters to be sent to the screen. They affect the terminal as described below.

Normal visible characters are ASCII characters with codes in the range ' ' – ' ~' (32 – 126), ' *c' (carriage return = 13) and ' *n' (new line = 10).

Other characters are sent directly to the device just as normal visible characters are, but the consequences are terminal dependent. Subsequent commands have unspecified consequences until a *reset* command is sent.

The effects of sending these characters are as follows:

  1 Characters in the range (' ') to (' ~') appear on the screen at the current position, moving the cursor one place to the right. The behaviour at the end of a line is undefined.

  2 ' *c' moves the cursor to the first character of the current line.

  3 ' *n' moves the cursor to the line beneath the current line, remaining at the same character position within the line. If the cursor position was the last line on the screen, then the contents of the screen are moved up a line, (the contents of the top line being lost) and the cursor is left on the last line, which is now blank.

Outputting characters to the screen

out.byte

The command *out.byte* contains a byte value. The ASCII character with that value is output at the current cursor position. The cursor position is moved one character position to the right.

out.string

The command *out.string* contains an integer length and then a byte array of that many bytes. ASCII characters with values of these bytes are output starting at the current cursor position. The maximum size of the array is **max.string.size** (currently 256). After each byte is output the cursor position is moved one character position to the right.

Any byte value sent to **screen** (using the *out.byte* and *out.string* commands) is sent to the screen hardware. If the byte values are normal visible characters, then **term.p** can track the cursor position and support commands such as *ins.char*, *ins.line*, etc. Values outside this range are also sent and allow the user to access more exotic features of the terminal hardware; however, **term.p** will probably not have tracked the screen state correctly and issuing commands that depend on the current cursor position after sending special control chars is not likely to have the desired effect.

Cursor movement

up, down, left, right

The commands *up*, *down*, *left* and *right* move the cursor one character position in the direction indicated.

goto

The command *goto* contains a column and row co-ordinate (x, y) and the cursor is moved to this position on the screen. The screen is addressed with the upper leftmost character position having the co-ordinates (0,0), with x going across and y going down the screen.

reset

The command *reset* causes **term.p** to reset the screen and the keyboard. This will perform any standard initialisations necessary to enable function keys for the particular type of terminal in use and will move the cursor to (0,0). It should be used after sending non-standard character codes to the screen if it is required to ensure that the user and system have the same idea as to where the cursor is.

Clearing the screen

clear.eol, clear.eos

The commands *clear.eol* and *clear.eos* clear the screen from the current cursor position to the end of the line or screen respectively. The cursor position remains unchanged.

Character operations

The commands *ins.char* and *del.char* can only be used if the terminal initialisation returns **0** after **ft.nocharops.prefix**. (see **tt.initialise**).

del.char

If supported *del.char* deletes the character at the current cursor position, shifting the characters to the right of the current position one place to the left and leaving the cursor unmoved.

ins.char

If supported *ins.char* inserts a character before the character at the current position, shifting the characters at and to the right of the current position one place to the right and moving the cursor one place to the right.

Line operations

The commands *ins.line* and *del.line* can only be used if the terminal initialisation returns 0 after `ft.noops.prefix`, (see `tt.initialise`).

del.line

If supported *del.line* deletes the contents of the current line, shifting the contents of all the lines below it up a line and making the last line on the screen blank.

ins.line

If supported *ins.line* shifts the contents of the current line and all lines below it down a line, losing the contents of the last line on the screen and making the current line blank.

In both cases the current cursor position remains unchanged.

Other operations

beep

The command *beep* makes a noise at the terminal without affecting the screen.

Initialising

Resetting the screen causes the driver to output codes to set the terminal modes to those required (e.g. setting the keypad to application mode) and sends the cursor to (0, 0). It can be used after non-standard values have been sent to the terminal to allow the virtual screen driver commands to be used again.

initialise

Initialising the the screen and keyboard is performed by sending *initialise* to **screen** and reading the initialisation information from **keyboard**. Initialising also causes the screen to be reset.

The initialisation information is returned as series of special values terminated by the value `ft.end.init`. The special values are:

```
VAL ft.lines.prefix     IS -1:
VAL ft.columns.prefix   IS -2:
VAL ft.nolineops.prefix IS -3:
VAL ft.end.init         IS -4:
VAL ft.table.error      IS -5:
VAL ft.noncom.table     IS -6:
VAL ft.nocharops.prefix IS -7:
```

The `ft.lines.prefix` and `ft.columns.prefix` values are followed by another non-negative integer that is the number of lines or columns on the screen.

The `ft.nolineops.prefix` and `ft.nocharops.prefix` value is followed by an integer that is 0 if the `term.p` process supports line insert/delete or character insert/delete respectively, otherwise it is 1.

All the above four values should be returned, along with their following values, terminated by `ft.end.init`.

The `ft.table.error` and `ft.noncom.table` values indicate that an error has occured on a table driven `term.p`, either the table cannot be read (`ft.table.error`) or the table has an invalid format (`ft.nocom.table`).

If there are characters read by the keyboard process into a typeahead buffer but not yet read by the user, they may have to be read by the user before the initialisation information is seen, as shown in the following example:

```
INT key :
SEQ
  screen ! tt.initialise
  keyboard ? key
  WHILE key >= 0
    keyboard ? key
  ... read initialisation
```

Changing the way keyboard input is processed

key.raw

The command *key.raw* causes future input from the keyboard to be passed exactly as received. Any characters already in the type-ahead buffer will still be read followed by the special value **ft.raw**. Subsequent characters will be passed raw.

key.cooked

The command *key.cooked* causes the input from the keyboard to be processed to recognise and 'bundle up' multi-character keys as single values. Any characters already in the type-ahead buffer will still be read followed by the special value **ft.cooked**. Subsequent characters will be passed cooked.

Other commands

set.poll

The integer sent to the TDS after this command is the maximum time interval (in transputer clock ticks) after which the TDS will poll the host keyboard during the execution of an EXE or UTIL. Keyboard polling may be suspended by sending the value (-1) and resumed by sending the value (-2).

endstream

This command has no effect.

16.7   The TDS user filer interface

This section describes in detail the user filer protocol and how it may be used to provide access to the folded filing system from an occam program running in the transputer development system.

The channel arrays **from.user.filer** and **to.user.filer** which are available within an executable procedure (**EXE**) allow user processes to communicate with a process called 'the user filer'. Through communication with this process, user processes may read and write data in the fold structure of the development system.

A running program accesses the fold structure in a similar manner to the TDS utilities. Just as a utility is given a portion of the fold structure on which it can operate by placing the cursor on a fold before running the utility, a user program may also be given some data by placing the cursor on a fold containing the data before the program is run. Any of these folds may be filed and these filed folds will correspond to files in the host operating system. Utility programs READ HOST and WRITE HOST are provided to convert files between the representation used by the fold system and the usual types of text files used by the host operating system.

The structure and representation of folds and files in the TDS is described fully in appendix G.

16.7.1   User filer protocol

The protocol used by all the folded filing system access channels **from.user.filer[i]** and **to.user. filer[i]**, i=0..3 is similar in style to screen stream protocol but more complicated. Each of the lines of the list below represents a command, a question, or a unit of data transfer to the process at the other end of the channel. The tag values are given in the library **filerhdr** listed in appendix D.

```
uf.number.of.folds
uf.test.filed; INT fold.number
uf.read.fold.string; INT fold.number
uf.read.fold.attr; INT fold.number
uf.read.file.id; INT file.number
uf.write.fold.string; INT fold.number; INT::[]BYTE record
uf.make.fold.set; INT fold.number
uf.unmake.fold.set; INT fold.number
uf.create.fold; [attr.size]INT attr
uf.delete.fold; INT fold.number
uf.make.filed; INT fold.number; INT::[]BYTE file.id
uf.unfile; INT file.number
uf.attach.file; INT fold.number; INT::[]BYTE file.id
uf.derive.file; INT fold.number
uf.delete.contents; INT fold.number
uf.open.data.read; INT file.number
uf.open.fold.read; INT file.number
uf.open.text.read; INT file.number
uf.open.data.write; INT file.number
uf.open.fold.write; INT file.number
uf.open.text.write; INT file.number

fsd.record; INT::[]BYTE record
fsd.attr; [attr.size]INT attr
fsd.file.id; INT::[]BYTE file.id
fsd.result; INT status
fsd.error; INT status
fsd.number.of.folds; INT fold.count
fsd.fold; INT::[]BYTE record
fsd.filed; INT::[]BYTE record
fsd.endfold
fsd.endfiled
fsd.endstream

fsc.read
fsc.close; INT status
fsc.read.file.id
fsc.read.attr
fsc.read.enc.attr
fsc.enter.fold
fsc.exit.fold
fsc.repeat.fold
```

16.7.2   Selecting a fold for access

Once the cursor has been placed on a fold (referred to as the top-level fold or fold bundle), and a user program has been started by pressing the [RUN EXE] key, the program may then do a number of things:

1 It may open the top-level fold and read its contents.

2 If the top-level fold is empty, it may open the fold and write data into it.

3 It may concurrently read and write a number of the folds directly nested inside the top-level fold.

4 It may read the attributes and the fold header of the top-level fold or any of the folds directly nested inside it.

5 It may write new fold headers and delete the contents of the top-level fold or any of the folds directly nested inside it.

6 It may delete folds or create new folds within the top-level fold.

If the top-level fold contains a sequence of fold lines and data lines, then the folds in the bundle are numbered from 1. Data lines (including blank lines) are ignored in the numbering.

The top-level fold is referred to as number 0.

For example, consider the following fold:

```
{{{   A bundle of folds
...F First member

...F Second member
any text
...F Third member
}}}
```

If the cursor were placed on the closed fold **A bundle of folds** before running a user program, then the program could open and read the top-level fold by referring to it as number 0, or it could concurrently open and read the **member** folds by referring to them as numbers 1, 2 and 3.

A program may access either the top-level fold or the folds directly contained within it; these two modes of access cannot be mixed. The user filer will return an error if the top-level fold is accessed while the inner folds are being accessed or vice versa.

16.7.3   User filer channels

A program may perform a number of independent sequences of communication with the user filer, possibly in parallel. Each sequence uses a pair of channels which must be corresponding elements (that is, the elements with the same subscript) from the arrays **from.user.filer** and **to.user.filer**. As their names indicate the **to.user.filer** channel is used for communications from the program to the environment, and the **from.user.filer** channel is used for communications in the other direction.

16.7.4   User filer modes

User filer channel pairs are used for two purposes:

1 They are used to communicate questions or commands to the user filer, and to receive answers or results corresponding to these questions or commands.

2 They are used to communicate a stream of data associated with reading or writing a file. A stream of data is communicated as a sequence of data items, with an acknowledgment on the other channel following each item.

When a user program is started by the TDS, the user filer is started in parallel with it. All the channel pairs are initially in *user filer command* mode. This means that valid communications consist of questions or commands to the user filer, followed immediately by the corresponding answer or result.

Once a successful open command has been issued on a particular pair of channels, that channel pair is then used to read or write a stream of data. The mode of the channel pair is then *file stream input* or *file stream output* according to the kind of open command which was used. Once a close operation has occurred on the stream, then the channel pair is once again available for commands.

There are two variants of the file stream modes:

   1 data stream modes,

   2 folded file stream modes.

In data stream modes only text lines within the fold structure are visible. In folded file stream modes the internal structure including embedded text and non-text folds is fully visible.

In the present implementation the user filer channel arrays each have 4 elements, numbered from 0 to 3. This is reflected by the system constant **max.files**. There are therefore four channel pairs, allowing up to four files to be open at the same time. The pairs may be used in parallel with each other or sequentially.

In user filer command mode the channel pairs are all connected to the one command-handling process in the TDS. This process only services one channel pair at a time; the process **ALT**s on all the **to.user.filer** channels until a command is received on one of them. The process then reads all the parameters associated with the command received, performs the required action and sends back the results on the corresponding **from.user.filer** channel. Only after these have been sent does the user filer return to the **ALT** and service commands on other **to.user.filer** channels.

However, once a file has been opened and the channel pair is being used to read or write a file stream, then the channel pair is in the appropriate file stream mode and is connected to a file streamer process started by the TDS for this purpose. The command handler and the file stream process can then proceed in parallel. Other channel pairs may be used to issue commands to the user filer, and become connected to further file stream processes, in parallel with and unaffected by the communications between the first channel pair and their file stream process.

16.7.5   Commands in user filer command mode

Commands for the following operations may be used in user filer command mode. All except the open commands leave the channel pair in user filer command mode, and so they may be issued in any order. Some commands require the relevant fold to be filed; in the case that it is not, an error result will be returned.

Count folds	Find out the number of folds in the bundle.
Read fold string	Read the fold header of any numbered fold.
Read fold attributes	Read the attributes of any numbered fold.
Test if filed	Test any numbered fold to see if it is filed.
Read file identifier	Read the file identifier of any numbered fold that is filed.
Write fold string	Write the fold header of any numbered fold.
Create fold	Create an empty fold in the bundle.
Delete fold	Delete an empty fold.
Delete contents	Delete the contents of a fold.
Make filed	Make a fold into a filed fold using a user supplied identifier.
Unfile	Make a filed fold into an unfiled fold.
Attach file	Make an empty fold into a filed fold containing a copy of a file already existing in the filing system. (Not available in unnamed filestore implementations.)
Derive file	Make a fold into a filed fold using the identifier of the first file in the bundle.
Make set valid	Change an invalid fold set to valid.
Make set invalid	Change a valid fold set to invalid.
Open to read	Open any numbered filed fold for reading.
Open to write	Open any empty numbered filed fold for writing.

Any numbered fold which is opened for reading or writing must be filed before opening (if it is not already filed).

Definitions of uf. commands

Questions and commands in user filer command mode are tagged by byte values conventionally associated with names beginning **uf.**. In the following definitions the pair of communications is first displayed as an occam fragment showing the error-free behaviour of the user filer. Procedures incorporating these fragments, and allowing for errors, are provided with the software.

In cases where possible errors are described the **tag** returned will have the value **fsd.error** and will be followed by an integer error status number. Filing system or hardware errors are possible for all commands which may involve disk hardware access. These numbers are all listed in appendix E.

The following variables are assumed to have been declared:

```
-- for commands
INT fold.number:--fold number within the bundle
INT file.number:--fold number within bundle
                --(must be filed)
INT len         :--length of a record or string

-- for answers
INT fold.count :   -- count of folds
INT status     :   -- error number
INT number     :   -- other number

-- for both
VAL attr.size IS 3:
[attr.size]INT attr:--array of fold attributes
VAL max.record.size IS 512:
[max.record.size]BYTE record:  -- data array
VAL max.string.size IS 256:
[max.string.size]BYTE file.id: -- file name
```

number.of.folds

```
-- question
to.uf   ! uf.number.of.folds

-- reply
from.uf ? tag        -- tag = fsd.number.of.folds
from.uf ? fold.count
```

The user filer responds to the question *number.of.folds* with a count of the number of folds within the bundle at the cursor position. If the command is issued when the cursor is not on such a bundle, a count of −1 will be returned.

read.fold.string

```
to.uf   ! uf.read.fold.string; fold.number
from.uf ? tag        -- tag = fsd.record
from.uf ? len:: record
```

The user filer responds to the question *read.fold.string* with the text of the comment on the top crease line of the fold indicated. An error will be signalled if the command is issued when the cursor is not on a fold or bundle of folds or the indicated fold does not exist.

This command, used to read the comment text on a root fold, must be distinguished from the command to read the crease comment of embedded folds, which can only be given when the channel pair is in folded stream input mode.

read.fold.attr

```
to.uf    ! uf.read.fold.attr; fold.number
from.uf ? tag         -- tag = fsd.attr
from.uf ? attr
```

The user filer responds to the question *read.fold.attr* with the array of attributes of the fold indicated. An error will be signalled if the command is issued when the cursor is not on a fold or bundle of folds or the indicated fold does not exist.

This command, used to read the attributes of a root fold, must be distinguished from the command to read the attributes of embedded folds, which can only be given when the channel pair is in folded stream input mode.

test.filed

```
to.uf    ! uf.test.filed; file.number
from.uf ? tag         -- tag = fsd.result
from.uf ? status      -- status = fi.ok or fi.not.filed
```

The user filer responds to the question *test.filed* with a result showing whether the indicated fold is filed or not. An error will be signalled if the command is issued when the cursor is not on a fold or bundle of folds or the indicated fold does not exist.

read.file.id

```
to.uf    ! uf.read.file.id; file.number
from.uf ? tag         -- tag = fsd.file.id
from.uf ? len:: file.id
```

The user filer responds to the question *read.file.id* with the name of the file corresponding to the fold indicated. On a system which does not support named files a zero length will be returned. An error will be signalled if the command is issued when the cursor is not on a filed fold or bundle of folds or the indicated fold does not exist.

This command, used to read the file id of a root fold, must be distinguished from the command to read the file id of embedded filed folds, which can only be given when the channel pair is in folded stream input mode.

write.fold.string

```
to.uf    ! uf.write.fold.string; fold.number;
              len:: record
from.uf ? tag         -- tag = fsd.result
from.uf ? status      -- status = fi.ok
```

The command *write.fold.string* causes the user filer to replace the fold comment on the indicated fold by the string given. An error will be signalled if the command is issued when the cursor is not on a fold or bundle of folds or the indicated fold does not exist.

create.fold

```
to.uf    ! uf.create.fold; attr
from.uf ? tag         -- tag = fsd.number.of.folds
from.uf ? fold.count
```

The command *create.fold* causes the user filer to create a new fold at the end of the bundle.
The **fold.count** returned includes the new fold and is therefore the number of the new fold. An error will be signalled if the command is issued when the cursor is not on a bundle of folds.

delete.fold

```
to.uf    ! uf.delete.fold; fold.number
from.uf ? tag       -- tag = fsd.result
from.uf ? status    -- status = fi.ok
```

The command *delete.fold* causes the user filer to delete the indicated fold in the bundle, which must be empty. The numbers used to access all subsequent folds are thereby decreased by 1. An error will be signalled if the command is issued when the cursor is not on a bundle of folds, or the indicated fold is not empty or does not exist.

delete.contents

```
to.uf    ! uf.delete.contents; fold.number
from.uf ? tag        -- tag = fsd.result
from.uf ? status     -- status = fi.ok
```

The command *delete.contents* causes the user filer to delete the contents of the indicated fold in the bundle. An error will be signalled if the command is issued when the cursor is not on a bundle of folds, or the indicated fold does not exist.

WARNING! This operation could, if used without care, cause loss of significant quantities of data. Programs including this command should provide adequate protection against being run with the cursor in an arbitrary position.

make.filed

```
to.uf    ! uf.make.filed; fold.number;
           len:: file.id
from.uf ? tag        -- tag = fsd.result
from.uf ? status     -- status = fi.ok
```

The command *make.filed* causes the user filer to file an unfiled fold. The `file.id` provided should be valid in the particular host system being used. For maximum portability it should consist of no more than 6 alphanumeric characters and should not include a filename extension which may be generated by the software from the fold attributes. An error will be signalled if the command is issued when the cursor is not on a bundle of folds, or the indicated fold does not exist. The server may create a random filename if an empty string is provided.

unfile

```
to.uf    ! uf.unfile; fold.number
from.uf ? tag        -- tag = fsd.result
from.uf ? status     -- status = fi.ok
```

The command *unfile* causes the user filer to unfile a filed fold. An error will be signalled if the command is issued when the cursor is not on a bundle of folds, or the indicated fold does not exist. An error will also be signalled if there is insufficient room in the system's fold manager buffer to read the contents of the file.

attach.file

```
to.uf    ! uf.attach.file; fold.number;
           len:: file.id
from.uf ? tag        -- tag = fsd.result
from.uf ? status     -- status = fi.ok
```

The command *attach.file* causes the user filer to create a filed fold from an existing empty fold by causing it to point to a copy of an existing file with the name `file.id`. The copy will have a name derived by a simple disambiguating algorithm from that of the previously existing file. An error will be signalled if the command is issued when the cursor is not on a bundle of folds, or the indicated fold or file does not exist.

derive.file

```
to.uf    ! uf.derive.file; fold.number;
from.uf ? tag        -- tag = fsd.result
from.uf ? status     -- status = fi.ok
```

The command *derive.file* causes the user filer to file an unfiled fold, giving it a name derived from the name of the first fold in the bundle and the attributes of the fold indicated. An error will be signalled if the command is issued when the cursor is not on a bundle of folds, the indicated fold does not exist, or the first fold in the bundle is not filed.

make.fold.set

```
to.uf    ! uf.make.fold.set; fold.number
from.uf ? tag        -- tag = fsd.result
from.uf ? status     -- status = fi.ok
```

The command *make.fold.set* is intended for use by compilers which need to keep control of the integrity of folds. It causes the user filer to change the `fold.type` attribute of the fold indicated to `ft.foldset`, implying that it contains corresponding source text and compiled code. An error will be signalled if the command is issued when the cursor is not on a bundle of folds, or the indicated fold does not exist or does not have appropriate attributes.

unmake.fold.set

```
to.uf    ! uf.unmake.fold.set; fold.number
from.uf ? tag        -- tag = fsd.result
from.uf ? status     -- status = fi.ok
```

The command *unmake.fold.set* is intended for use by compilers which need to keep control of the integrity of folds. It causes the user filer to change the `fold.type` attribute of the fold indicated to `ft.voidset`, implying that it requires recompilation. An error will be signalled if the command is issued when the cursor is not on a bundle of folds, or the indicated fold does not exist or does not have appropriate attributes.

Example showing use of a `uf.` command

```
PROC read.fold.attr(CHAN OF ANY from.uf, to.uf,
                    VAL INT seq.no,
                    [attr.size]INT attr,
                    INT result)
  BYTE    tag :
  SEQ
    IF
      result = fi.ok
        SEQ
          to.uf ! uf.read.fold.attr; seq.no
          from.uf ? tag
          IF
            tag = fsd.error
              from.uf ? result
            tag = fsd.attr
              from.uf ? attr
      TRUE
        SKIP
:
```

Procedures in this style which test the value of `result` on entry and only perform the operation if `result = fi.ok` may be written for all the user filer control commands. They have the advantage that a sequence of different commands may be programmed without the need to test the value of the `result` parameter after each call. A collection of such procedures is supplied in the library `ufiler` and user level procedures calling these are supplied in the library `userio`.

Opening a fold for reading

A numbered fold, if it is already filed, may be opened for reading. If fold 0, the top-level fold, is opened either for reading or for writing then no other fold may be opened until it has been closed. Fold 0 may not be opened if any other fold is open.

Before being opened a fold must be filed, and so in subsequent discussion the terms fold and file are used interchangably. A file may be opened either as a data stream or as a folded stream.

When a file is opened as a folded stream, all the information in the fold is sent, including where folds begin and end, and the header and attributes of each fold.

When a file is opened as a data stream the user filer outputs a sequence of data records which are the data stored in the file. The contents of internal text folds are sent, but the information associated with the internal fold itself (the fold attributes and the header) is not.

From the point of view of the protocol which must be used for channel communications, data stream operations are a subset of folded stream operations, but they do differ in their handling of indentation (implicit leading spaces in text within folds).

open.fold.read, open.data.read

The **uf.** commands for opening a fold for reading are:

```
to.uf    ! uf.open.fold.read; fold.number
from.uf ? tag        -- tag = fsd.result
from.uf ? status     -- status = fi.ok

to.uf    ! uf.open.data.read; fold.number
from.uf ? tag        -- tag = fsd.result
from.uf ? status     -- status = fi.ok
```

If the status value returned is **fi.ok** then the open was successful.

The channel pair used to open the file is then in file stream input mode and must then be used to read the resulting stream. If the open fails, one of the error results listed later is returned as the status value and the channel pair is still in user filer command mode.

The operations needed to read a fold or data stream are described in the next section.

Opening a fold for writing

A numbered fold, if it is filed, may be opened for writing. If fold 0, the top-level fold, is opened then no other fold may be opened until it has been closed.

Only an empty filed fold may be opened for writing.

A file may be opened as a data stream or as a file stream. Data stream mode allows the user program to write a sequence of text or data records into the file. Folded stream mode allows the user to write an arbitrarily complex nested folded structure into the file.

open.fold.write, open.data.write

The **uf.** commands for opening a fold for writing are:

```
to.uf    ! uf.open.fold.write; fold.number
from.uf ? tag      -- tag = fsd.result
from.uf ? status   -- status = fi.ok

to.uf    ! uf.open.data.write; fold.number
from.uf ? tag      -- tag = fsd.result
from.uf ? status   -- status = fi.ok
```

If the status value is **fi.ok** then the open was successful.

The channel pair used to open the file is then in file stream output mode and must then be used to output data to the file.

If the open fails, one of the error results listed later is returned as the status integer value following the *result* and the channel pair is then still in user filer command mode.

The operations needed to write a fold or data stream are described in the next section.

16.7.6   Communications in file stream modes

Introduction to file stream modes

The way the channel pair is used in file stream modes is strictly symmetrical. One channel of the pair sends requests from a receiver process to a sender process, the other returns data, results, or errors from the sender process to the receiver. This symmetry has been provided to allow user processes to use the same protocol when transferring folded data streams amongst themselves as they do when communicating with the user filer in the TDS.

Communications in file stream input mode are between a system sender process in the user filer and the user program acting as a receiver.

Communications in file stream output mode are between the user program acting as a sender and a system receiver process in the user filer.

The sequence of communications in file stream modes is strictly determined by a sequential pass (possibly with skips and/or repeats of parts of the structure) through a properly nested fold structure by the sender process.

In folded file stream modes, this sequential pass includes the option to enter, or not to enter any embedded folds, and to provide additional information before entering folds. These options are exercised by the receiver process.

The valid communications by a sender process depend on the mode of opening and the current position in the folded data structure. A receiver process must base its actions on the tags and data it receives from the sender.

It is important to note that all the facilities provided in the protocol for file stream communication are not necessarily applicable in all programs. In particular it is important for user programs to know what the system sender and receiver processes do when they are in a state where the protocol allows options. This is described in detail below.

Syntax of valid sequences of communications

In order to define the permitted sequences of operations in a syntactic notation it is necessary to define the tagged commands and their data as 'terminal symbols' for the syntax. Tags output by a sending process all have names beginning **fsd.**, those output by a receiving process all have names beginning **fsc.**. The values of these tags are defined in the library **filerhdr** and are listed in appendix D.

In subsequent discussion in this chapter one of these words in *italics* always means the communication of the appropriate tag followed, if necessary, by its data in the form indicated.

Communications from the sender to the receiver:

record	=	`fsd.record;`
		`INT::[]BYTE record`
number	=	`fsd.number;`
		`INT value`
attr	=	`fsd.attr;`
		`[attr.size]INT attr`
file.id	=	`fsd.file.id;`
		`INT::[]BYTE file.id`
result	=	`fsd.result;`
		`INT status`
error	=	`fsd.error;`
		`INT status`
fold	=	`fsd.fold;`
		`INT::[]BYTE record`
filed	=	`fsd.filed;`
		`INT::[]BYTE record`
endfold	=	`fsd.endfold`
endfiled	=	`fsd.endfiled`
endstream	=	`fsd.endstream`

Communications from the receiver to the sender:

read	=	`fsc.read`
close	=	`fsc.close;`
		`INT status`
read.file.id	=	`fsc.read.file.id`
read.attr	=	`fsc.read.attr`
read.enc.attr	=	`fsc.read.enc.attr`
enter.fold	=	`fsc.enter.fold`
exit.fold	=	`fsc.exit.fold`
repeat.fold	=	`fsc.repeat.fold`

The syntax defining permitted sequences of communications is displayed in two columns representing the sender and receiver respectively. Time advances downwards and the syntactic metasymbols { , } and | have their normal meanings. Ordinary parentheses are used for bracketting.

Data stream modes

Data stream syntax

```
      Sender                    ....  Receiver

                                      read
{    (record| number)
                                      read            }

     (record| number| endstream)
                                      close
     result
```

This syntax represents a sequence of data transfers from the sender to the receiver. Each transfer is of a *record* item (an array of up to 512 bytes) or of a *number* item (a single non-negative integer) and is sent as a response to a *read* from the receiver. At any time the receiver may terminate the stream transfer by sending a *close* instead of a *read*. If the sender has no more data to send it will send an *endstream* in response to each subsequent *read*. Note that a *close* includes a status value which should normally be `fi.ok`.

At any time (not shown in the syntax) the sender may send an *error* instead of a data item or *endstream*. If an *error* is sent the receiver may then send another request.

The syntax shows the permissible temporal sequences of communications using a user filer channel pair. From the point of view of a sender process the left hand column defines outputs and the right hand column inputs. From the point of view of a receiver process the left hand column defines inputs and the right hand column outputs.

The system sender process, which communicates with a user program as receiver, using a user filer channel pair in input file stream mode, produces a stream of records from the fold specified in the *open.data.read* operation which defined the stream. Any embedded told creases are ignored and the records within folds are communicated in sequence.

In a text fold a record corresponds to a text line as seen by the editor. Each record is preceded by a number of ASCII space characters corresponding to the cumulated sum of the indent attributes of the folds entered within the stream. The system sender sends *endstream* at the bottom of the fold structure. It may send an *error* at any time if a hardware or low-level software problem arises. The user may send a *close* at any time, and must then read the corresponding *result*.

The system receiver process, which communicates with a user program as sender, using a user filer channel pair in file stream output mode, receives *record* or *number* items from the user program and inserts them sequentially into an initially empty file. After opening and after each item it will normally return a *read* but may return a *close* (including an error number) if there are hardware or low-level software problems inhibiting progress. Normal termination is by the user program sending an *endstream* to the system receiver. This will be acknowledged by a *close*, after which the user program must send the final *result* (in which the status value should be **fi.ok**).

After a stream has been transferred in either direction and a close sequence has been completed the user filer channel pair returns to user filer command mode, and is available for reuse.

Examples of the user filer being used in data stream mode may be found in the implementation of the interface procedures **keystream.from.file** and **scrstream.to.file** (described in section 14.5.13, interface procedure library **interf**), which are provided as occam source with the software.

Folded stream modes

The structure and representation of folded data is described in full in appendix G. For the purpose of the present description it is only necessary to think in terms of folds as displayed on the terminal screen by the TDS editor.

A fold is a structure consisting of a sequence of items. An item may be a data item or a fold item, where a fold item in turn consists of a top crease item, a sequence of items and a bottom crease item. A data item is either a record item or a number item. A fold item (or top crease item) has associated with it a record item which is the text displayed on the fold line by the editor, and an array of three attributes, defining certain properties of the contents of the fold.

Some values of attributes define folds whose contents are not suitable for display on the screen by the editor. This is not a property that concerns access to folds across the user filer interface. The principal constraint imposed by the implementation of folded files is a maximum size for records stored in the folds. This is 512 bytes. Each byte may contain arbitrary data and is not restricted to displayable ASCII characters.

As a fold structure is traversed there is always an item which is deemed to be the 'current item'. Immediately after opening a fold stream the current item is undefined. Thereafter the current item is that item which was most recently transmitted from the sending process to the receiving process. The fold most closely enclosing the current item is called 'the current enclosing fold'. The identities of the current fold and the current enclosing fold constitute the state of the sending process.

Folded stream protocol makes the folds and their attributes visible to the user program and gives the programmer the ability to control his navigation of an existing fold structure by deciding at each top crease whether or not to enter the fold. It is also possible to abandon the sequential traverse of the current enclosing fold or to return to its first item for a repeated traverse. The table below defines the general form of the syntax of a valid sequence of communications between two processes using folded stream protocol.

Note that the alternative read commands allowed at certain points imply that full implementations must allow the receiver process to send any one of these. According to the way the sender is creating its stream it is not always possible to perform all of the possible operations that may be requested.

It may sometimes be desirable for user programs communicating with a system receiver process to take advantage of knowledge of the particular options which will be taken at various points.

Specialised syntaxes for communicating with the system sender and system receiver are given later.

Folded stream syntax

To save space in the tabulation we define:

read.command	=	*read*	*enter.fold*	*exit.fold*	*repeat.fold*
data.item	=	*record*	*number*
top.crease	=	*fold*	*filed*
bottom.crease	=	*endfold*	*endfiled*
item	=	*data.item*	*top.crease*	*bottom.crease*

Sender	. . . .	**Receiver**
		read
{ *item*		
{		(*read.attr*| *read.enc.attr*) }
attr		
{		*read.file.id* }
file.id		
		read.command }
(*item*| *endstream*)		
		close
result		

As in the case of data stream mode communications (which are a strict subset of these) the sender may at any time send an *error* as an alternative to what the syntax shows. If an *error* is sent the receiver may then send another request.

fsc commands

The **fsc.** commands used by a receiver process are defined as follows:

close

```
to.sender    ! fsc.close; status
from.sender  ? tag; status
                  -- tag = fsd.result or fsd.error
```

The *close* command requests the sender to stop sending data and to terminate. Before doing so the sender should (and the system sender will) return a *result* or an *error*.

read

```
to.sender    ! fsc.read
from.sender ? tag --tag = fsd.record,fsd.number,
              --          fsd.fold,  fsd.filed,
              -- fsd.endfold, fsd.endfiled,
              -- fsd.endstream or fsd.error
-- act according to tag value
-- (in data stream mode only the first and last
--  three are possible, in folded stream mode
--  all are)
IF
  tag = fsd.record                 -- data record
    from.sender ? len:: record
  tag = fsd.number                 -- number item
    from.sender ? number
  (tag = fsd.fold) OR (tag = fsd.filed)
                                   -- crease comment
    from.sender ? len:: record
  (tag = fsd.endfold) OR (tag = fsd.endfiled) OR
  (tag = fsd.endstream)
    SKIP
  tag = fsd.error
    from.sender ? status
```

The *read* command requests the sender to return the next item. This is the item immediately following the current item in the fold stream. When the current item is a fold or filed item, the next item is the item after the fold, not the first item within it.

If a *read* command is issued when the current item is an *endfold* or an *endfiled* the sender should (and the system sender will) return that item again. The receiver should use *exit.fold* in this situation.

The sender should (and in the absence of errors or a premature *close* the system sender will) ensure that the sequence of items represents a properly nested fold structure.

The system receiver will accept either an *endfold* or an *endfiled* at the bottom of any fold, and so a user program when sending to it does not need to keep track of whether or not its folds are filed.

enter.fold

```
to.sender    ! fsc.enter.fold
from.sender ? tag   -- tag = fsd.record,
              --          fsd.number,
              --          fsd.fold, fsd.filed,
              -- fsd.endfold, fsd.endfiled,
              -- fsd.endstream or fsd.error
...   act according to tag value
```

The *enter.fold* command should only be used in folded stream input mode when the current item is a *fold* or a *filed*. It requests the sender to return the first item within the fold, which becomes the current item.

The system receiver will always send an *enter.fold* after receiving a *fold* or a *filed* from a user program and requesting and receiving the attributes (and possibly the file name) of the fold (see *read.attr* and *read.file.id*).

exit.fold

```
to.sender    ! fsc.exit.fold
from.sender ? tag  -- tag = fsd.record,
                   --         fsd.number,
                   --         fsd.fold, fsd.filed,
                   -- fsd.endfold, fsd.endfiled,
                   -- fsd.endstream or fsd.error
...  act according to tag value
```

The *exit.fold* command, applicable in folded stream input mode only, requests the sender to cease sending the items of the current enclosing fold and to return the first item after this fold, which becomes the current item.

The system sender will accept *exit.fold* commands at any time, thereby allowing a user to skip the remaining items in any fold.

The system receiver will only send an *exit.fold* after it has received an *endfold* or an *endfiled*.

repeat.fold

```
to.sender    ! fsc.repeat.fold
from.sender ? tag  -- tag = fsd.record,
                   --         fsd.number,
                   --         fsd.fold, fsd.filed,
                   -- fsd.endfold, fsd.endfiled,
                   -- fsd.endstream or fsd.error
...  act according to tag value
```

The *repeat.fold* command, applicable only in folded stream input mode, requests the sender to cease sending the items following the current item and to return again the first item within the current enclosing fold, which becomes the current item.

The system sender will accept *repeat.fold* commands at any time, thereby allowing a user to repeat the reading of any fold.

The system receiver will never send a *repeat.fold*.

read.attr

```
to.sender    ! fsc.read.attr
from.sender ? tag  -- tag = fsd.attr or fsd.error
from.sender ? attr -- assuming tag = fsd.attr
```

The *read.attr* command, applicable only in folded stream input mode, should only be used when the current item is a *fold* or a *filed*. The sender should (and the system sender will) respond by returning an array of attributes for the fold which is the current item. The current item does not change.

The system receiver will always send a *read.attr* after receiving a *fold* or a *filed*, and before sending an *enter.fold*.

read.enc.attr

```
to.sender    ! fsc.read.enc.attr
from.sender ? tag  -- tag = fsd.attr or fsd.error
from.sender ? attr -- assuming tag = fsd.attr
```

The *read.enc.attr* command, applicable only in folded stream input mode, requests the sender to return the attributes of the current enclosing fold.

The system sender will respond to this command independently of the nature of the current item which does not change. User programs may wish to use this command before leaving a fold to determine its relative indentation, or after doing so to reestablish knowledge about the type or contents of the enclosing fold.

The system receiver will never send *read.enc.attr.*

read.file.id

```
to.sender    ! fsc.read.file.id
from.sender  ? tag
               -- tag = fsd.file.id or fsd.error
from.sender  ? len:: file.id
                        -- assuming no error
```

The *read.file.id* command, applicable only in folded stream input mode, should only be used in an environment where named files are being used, and should only be used when the current item is a *filed*. It requests the sender to return the name of the file in which the contents of the current item, a filed fold, are stored.

The system sender in named filestore implementations will respond to this command at any time when the current item is a *filed*. This may be before or after supplying the attributes but before entering the fold.

The system receiver in named filestore implementations will always send this command after requesting the attributes of a filed fold.

Reading a fold stream from the system sender

This section summarises the application of the details of the protocol already defined to the specific task of writing a program which reads a folded file from the filing system of the transputer development system host.

Wherever possible such communications should be coded using the procedures described in the section on the i/o library **userio**.

If the folds in the file are irrelevant then the simple user procedures may be used and access to the file obtained by running the interface procedure **keystream.from.file** in parallel with the application.

If the fold structure is to be traversed sequentially with the folds having significance then the fold access procedures may be used.

Examples of both these styles are provided with the software and the user may extend them as appropriate to support additional features of the interface as necessary.

Any sequence of communications with the files of the development system must use the channels to and from the user filer provided as parameters of the executable procedure (**EXE**).

A channel pair will start in user filer command mode. Any sequence of commands meaningful in that mode may then be used. An open command may then be used to put the channel pair into data stream input mode or folded stream input mode. In this mode the channel pair connect a system sender process to a user process as a receiver. Operations in data stream modes are a subset of those in the corresponding folded stream modes.

In these modes communications must obey the bidirectional syntax presented above. That syntax is repeated below:

Data stream input mode

Sender (system)	....	Receiver (user)	
		read	
{ (*record| number*)			
		read	}
(*record| number| endstream*)			
		close	
result			

Folded stream input mode

	Sender	....	Receiver		
			read		
{	*item*				
	{		(*read.attr* | *read.enc.attr*)		
	attr			**}**	
	{		*read.file.id*		
	file.id			**}**	
			read.command		**}**
	(*item* | *endstream*)				
			close		
	result				

Where:

read.command	=	*read* | *enter.fold* | *exit.fold* | *repeat.fold*
data.item	=	*record* | *number*
top.crease	=	*fold* | *filed*
bottom.crease	=	*endfold* | *endfiled*
item	=	*data.item* | *top.crease* | *bottom.crease*

The type of each item received determines the valid commands which may be sent back. An *enter.fold* or a *read.attr* may only be sent after receiving a *fold* or a *filed*. A *read.file.id* may only be sent after receiving a *filed*.

Writing a fold stream to the system receiver

This section summarises the application of the details of the protocol already defined to the specific task of writing a program which writes a folded file into the filing system of the transputer development system host.

Wherever possible such communications should be coded using the procedures described in the section on the i/o library **userio**. If the folds in the file are irrelevant then the simple user procedures may be used and access to the file obtained by running the interface procedure **scrstream.to.file** in parallel with the application.

If it is required to generate a fold structure with nested folds then the fold access procedures may be used.

Examples of both these styles are provided and the user may extend them as appropriate to support additional features of the interface as necessary.

Any sequence of communications with the files of the development system must use the channels to and from the user filer provided as parameters of the executable procedure (**EXE**).

A channel pair will start in user filer command mode. Any sequence of commands meaningful in that mode may then be used. An open command may then be used to put the channel pair into data stream output mode or folded stream output mode. In this mode the channel pair connect a user process as sender to a system receiver process. Operations in data stream modes are a subset of those in the corresponding folded stream modes.

In these modes communications must obey the bidirectional syntax presented above. An alternative presentation of this syntax applicable when the sender is a user process and the receiver is a system receiver process whose particular behaviour is defined, is as follows:

Data stream output mode

Sender (user)	. . . .	**Receiver** (system)
		read
{ (*record* | *number*)		
		read **}**
(*record* | *number* | *endstream*)		
		close
result		

Folded stream output mode

Sender	. . . .	**Receiver**
		read
{ (*record*		
		read)
| (*number*		
		read)
| (*fold*		
		read.attr
attr		
		enter.fold)
| (*filed*		
		read.attr
attr		
		read.file.id
file.id		
		enter.fold)
| ((*endfold* | *endfiled*)		
		exit.fold) **}**
endstream		
		close
result		

The user process may send an *error* at any time which will cause the file to be terminated with no contents.

16.8   Other TDS interfaces

In earlier versions of the TDS there was an additional documented interface known as the kernel filer interface, using **tkf.** commands. Whereas this interface is still implemented in TDS3 so that old programs may run unchanged, it has been reduced to the level of an undocumented and unsupported interface as its limitation to one file open at any one time has made it unattractive to the writers of **EXE**s. Most of its facilities are available across the **iserver** channels (see section 16.5) which do not impose such a limitation.

The undocumented and unsupported fold manager interface is also implemented.

The kernel channel interface is only implemented to the extent needed by TDS tools. Any user programs which used undocumented additional features of this interface may no longer work.

The **k.get.abort.state** command is used to implement the ⌐SET ABORT FLAG⌐ feature of the TDS. It returns a result indicating whether the Control-A key has been pressed on the keyboard, since the last time the flag was tested.

Usage is:

```
INT result:
SEQ
  to.kernel ! k.get.abort.state
  from.kernel ? result
```

If result is non-zero, Control-A has been pressed, if zero it has not.

Appendices

A Keyboard layouts

A.1   IBM PC function keys

	F1	F2
Ctrl		
Shift	Browse	File/Unfile
Alt	Toolkit	Code Info
	Help	Fold Info
Ctrl	Get Macro	Save Macro
Shift	Put	
Alt	Pick Line	Copy Pick
	Move Line	Copy Line
Ctrl	Clear UTIL	Clear EXE
Shift	Autoload	Clear All
Alt	Next UTIL	Next EXE
	Get Code	Run EXE
Ctrl	◄Del Word	Del Word ►
Shift	◄─ Word	Word ─►
Alt	Delete Line	Restore Line
	◄─ Line	Line ─►
Ctrl	Define Macro	Call Macro
Shift	Top of fold	Bottom of fold
Alt	Page Up	Page Down
	Line Up	Line Down
	F9	F10

A.2   IBM PC keyboard layout

		F1	F2	F3	F4		F5	F6	F7	F8
Esc	Ctrl			Get Macro	SaveMacro	Ctrl	Clear UTIL	Clear EXE	◄ Del Word	Del Word ►
Refresh	Shift	Browse	File/Unfile	Put		Shift	Autoload	Clear All	◄ Word	Word ►
	Alt	Toolkit	Code Info	Pick Line	Copy Pick	Alt	Next UTIL	Next EXE	Delete Line	Restore Line
		Help	Fold Info	Move Line	Copy Line		Get Code	Run EXE	◄ Line	Line ►

	Alt	1	2	3	4	5	6	7	8	9
File Handler		Attach/ Detach	Copy Attach	Compact Libs	Rename File	Write Protect	Write Enable	Copy In	Copy Out	Read Host
Debugger		Inspect	Channel	Top	Retrace	Relocate	Info		Links	Monitor
Compiler		Check	Compile	Extract	Load	Recompile	Compile Info	Make Foldset	Search	Replace

	Q	W	E	R	T	Y*	U	I	O
Select Parameter						Delete Line			

	A*	S	D	F	G	H	J	K	L
	Set abort flag								

Z	X	C	V	B	N	M

∗ Ctrl + key

Note that additional keys or combinations may be defined by modifying the ITERM file (see section 16.3).

F9	F10	F11	F12
Define Macro	Call Macro			Ctrl
Top of fold	Bottom of fold			Shift
Page Up	Page Down			Alt
Line Up	Line Down			

0   –   Alt

Write			Delete
Host	Make Comment		←
Backtrace			
List Fold			
P			Return
			Enter

	Enter Fold	↑	Exit Fold
Ctrl	Delete to EOL		Remove Fold
	←		→
Ctrl	◄ Word		Word ►
	Open Fold	↓	Close Fold
Ctrl	Finish		Suspend TDS
	Create Fold		Delete →

A.3   NEC PC keyboard layout

	F1	F2	F3	F4	F5	F6
Shift	Toolkit	Code Info	Pick Line	Copy Pick	Next UTIL	Next EXE
	Help	Fold Info	Move Line	Copy Line	Get Code	Run EXE

	Esc 1	2	3	4	5	6	7	8	9
File Handler	Attach/ Detatch	Copy Attach	Compact Libs	Rename File	Write Protect	Write Enable	Copy In	Copy Out	Read Host
Debugger	Inspect	Channel	Top	Retrace	Relocate	Info		Links	Monitor
Compiler	Check	Compile	Extract	Load	Recompile	Compile Info	Make Foldset	Search	Replace

Esc — Select Parameter

Q	W* Save Macro	E* Delete to EOL	R* Remove Fold	T* Top of Fold	Y* Delete Line	U* Put	I	O* Browse

A* Set abort flag	S	D* Define Macro	F* Word ←	G* Word →	H	J* Call Macro	K* Del Word ←	L* Del Word →

Z* Suspend TDS	X* Finish	C	V* Get Macro	B* Bottom of fold	N	M

* Ctrl + key

Autoload   = Esc GET CODE   Clear UTIL = Esc NEXT UTIL
Refresh   = Esc Esc   Clear EXE = Esc NEXT EXE
                       Clear ALL = Esc RUN EXE

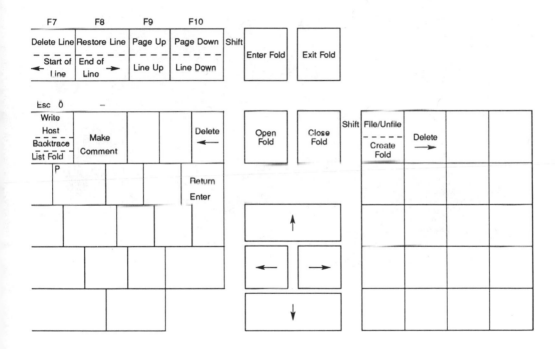

B Summary of standard utilities

These are the standard utility packages supplied with the TDS:

occam **program development package**

1	`CHECK`	Check program syntax
2	`COMPILE`	Compile and/or configure program
3	`EXTRACT`	Extract and link code for network
4	`LOAD NETWORK`	Load program onto network
5	`RECOMPILE`	Recompile program with old parameters
6	`COMPILATION INFO`	Show compilation or configuration info
7	`MAKE FOLDSET`	Make foldset around the current line
8	`SEARCH`	Search for string
9	`REPLACE`	Replace string at current cursor position
0	`LIST FOLD`	List utility

File handling package

1	`ATTACH/DETACH`	Attach or detach a file in the fold structure
2	`COPY ATTACH`	Copy TDS file(s) into the fold structure
3	`COMPACT LIBRARIES`	Compact libraries into another directory
4	`RENAME FILE`	Change the name of the file for a fold
7	`COPY IN`	Copy a TDS file structure from another directory
8	`COPY OUT`	Copy a TDS file structure to another directory
9	`READ HOST`	Read a host format file into a fold
0	`WRITE HOST`	Write a fold out into a host format file

C Names defined by the software

All names which may appear in occam source text and which are defined either by the language, the compiler, the libraries or the TDS are given below in alphabetical order. Library constants are not included, but see Appendix D for tables of these.

The names in this table are classified into the following different classes:

1 language keyword. Keyword defined in the language reference manual.

2 compiler keyword. Keyword defined by the current compiler implementation.

3 compiler predefine. A procedure or function which is predeclared by the compiler and implemented by in line code.

4 compiler library. A library procedure or function which is used by compiler generated code. On some processors these are implemented by a function in a library with the name indicated, on others they are implemented as in line code.

5 system library. A library procedure for special transputer system operations.

6 maths library. A function in the elementary function library. Logical library name depends on version required.

7 maths utility library. Supporting function for maths library

8 io library. A procedure or function in the input/output library. The library logical name which must be used to access it is also shown.

9 host io library. A procedure in the host input/output library. The library logical name which must be used to access it is also shown.

10 compiler input directive. Word in occam source code recognised by compilation system for special action at compile time.

11 TDS foldtype name. Name written by TDS editor or compilation utility on certain fold lines.

Any name which is not a language keyword or a compiler keyword may be redeclared as an identifier in an occam program. However, redefining a name of a compiler library procedure or function can have unexpected consequences; it is strongly recommended that all the names below are reserved for the use specified in this table.

Name	Class	Library	Notes
ABS	compiler library	realpds	
ACOS	maths library		
add.char	io library	strings	
add.hex.int32	io library	strings	
add.hex.int64	io library	strings	
add.hex.int	io library	strings	
add.int32	io library	strings	
add.int64	io library	strings	
add.int	io library	strings	
add.real32	io library	strings	
add.real64	io library	strings	
add.text	io library	strings	
af.buffer	io library	afinterf	
af.multiplexor	io library	afinterf	
af.read.integer	io library	afio	
af.read.record	io library	afio	
af.to.sp	host io library	afsp	
af.write.integer	io library	afio	
af.write.record	io library	afio	
AFTER	language keyword		
ALOG	maths library		
ALOG10	maths library		
ALT	language keyword		
AND	language keyword		
ANY	language keyword		
append.char	io library	strings	
append.hex.int32	io library	strings	
append.hex.int64	io library	strings	
append.hex.int	io library	strings	
append.int32	io library	strings	
append.int64	io library	strings	
append.int	io library	strings	
append.real32	io library	strings	
append.real64	io library	strings	
append.text	io library	strings	
ARGUMENT.REDUCE	compiler library	realpds	
ASHIFTLEFT	compiler predefine		
ASHIFTRIGHT	compiler predefine		
ASIN	maths library		
assign.bslice	io library	slice	
AT	language keyword		
ATAN	maths library		
ATAN2	maths library		
B00x.term.p.driver	io library	t2board	
B00x.term.p.driver	io library	t4board	
beep	io library	userio	
BITAND	language keyword		
BITCOUNT	compiler library	intpds	
BITNOT	language keyword		
BITOR	language keyword		
BITREVNBITS	compiler library	intpds	

Name	Class	Library	Notes
BITREVWORD	compiler library	intpds	
BOOL	language keyword		
BOOLREAD	io library	derivio	
BOOLTOSTRING	io library	ioconv	
BOOLWRITE	io library	derivio	
BYTE	language keyword		
call.interrupt	io library	afiler	
CASE	language keyword		
CAUSEERROR	compiler predefine		
CHAN	language keyword		
char.pos	io library	strings	
clean.string	io library	ufiler	
clear.eol	io library	userio	
clear.eos	io library	userio	
CLIP2D	compiler library	intpds	
close.folded.stream	io library	userio	
close.stream	io library	afiler	
close.uf.stream	io library	ufiler	
close.tkf.file	io library	msdos	
CODE EXE	TD3 foldtype name		
CODE PROGRAM	TDS foldtype name		
CODE SC	TDS foldtype name		
CODE UTIL	TDS foldtypo name		
COMMENT	TDS foldtype name		
compare.strings	lo library	strings	
CONFIG INFO	TDS foldtype name		
COPYSIGN	compiler library	realpds	
COS	maths library		
COSH	maths library		
CRCBYTE	compiler library	intpds	
CRCFROMLSB	system library	blockcrc	
CRCFROMMSB	system library	blockcrc	
CRCWORD	compiler library	intpds	
create.fold	io library	ufiler	
create.new.fold	io library	userio	
DABS	compiler library	realpds	
DACOS	maths library		
DALOG	maths library		
DALOG10	maths library		
DARGUMENT.REDUCE	compiler library	realpds	
DASIN	maths library		
DATAN	maths library		
DATAN2	maths library		
DCOPYSIGN	compiler library	realpds	
DCOS	maths library		
DCOSH	maths library		
DDIVBY2	compiler library	realpds	
del.line	io library	userio	
delete.chl	io library	uscrio	
delete.chr	io library	userio	
delete.string	io library	strings	

Name	Class	Library	Notes
DEXP	maths library		
Dexp	maths library		
DFLOATING.UNPACK	compiler library	realpds	
DFPINT	compiler library	realpds	
DIEEECOMPARE	compiler library	realpds	
DISNAN	compiler library	realpds	
DIVBY2	compiler library	realpds	
DLOGB	compiler library	realpds	
DMINUSX	compiler library	realpds	
DMULBY2	compiler library	realpds	
DNEXTAFTER	compiler library	realpds	
DNOTFINITE	compiler library	realpds	
DORDERED	compiler library	realpds	
down	io library	userio	
DPOWER	maths library		
DRAN	maths library		
DRAW2D	compiler library	intpds	
DSCALEB	compiler library	realpds	
DSIN	maths library		
DSINH	maths library		
DSQRT	compiler library	realpds	
DTAN	maths library		
DTANH	maths library		
ELSE	language keyword		
enter.fold	io library	userio	
eqstr	io library	strings	
EXE	TDS foldtype name		
exit.fold	io library	userio	
EXP	maths library		
exp	maths library		
FALSE	language keyword		
file.lock	io library	msdos	
file.release	io library	msdos	
finish.new.fold	io library	userio	
FIX	maths utility library		
FIX64	maths utility library		
FLOATING.UNPACK	compiler library	realpds	
FOR	language keyword		
FPINT	compiler library	realpds	
FracDiv	maths utility library		
FracDiv64	maths utility library		
FRACMUL	compiler predefine		T4,T8 only
FracMult64	maths utility library		
FROM	language keyword		
FUNCTION	language keyword		
get.real.string	io library	userio	
get.real.with.del	io library	userio	
get.stream.result	io library	ufiler	
GETSTRING	io library	derivio	
goto.xy	io library	userio	
GUY	compiler keyword		

Name	Class	Library	Notes
handle.af.transaction	io library	afiler	
HEX16READ	io library	derivio	
HEX16TOSTRING	io library	extrio	
HEX16WRITE	io library	derivio	
HEX32READ	io library	derivio	
HEX32TOSTRING	io library	extrio	
HEX32WRITE	io library	derivio	
HEX64READ	io library	derivio	
HEX64TOSTRING	io library	extrio	
HEX64WRITE	io library	derivio	
HEXREAD	io library	derivio	
HEXTOSTRING	io library	ioconv	
HEXWRITE	io library	derivio	
IEEE32OP	compiler library	reals	
IEEE32REM	compiler library	reals	
IEEE64OP	compiler library	dreals	
IEEE64REM	compiler library	dreals	
IEEECOMPARE	compiler library	realpds	
IF	language keyword		
IN	compilor keyword		
IncExp	maths utility library		
IncExp64	maths utility library		
input.error.item	io library	uscrio	
input.len.bslice	io library	slice	
input.number.item	io library	userio	
input.record.item	io library	userio	
input.top.crease	io library	userio	
InputOrFail.c	system library	reinit	
InputOrFail.t	system library	reinit	
ins.line	io library	userio	
insert.char	io library	userio	
insert.string	io library	strings	
INT	language keyword		
INT16	language keyword		
INT16ADD	compiler library	ints	
INT16BITAND	compiler library	ints	
INT16BITNOT	compiler library	ints	
INT16BITOR	compiler library	ints	
INT16DIV	compiler library	ints	
INT16EQ	compiler library	ints	
INT16GT	compiler library	ints	
INT16LSHIFT	compiler library	ints	
INT16MINUS	compiler library	ints	
INT16MUL	compiler library	ints	
INT16PLUS	compiler library	ints	
INT16READ	io library	derivio	
INT16REM	compiler library	ints	
INT16RSHIFT	compiler library	ints	
INT16SUB	compiler library	ints	
INT16TIMES	compiler library	ints	
INT16TOINT32	compiler library	ints	

Name	Class	Library	Notes
INT16TOINT64	compiler library	ints	
INT16TOREAL32	compiler library	reals	
INT16TOREAL64	compiler library	dreals	
INT16TOSTRING	io library	extrio	
INT16WRITE	io library	derivio	
INT16XOR	compiler library	ints	
INT32	language keyword		
INT32ADD	compiler library	ints	
INT32BITAND	compiler library	ints	
INT32BITNOT	compiler library	ints	
INT32BITOR	compiler library	ints	
INT32DIV	compiler library	ints	
INT32DIVREM	compiler library	ints	
INT32EQ	compiler library	ints	
INT32GT	compiler library	ints	
INT32LSHIFT	compiler library	ints	
INT32MINUS	compiler library	ints	
INT32MUL	compiler library	ints	
INT32PLUS	compiler library	ints	
INT32READ	io library	derivio	
INT32REM	compiler library	ints	
INT32RSHIFT	compiler library	ints	
INT32SUB	compiler library	ints	
INT32TIMES	compiler library	ints	
INT32TOINT16	compiler library	ints	
INT32TOINT64	compiler library	ints	
INT32TOINT64	compiler library	ints	
INT32TOREAL32	compiler library	reals	
INT32TOREAL64	compiler library	dreals	
INT32TOSTRING	io library	extrio	
INT32WRITE	io library	derivio	
INT32XOR	compiler library	ints	
INT64	language keyword		
INT64ADD	compiler library	ints	
INT64BITAND	compiler library	ints	
INT64BITNOT	compiler library	ints	
INT64BITOR	compiler library	ints	
INT64DIV	compiler library	ints	
INT64DIVREM	compiler library	ints	
INT64EQ	compiler library	ints	
INT64GT	compiler library	ints	
INT64LSHIFT	compiler library	ints	
INT64MINUS	compiler library	ints	
INT64MUL	compiler library	ints	
INT64PLUS	compiler library	ints	
INT64READ	io library	derivio	
INT64REM	compiler library	ints	
INT64RSHIFT	compiler library	ints	
INT64SUB	compiler library	ints	
INT64TIMES	compiler library	ints	
INT64TOINT16	compiler library	ints	

Name	Class	Library	Notes
INT64TOINT32	compiler library	ints	
INT64TOREAL32	compiler library	reals	
INT64TOREAL64	compiler library	dreals	
INT64TOSTRING	io library	extrio	
INT64WRITE	io library	derivio	
INT64XOR	compiler library	ints	
INTREAD	io library	derivio	
INTTOSTRING	io library	ioconv	
INTWRITE	io library	derivio	
IS	language keyword		
is.digit	io library	strings	
is.hex.digit	io library	strings	
is.id.char	io library	strings	
is.in.range	io library	strings	
is.lower	io library	strings	
is.upper	io library	strings	
ISNAN	compiler library	realpds	
KERNEL.RUN	compiler predefine		
keystream.from.afserver	io library	afinterf	
keystream.from.B004.link	io library	t4board	
keystream.from.file	io library	interf	
keystream.from.server	io library	interf	
keystream.sink	io library	interf	
keystream.to.screen	io library	interf	
ks.get.real.string	io library	streamio	
ks.get.real.with.del	io library	streamio	
ks.keystream.sink	io library	streamio	
ks.keystream.to.screen	io library	streamio	
ks.read.char	io library	streamio	
ks.read.echo.char	io library	streamio	
ks.read.echo.hex.int	io library	streamio	
ks.read.echo.hex.int32	io library	streamio	
ks.read.echo.hex.int64	io library	streamio	
ks.read.echo.int	io library	streamio	
ks.read.echo.int32	io library	streamio	
ks.read.echo.int64	io library	streamio	
ks.read.echo.real32	io library	streamio	
ks.read.echo.real64	io library	streamio	
ks.read.echo.text.line	io library	streamio	
ks.read.hex.int	io library	streamio	
ks.read.hex.int32	io library	streamio	
ks.read.hex.int64	io library	streamio	
ks.read.int	io library	streamio	
ks.read.int32	io library	streamio	
ks.read.int64	io library	streamio	
ks.read.real32	io library	streamio	
ks.read.real64	io library	streamio	
ks.read.text.line	io library	streamio	
left	io library	userio	
LIB	TDS foldtype name		
LOAD.BYTE.VECTOR	compiler predefine		

Name	Class	Library	Notes
LOAD.INPUT.CHANNEL	compiler predefine		
LOAD.INPUT.CHANNEL.VECTOR	compiler predefine		
LOAD.OUTPUT.CHANNEL	compiler predefine		
LOAD.OUTPUT.CHANNEL.VECTOR	compiler predefine		
LOGB	compiler library	realpds	
LONGADD	compiler predefine		
LONGDIFF	compiler predefine		
LONGDIV	compiler predefine		
LONGPROD	compiler predefine		
LONGSUB	compiler predefine		
LONGSUM	compiler predefine		
make.filed	io library	ufiler	
make.id	io library	msdos	
MINUS	language keyword		
MINUSX	compiler library	realpds	
MOSTNEG	language keyword		
MOSTPOS	language keyword		
MOVE2D	compiler library	intpds	
MULBY2	compiler library	realpds	
newline	io library	userio	
next.int.from.line	io library	strings	
next.word.from.line	io library	strings	
NEXTAFTER	compiler library	realpds	
NORMALISE	compiler predefine		
NORMALISE64	maths utility library		
NOT	language keyword		
NOTFINITE	compiler library	realpds	
number.of.folds	io library	ufiler	
OF	language keyword		
open.data.stream	io library	ufiler	
open.file	io library	afiler	
open.folded.stream	io library	userio	
open.input.stream	io library	afiler	
open.output.stream	io library	afiler	
open.stream	io library	ufiler	
open.temp.file	io library	afiler	
open.tkf.file	io library	msdos	
OR	language keyword		
ORDERED	compiler library	realpds	
output.len.bslice	io library	slice	
OutputOrFail.c	system library	reinit	
OutputOrFail.t	system library	reinit	
PAR	language keyword		
PLACE	language keyword		
PLACED	language keyword		
PLUS	language keyword		
PORT	language keyword		
port.read	io library	afiler	
port.write	io library	afiler	
POWER	maths library		
PRI	language keyword		

Name	Class	Library	Notes
PROC	language keyword		
PROCESSOR	language keyword		
PROGRAM	TDS foldtype name		
PROTOCOL	language keyword		
QRealIDiv	compiler library	t2utils	
QRealIMul	compiler library	t2utils	
QUADNORMALISE	compiler library	t2utils	
QUADSHIFTLEFT	compiler library	t2utils	
QUADSHIFTRIGHT	compiler library	t2utils	
RAN	maths library		
read.block	io library	afiler	
read.bottom.crease	io library	userio	
read.char	io library	userio	
read.core.dump	io library	afiler	
read.data.record	io library	ufiler	
read.echo.char	io library	userio	
read.echo.hex.int64	io library	userio	
read.echo.hex.int	io library	userio	
read.echo.int64	io library	userio	
read.echo.int	io library	userio	
read.echo.real32	io library	userio	
read.echo.real64	io library	userio	
read.echo.text.line	io library	userio	
read.environment	io library	afiler	
read.error.item	io library	userio	
read.file.name	io library	userio	
read.filed.top.crease	io library	userio	
read.fold.attr	io library	ufiler	
read.fold.heading	io library	userio	
read.fold.string	io library	ufiler	
read.fold.top.crease	io library	userio	
read.hex.int64	io library	userio	
read.hex.int	io library	userio	
read.int64	io library	userio	
read.int	io library	userio	
read.key.wait	io library	afiler	
read.key	io library	afiler	
read.number.item	io library	userio	
read.real32	io library	userio	
read.real64	io library	userio	
read.record.item	io library	userio	
read.regs	io library	afiler	
read.text.line	io library	userio	
read.time	io library	afiler	
read.tkf.block	io library	msdos	
read.tkf.line	io library	msdos	
REAL32	language keyword		
REAL32EQ	compiler library	reals	
REAL32EQERR	compiler library	reals	
REAL32GT	compiler library	reals	
REAL32GTERR	compiler library	reals	

Name	Class	Library	Notes
REAL32OP	compiler library	reals	
REAL32OPERR	compiler library	reals	
REAL32READ	io library	derivio	
REAL32REM	compiler library	reals	
REAL32REMERR	compiler library	reals	
REAL32TOINT16	compiler library	reals	
REAL32TOINT32	compiler library	reals	
REAL32TOINT64	compiler library	reals	
REAL32TOREAL64	compiler library	reals	
REAL32TOSTRING	io library	extrio	
REAL32WRITE	io library	derivio	
REAL64	language keyword		
REAL64EQ	compiler library	dreals	
REAL64EQERR	compiler library	dreals	
REAL64GT	compiler library	dreals	
REAL64GTERR	compiler library	dreals	
REAL64OP	compiler library	dreals	
REAL64OPERR	compiler library	dreals	
REAL64READ	io library	derivio	
REAL64REM	compiler library	dreals	
REAL64REMERR	compiler library	dreals	
REAL64TOINT16	compiler library	dreals	
REAL64TOINT32	compiler library	dreals	
REAL64TOINT64	compiler library	dreals	
REAL64TOREAL32	compiler library	reals	
REAL64TOSTRING	io library	extrio	
REAL64WRITE	io library	derivio	
RealIDiv	compiler library	t2utils,r64util	
RealIMul	compiler library	t2utils,r64util	
receive.block	io library	afiler	
ReFloat	maths utility library		
ReFloat64	maths utility library		
Reinitialise	system library	reinit	
REM	language keyword		
rename.file	io library	afiler	
repeat.fold	io library	userio	
RESULT	language keyword		
RETYPES	language keyword		
right	io library	userio	
ROTATELEFT	compiler predefine		
ROTATERIGHT	compiler predefine		
ROUND	language keyword		
ROUNDSN	compiler predefine		*T4 only*
run.command	io library	afiler	
runtime.data	io library	afiler	
SC	TDS foldtype name		
SCALEB	compiler library	realpds	
scrstream.copy	io library	interf	
scrstream.fan.out	io library	interf	
scrstream.from.array	io library	interf	
scrstream.multiplexor	io library	interf	

Name	Class	Library	Notes
`scrstream.sink`	io library	interf	
`scrstream.to.afserver`	io library	afinterf	
`scrstream.to.ANSI`	io library	interf	
`scrstream.to.array`	io library	interf	
`scrstream.to.B004.link`	io library	t4board	
`scrstream.to.file`	io library	interf	
`scrstream.to.server`	io library	interf	
`scrstream.to.TVI920`	io library	interf	
`search.match`	Io library	strings	
`search.no.match`	io library	strings	
`seek`	io library	afiler	
`send.block`	io library	afiler	
`send.command`	io library	ufiler	
`SEQ`	language keyword		
`server.version`	io library	afiler	
`set.return.result`	io library	afiler	
`SHIFTLEFT`	compiler predefine		
`SHIFTRIGHT`	compiler predefine		
`SHIFTRIGHT64`	maths utility library		
`SIN`	maths library		
`SINH`	maths library		
`SIZE`	language keyword		
`SKIP`	language keyword		
`skip.item`	io library	userio	
`so.ask`	host io library	sklib	
`so.buffer`	host io library	spinterf	
`so.close`	host io library	splib	
`so.commandline`	host io library	splib	
`so.core`	host io library	splib	
`so.date.to.ascii`	host io library	solib	
`so.eof`	host io library	splib	
`so.exit`	host io library	splib	
`so.ferror`	host io library	splib	
`so.flush`	host io library	splib	
`so.fwrite.char`	host io library	solib	
`so.fwrite.nl`	host io library	solib	
`so.fwrite.hex.int`	host io library	solib	
`so.fwrite.hex.int64`	host io library	solib	
`so.fwrite.int`	host io library	solib	
`so.fwrite.int64`	host io library	solib	
`so.fwrite.real32`	host io library	solib	
`so.fwrite.real64`	host io library	solib	
`so.fwrite.string`	host io library	solib	
`so.fwrite.string.nl`	host io library	solib	
`so.getenv`	host io library	splib	
`so.getkey`	host io library	splib	
`so.gets`	host io library	splib	
`so.keystream.from.file`	host io library	spinterf	
`so.keystream.from.kbd`	host io library	spinterf	
`so.keystream.from.stdin`	host io library	spinterf	
`so.multiplexor`	host io library	spinterf	

Name	Class	Library	Notes
so.open	host io library	splib	
so.open.temp	host io library	solib	
so.overlapped.buffer	host io library	spinterf	
so.overlapped.multiplexor	host io library	spinterf	
so.parse.command.line	host io library	solib	
so.pollkey	host io library	splib	
so.popen.read	host io library	solib	
so.puts	host io library	splib	
so.read	host io library	splib	
so.read.echo.any.int	host io library	sklib	
so.read.echo.hex.int	host io library	sklib	
so.read.echo.hex.int64	host io library	sklib	
so.read.echo.int	host io library	sklib	
so.read.echo.int64	host io library	sklib	
so.read.echo.line	host io library	solib	
so.read.echo.real32	host io library	sklib	
so.read.echo.real64	host io library	sklib	
so.read.line	host io library	solib	
so.remove	host io library	splib	
so.rename	host io library	splib	
so.scrstream.to.ANSI	host io library	spinterf	
so.scrstream.to.file	host io library	spinterf	
so.scrstream.to.stdout	host io library	spinterf	
so.keystream.to.TVI920	host io library	spinterf	
so.seek	host io library	splib	
so.system	host io library	splib	
so.tell	host io library	splib	
so.test.exists	host io library	solib	
so.time	host io library	splib	
so.time.to.ascii	host io library	solib	
so.time.to.date	host io library	solib	
so.today.ascii	host io library	solib	
so.today.date	host io library	solib	
so.version	host io library	splib	
so.write	host io library	splib	
so.write.char	host io library	solib	
so.write.hex.int	host io library	solib	
so.write.hex.int64	host io library	solib	
so.write.int	host io library	solib	
so.write.int64	host io library	solib	
so.write.nl	host io library	solib	
so.write.real32	host io library	solib	
so.write.real64	host io library	solib	
so.write.string	host io library	solib	
so.write.string.nl	host io library	solib	
SQRT	compiler library	realpds	
ss.beep	io library	streamio	
ss.clear.eol	io library	streamio	
ss.clear.eos	io library	streamio	
ss.delete.chl	io library	streamio	
ss.delete.chr	io library	streamio	

Name	Class	Library	Notes
`ss.del.line`	io library	streamio	
`ss.down`	io library	streamio	
`ss.goto.xy`	io library	streamio	
`ss.insert.char`	io library	streamio	
`ss.ins.line`	io library	streamio	
`ss.left`	io library	streamio	
`ss.right`	io library	streamio	
`ss.scrstream.copy`	io library	streamio	
`ss.scrstream.fan.out`	io library	streamio	
`ss.scrstream.from.array`	io library	streamio	
`ss.scrstream.from.fold`	io library	ssinterf	
`ss.scrstream.sink`	io library	streamio	
`ss.scrstream.to.ANSI.bytes`	io library	ssinterf	
`ss.scrstream.to.array`	io library	streamio	
`ss.scrstream.to.fold`	io library	ssinterf	
`ss.scrstream.to.TVI920.bytes`	io library	ssinterf	
`ss.up`	io library	streamio	
`ss.write.char`	io library	streamio	
`ss.write.endstream`	io library	streamio	
`ss.write.hex.int`	io library	streamio	
`ss.write.hex.int32`	io library	streamio	
`ss.write.hex.int64`	io library	streamio	
`ss.write.int`	io library	streamio	
`ss.write.int32`	io library	streamio	
`ss.write.int64`	io library	streamio	
`ss.write.nl`	io library	streamio	
`ss.write.real32`	io library	streamio	
`ss.write.real64`	io library	streamio	
`ss.write.string`	io library	streamio	
`ss.write.text.line`	io library	streamio	
`STOP`	language keyword		
`str.shift`	io library	strings	
`stream.access`	io library	afiler	
`stream.connect`	io library	afiler	
`stream.file`	io library	afiler	
`stream.length`	io library	afiler	
`stream.status`	io library	afiler	
`string.pos`	io library	strings	
`STRINGTOBOOL`	io library	ioconv	
`STRINGTOHEX16`	io library	extrio	
`STRINGTOHEX32`	io library	extrio	
`STRINGTOHEX64`	io library	extrio	
`STRINGTOHEX`	io library	ioconv	
`STRINGTOINT16`	io library	extrio	
`STRINGTOINT32`	io library	extrio	
`STRINGTOINT64`	io library	extrio	
`STRINGTOINT`	io library	ioconv	
`STRINGTOREAL32`	io library	extrio	
`STRINGTOREAL64`	io library	extrio	
`T2`	compiler keyword		

Name	Class	Library	Notes
T4	compiler keyword		
T8	compiler keyword		
TAN	maths library		
TANH	maths library		
terminate.filer	io library	afiler	
terminate.server	io library	t4board	
test.exists	io library	msdos	
TIMER	language keyword		
TIMES	language keyword		
to.lower.case	io library	strings	
to.upper.case	io library	strings	
TRUE	language keyword		
TRUNC	language keyword		
truncate.file.id	io library	ufiler	
UNPACKSN	compiler predefine		*T4 only*
up	io library	userio	
USE	compiler input directive		
UTIL	TDS foldtype name		
VAL	language keyword		
VALOF	language keyword		
VECSPACE	compiler keyword		
WHILE	language keyword		
WORKSPACE	compiler keyword		
write.block	io library	afiler	
write.bottom.crease	io library	userio	
write.char	io library	userio	
write.endstream	io library	userio	
write.filed.top.crease	io library	userio	
write.fold.string	io library	ufiler	
write.fold.top.crease	io library	userio	
write.full.string	io library	userio	
write.hex.int64	io library	userio	
write.hex.int	io library	userio	
write.int64	io library	userio	
write.int	io library	userio	
write.len.string	io library	userio	
write.number.item	io library	userio	
write.real32	io library	userio	
write.real64	io library	userio	
write.record.item	io library	userio	
write.text.line	io library	userio	
write.tkf.block	io library	msdos	
write.top.crease	io library	userio	

D System constant definitions

The TDS libraries include several groups of constants as indicated in the table in chapter 14. Some of these libraries are provided for compatibility with earlier releases only and are not tabulated here. The libraries tabulated here are:

complibs :	linkaddr	transputer link addresses
mathlibs :	mathvals	important REAL numbers
hostlibs :	sphdr	constants used in the iserver interface
iolibs :	strmhdr	protocols SP, SS and KS
	userhdr	constants used with KS and SS protocol
	filerhdr	constants used in the user filer interface

D.1 LINKADDR

To declare these addresses, used in **PLACE** allocations for channels on hard transputer links, insert the following line in a compilation unit:

```
#USE linkaddr

{{{   link addresses
-- Transputer link addresses

VAL link0.in  IS 4:
VAL link0.out IS 0:

VAL link1.in  IS 5:
VAL link1.out IS 1:

VAL link2.in  IS 6:
VAL link2.out IS 2:

VAL link3.in  IS 7:
VAL link3.out IS 3:

VAL event.in  IS 8:

}}}
```

D.2   MATHVALS

This library includes specifications of a variety of general purpose real constants.

To declare these values, insert the following line in a compilation unit:

```
#USE mathvals

{{{   REAL32 Constants
VAL REAL32 INFINITY RETYPES #7F800000(INT32) :
VAL REAL32 MINREAL  RETYPES #00000001(INT32) :
VAL REAL32 MAXREAL  RETYPES #7F7FFFFF(INT32) :   -- 3.40282347 E+38
VAL REAL32 E        RETYPES #402DF854(INT32) :   -- 2.71828174 E+00
VAL REAL32 PI       RETYPES #40490FDB(INT32) :   -- 3.14159274 E+00
VAL REAL32 LOGE2    RETYPES #3F317218(INT32) :   -- 6.93147182 E-01
VAL REAL32 LOG10E   RETYPES #3EDE5BD9(INT32) :   -- 4.34294492 E-01
VAL REAL32 ROOT2    RETYPES #3FB504F3(INT32) :   -- 1.41421354 E+00
VAL LOGEPI          IS       1.1447298858(REAL32) :
VAL RADIAN          IS      57.295779513(REAL32) :
VAL DEGREE          IS       1.74532925199E-2(REAL32) :
VAL GAMMA           IS       0.5772156649(REAL32) :
}}}
{{{   REAL64 Constants
VAL REAL64 DINFINITY RETYPES #7FF0000000000000(INT64):
VAL REAL64 DMINREAL  RETYPES #0000000000000001(INT64):
VAL REAL64 DMAXREAL  RETYPES #7FEFFFFFFFFFFFFF(INT64):
                             -- 1.7976931348623157E+308
VAL REAL64 DE        RETYPES #4005BF0A8B145769(INT64) :
                             -- 2.7182818284590451E+000
VAL REAL64 DPI       RETYPES #400921FB54442D18(INT64) :
                             -- 3.1415926535897931E+000
VAL REAL64 DLOGE2    RETYPES #3FE62E42FEFA39EF(INT64) :
                             -- 6.9314718055994529E-001
VAL REAL64 DLOG10E   RETYPES #3FDBCB7B1526E50E(INT64) :
                             --4.3429448190325182E-001
VAL REAL64 DROOT2    RETYPES #3FF6A09E667F3BCD(INT64) :
                             -- 1.4142135623730951E+000
VAL DLOGEPI          IS       1.1447298858494001741(REAL64) :
VAL DRADIAN          IS      57.295779513082320877(REAL64) :
VAL DDEGREE          IS       1.745329251994329576E-2(REAL64) :
VAL DGAMMA           IS       0.57721566490153286061(REAL64) :
}}}
```

D.3 SPHDR

The constants in this library are reserved for use in communications between a program and the **iserver** or any other process which communicates using **SP** protocol. Similar declarations are included in all INMOS occam implementations which allow the creation of programs supported by **iserver**.

To declare these constants include the following line in a compilation unit:

#USE sphdr

```
{{{   command tags
-- values up to 127 are reserved for use by INMOS
{{{   file command tags
VAL sp.open.tag    IS 10(BYTE) :
VAL sp.close.tag   IS 11(BYTE) :
VAL sp.read.tag    IS 12(BYTE) :
VAL sp.write.tag   IS 13(BYTE) :
VAL sp.gets.tag    IS 14(BYTE) :
VAL sp.puts.tag    IS 15(BYTE) :
VAL sp.flush.tag   IS 16(BYTE) :
VAL sp.seek.tag    IS 17(BYTE) :
VAL sp.tell.tag    IS 18(BYTE) :
VAL sp.eof.tag     IS 19(BYTE) :
VAL sp.ferror.tag  IS 20(BYTE) :
VAL sp.remove.tag  IS 21(BYTE) ;
VAL sp.rename.tag  IS 22(BYTE) :
}}}
{{{   host command tags
VAL sp.getkey.tag  IS 30(BYTE) :
VAL sp.pollkey.tag IS 31(BYTE) :
VAL sp.getenv.tag  IS 32(BYTE) :
VAL sp.time.tag    IS 33(BYTE) :
VAL sp.system.tag  IS 34(BYTE) :
VAL sp.exit.tag    IS 35(BYTE) :
}}}
{{{   server command tags
VAL sp.commandline.tag IS 40(BYTE) :
VAL sp.core.tag        IS 41(BYTE) :
VAL sp.version.tag     IS 42(BYTE) :
}}}
{{{   OS specific command tags
-- These OS specific tags will be followed by another tag indicating
-- which OS specific function is required

VAL sp.DOS.tag    IS 50(BYTE) :
VAL sp.HELIOS.tag IS 51(BYTE) :
VAL sp.VMS.tag    IS 52(BYTE) :
VAL sp.SUNOS.tag  IS 53(BYTE) :
VAL sp.TDS.tag    IS 65(BYTE) :
}}}
}}}
```

```
{{{
packet and buffer Sizes
VAL sp.max.packet.size IS 512 :
                            -- bytes transferred, includes length & data
VAL sp.min.packet.size IS   8 :
                            -- bytes transferred, includes length & data

VAL sp.max.packet.data.size IS sp.max.packet.size - 2 :  -- INT16 length
VAL sp.min.packet.data.size IS sp.min.packet.size - 2 :  -- INT16 length

{{{  Individual command maxima
VAL sp.max.openname.size     IS sp.max.packet.data.size - 5 :
                                         -- 5 bytes extra
VAL sp.max.readbuffer.size   IS sp.max.packet.data.size - 3 :
                                         -- 3 bytes extra
-- ditto for gets
VAL sp.max.writebuffer.size  IS sp.max.packet.data.size - 7 :
                                         -- 7 bytes extra
-- ditto for puts
VAL sp.max.removename.size   IS sp.max.packet.data.size - 3 :
                                         -- 3 bytes extra
VAL sp.max.renamename.size   IS sp.max.packet.data.size - 5 :
                                         -- 5 bytes extra
VAL sp.max.getenvname.size   IS sp.max.packet.data.size - 3 :
                                         -- 3 bytes extra
VAL sp.max.systemcommand.size IS sp.max.packet.data.size - 3 :
                                         -- 3 bytes extra
VAL sp.max.corerequest.size  IS sp.max.packet.data.size - 3 :
                                         -- 3 bytes extra

VAL sp.max.buffer.size IS sp.max.writebuffer.size :
                                    -- smaller of read & write
}}}
}}}
{{{  result values       (spr.)
VAL spr.ok              IS   0(BYTE) :

VAL spr.not.implemented IS   1(BYTE) :
VAL spr.bad.name        IS   2(BYTE) : -- filename is null
VAL spr.bad.type        IS   3(BYTE) : -- open file type is incorrect
VAL spr.bad.mode        IS   4(BYTE) : -- open file mode is incorrect
VAL spr.invalid.streamid IS  5(BYTE) : -- never opened that streamid
VAL spr.bad.stream.use  IS   6(BYTE) :
                            -- reading an output file, or vice versa
VAL spr.buffer.overflow IS   7(BYTE) :
                                -- buffer too small for required data
VAL spr.bad.packet.size IS   8(BYTE) :
                                -- data too big or small for packet
VAL spr.bad.origin      IS   9(BYTE) : -- seek origin is incorrect
VAL spr.notok           IS 127(BYTE) : -- a general fail result

-- anything 128 or above is a server dependent 'failure' result
VAL spr.operation.failed IS 128(BYTE) :
}}}
{{{  predefined streams    (spid.)
VAL spid.stdin  IS 0(INT32) :
VAL spid.stdout IS 1(INT32) :
VAL spid.stderr IS 2(INT32) :
}}}
```

```
{{{   open types              (spt.)
VAL spt.binary IS 1(BYTE) :
VAL spt.text   IS 2(BYTE) :
}}}
{{{   open modes              (spm.)
VAL spm.input            IS 1(BYTE) :
VAL spm.output           IS 2(BYTE) :
VAL spm.append           IS 3(BYTE) :
VAL spm.existing.update  IS 4(BYTE) :
VAL spm.new.update       IS 5(BYTE) :
VAL spm.append.update    IS 6(BYTE) :
}}}
{{{   status values           (sps.)
VAL sps.success IS  999999999(INT32) :
VAL sps.failure IS -999999999(INT32) :
}}}
{{{   seek origins            (spo.)
VAL spo.start   IS 1(INT32) :
VAL spo.current IS 2(INT32) :
VAL spo.end     IS 3(INT32) :
}}}
{{{   version information    (sph., spo., spb.)
{{{   host types            (sph.)
-- values up to 127 are reserved for use by INMOS
VAL sph.PC    IS 1(BYTE) :
VAL sph.NECPC IS 2(BYTE) :
VAL sph.VAX   IS 3(BYTE) :
VAL sph.SUN3  IS 4(BYTE) :
VAL sph.SUN4  IS 5(BYTE) :
}}}
{{{   OS types                (spo.)
VAL spo.DOS    IS 1(BYTE) :
VAL spo.HELIOS IS 2(BYTE) :
VAL spo.VMS    IS 3(BYTE) :
VAL spo.SUNOS  IS 4(BYTE) :
-- values up to 127 are reserved for use by INMOS
}}}
{{{   interface Board types (spb.)
-- This determines the interface between the link and the host
VAL spb.B004  IS 1(BYTE) :
VAL spb.B008  IS 2(BYTE) :
VAL spb.B010  IS 3(BYTE) :
VAL spb.B011  IS 4(BYTE) :
VAL spb.B014  IS 5(BYTE) :
VAL spb.DRX11 IS 6(BYTE) :
VAL spb.QT0   IS 7(BYTE) :
-- values up to 127 are reserved for use by INMOS
}}}
}}}
{{{   command line
VAL sp.short.commandline IS BYTE 0 :   -- remove  server's own arguments
VAL sp.whole.commandline IS BYTE 1 :   -- include server's own arguments

VAL spopt.never  IS 0 : -- values for so.parse.commandline
VAL spopt.maybe  IS 1 : -- indicate whether an option requires
VAL spopt.always IS 2 : -- a following parameter
}}}
```

```
{{{   time string and date lengths
VAL so.time.string.len IS 19 :  -- enough for "HH:MM:SS DD/MM/YYYY"
VAL so.date.len        IS  6 :  -- enough for DDMMYY (as integers)
}}}
{{{   temp filename length
VAL so.temp.filename.length IS 6 :  -- six chars will work on anything!
}}}
```

D.4   STRMHDR

The occam protocols, and supporting constants, in this library are used for communication with the channels **from.isv**, **to.isv**, **keyboard** and **screen** in an **EXE** in **TDS3**.

To declare these protocols include the following line in a compilation unit:

#USE strmhdr

```
{{{   SP protocol
PROTOCOL SP IS INT16::[]BYTE :
}}}
{{{   streamio constants and protocols
VAL st.max.string.size IS 256 :
VAL ft.terminated   IS  -8 : -- used to terminate a keystream
VAL ft.number.error IS -11 :
PROTOCOL KS IS INT:
PROTOCOL SS
  CASE
    st.reset
    st.up
    st.down
    st.left
    st.right
    st.goto; INT32; INT32
    st.ins.char; BYTE
    st.del.char
    st.out.string; INT32::[]BYTE
    st.clear.eol
    st.clear.eos
    st.ins.line
    st.del.line
    st.beep
    st.spare
    st.terminate
    st.help
    st.initialise
    st.out.byte; BYTE
    st.out.int; INT32
    st.key.raw
    st.key.cooked
    st.release
    st.claim
    st.endstream
    st.set.poll; INT32
  :
}}}
```

D.5   USERHDR

In this library are the specifications of names for constants whose values are defined by the TDS or underlying hardware. Note that the user may change the names but not the values. A further set of constants for filing system access, etc are given elsewhere

To declare these constants include the following line in a compilation unit:

```
#USE userhdr

{{{   specifications of all the system defined constants
{{{   common array sizes
VAL max.record.size IS 512 :
VAL max.string.size IS 256 :
VAL abs.id.size IS 63:                    -- used by obsolete msdos library
}}}
{{{   clock ticks
VAL tptr.h.ticks.per.second IS 1000000(INT32) : -- high priority process
VAL tptr.l.ticks.per.second IS 15625(INT32) :   -- low priority process
VAL tptr.a.ticks.per.second IS 625000(INT32) :  -- rev A silicon
}}}

--    term.p protocol (keyboard and screen)
{{{   terminal definitions
VAL VT220 IS 1:
VAL VT100 IS 2:
VAL TVi920 IS 3:
VAL WYS50 IS 4:
VAL UNKNOWN IS 5 :
}}}
{{{   ITERM table sizes and values
VAL max.entries.per.screen.line   IS 10 :
VAL max.entries.per.keyboard.line IS 10 :
{{{   ITERM screen codes
VAL iterm.up.code        IS  1 :
VAL iterm.down.code      IS  2 :
VAL iterm.left.code      IS  3 :
VAL iterm.right.code     IS  4 :
VAL iterm.goto.code      IS  5 :
VAL iterm.ins.char.code  IS  6 :
VAL iterm.del.char.code  IS  7 :
VAL iterm.clear.eol.code IS  8 :
VAL iterm.clear.eos.code IS  9 :
VAL iterm.ins.line.code  IS 10 :
VAL iterm.del.line.code  IS 11 :
VAL iterm.beep.code      IS 12 :
VAL iterm.cls.code       IS 13 :
}}}
}}}
{{{   screen cursor addressing
VAL first.screen.col  IS 0 :
VAL first.screen.line IS 0 :
VAL first.text.col    IS first.screen.col :
{{{   screen size dependent non-constants (debugger assumes constant width)
-- avoid these by using tt.initialise
VAL last.screen.col    IS 79 :
VAL last.screen.line   IS 23 :
VAL last.text.col      IS last.screen.col :
}}}
}}}
```

```
{{{  ascii visible characters
VAL min.visible.char  IS 32 :          -- space
VAL max.visible.char  IS 126 :         -- '~'
}}}
{{{  message tags for screen protocol (to.term.p)
-- these correspond to the tags of the SS protocol declared in strmhdr
VAL tt.reset       IS BYTE 0 :
VAL tt.up          IS BYTE 1 :
VAL tt.down        IS BYTE 2 :
VAL tt.left        IS BYTE 3 :
VAL tt.right       IS BYTE 4 :
VAL tt.goto        IS BYTE 5 :
VAL tt.ins.char    IS BYTE 6 :
VAL tt.del.char    IS BYTE 7 :
VAL tt.out.string  IS BYTE 8 :
VAL tt.clear.eol   IS BYTE 9 :
VAL tt.clear.eos   IS BYTE 10 :
VAL tt.ins.line    IS BYTE 11 :
VAL tt.del.line    IS BYTE 12 :
VAL tt.beep        IS BYTE 13 :
VAL tt.terminate   IS BYTE 15 :
VAL tt.help        IS BYTE 16 :
VAL tt.initialise  IS BYTE 17 :
VAL tt.out.byte    IS BYTE 18 :
VAL tt.out.int     IS BYTE 19 :
VAL tt.key.raw     IS BYTE 20 :
VAL tt.key.cooked  IS BYTE 21 :
VAL tt.release     IS BYTE 22 :   -- not for general use
VAL tt.claim       IS BYTE 23 :   -- not for general use
VAL tt.endstream   IS BYTE 24 :
VAL tt.set.poll    IS BYTE 25 :   -- used inside TDS only
}}}
{{{  values for "cooked" keys
VAL tab            IS '*t' :
VAL return         IS INT '*c' :
VAL delete         IS 127 :          -- may also receive tt.del.chl
VAL ft.tag         IS 200 :
VAL ft.return      IS ft.tag + 0 :
VAL ft.up          IS ft.tag + 1 :
VAL ft.down        IS ft.tag + 2 :
VAL ft.left        IS ft.tag + 3 :
VAL ft.right       IS ft.tag + 4 :
VAL ft.del.chl     IS ft.tag + 5 :
VAL ft.del.chr     IS ft.tag + 6 :
VAL ft.del.line    IS ft.tag + 7 :
VAL ft.undel.line  IS ft.tag + 8 :
VAL ft.sol         IS ft.tag + 9 :
VAL ft.eol         IS ft.tag + 10 :
VAL ft.move        IS ft.tag + 11 :
VAL ft.copy        IS ft.tag + 12 :
VAL ft.line.up     IS ft.tag + 13 :
VAL ft.line.down   IS ft.tag + 14 :
VAL ft.page.up     IS ft.tag + 15 :
VAL ft.page.down   IS ft.tag + 16 :
VAL ft.create.fold IS ft.tag + 17 :
VAL ft.remove.fold IS ft.tag + 18 :
VAL ft.open.fold   IS ft.tag + 19 :
VAL ft.close.fold  IS ft.tag + 20 :
VAL ft.enter.fold  IS ft.tag + 21 :
VAL ft.exit.fold   IS ft.tag + 22 :
```

```
VAL ft.refresh      IS ft.tag + 23 :
VAL ft.file.fold    IS ft.tag + 24 :
VAL ft.finish       IS ft.tag + 26 :
VAL ft.edit.parms   IS ft.tag + 27 :    -- enter toolkit
VAL ft.fold.info    IS ft.tag + 28 :
VAL ft.help         IS ft.tag + 29 :
VAL ft.get.code     IS ft.tag + 31 :
VAL ft.save.macro   IS ft.tag + 32 :
VAL ft.get.macro    IS ft.tag + 33 :
VAL ft.run          IS ft.tag + 34 :
VAL ft.tool0        IS ft.tag + 35 :
VAL ft.tool1        IS ft.tag + 36 :
VAL ft.tool2        IS ft.tag + 37 :
VAL ft.tool3        IS ft.tag + 38 :
VAL ft.tool4        IS ft.tag + 39 :
VAL ft.tool5        IS ft.tag + 40 :
VAL ft.tool6        IS ft.tag + 41 :
VAL ft.tool7        IS ft.tag + 42 :
VAL ft.tool8        IS ft.tag + 43 :
VAL ft.tool9        IS ft.tag + 44 :
VAL ft.word.left    IS ft.tag + 45 :
VAL ft.word.right   IS ft.tag + 46 :
VAL ft.del.wordl    IS ft.tag + 47 :
VAL ft.del.wordr    IS ft.tag + 48 :
VAL ft.delto.eol    IS ft.tag + 49 :
VAL ft.top.of.fold  IS ft.tag + 50 :
VAL ft.bottom.of.fold   IS ft.tag + 51 :
VAL ft.select.param IS ft.tag + 52 :
VAL ft.code.info    IS ft.tag + 53 :
VAL ft.pick         IS ft.tag + 54 :
VAL ft.copy.pick    IS ft.tag + 55 :
VAL ft.put          IS ft.tag + 56 :
VAL ft.next.util    IS ft.tag + 57 :
VAL ft.clear.util   IS ft.tag + 58 :
VAL ft.autoload     IS ft.tag + 59 :
VAL ft.next.exe     IS ft.tag + 60 :
VAL ft.clear.exe    IS ft.tag + 61 :
VAL ft.clear.all    IS ft.tag + 62 :
VAL ft.browse       IS ft.tag + 63 :
VAL ft.suspend.tds  IS ft.tag + 64 :
VAL ft.define.macro IS ft.tag + 65 :
VAL ft.call.macro   IS ft.tag + 66 :
VAL ft.make.comment IS ft.tag + 67 :
VAL ft.bad          IS ft.tag + 70 :
}}}
{{{   special codes from the keyboard channel (from.term.p)
VAL ft.lines.prefix     IS -1 :
VAL ft.columns.prefix   IS -2 :
VAL ft.nolineops.prefix IS -3 :    -- TRUE if no line insert/delete ops
VAL ft.end.init         IS -4 :
VAL ft.table.error      IS -5 :    -- read error
VAL ft.noncom.table     IS -6 :    -- non-compatable table
VAL ft.nocharops.prefix IS -7 :    -- TRUE if no char insert/delete ops
VAL ft.terminated       IS -8 :
VAL ft.released         IS -9 :
VAL ft.claimed          IS -10 :
VAL ft.number.error     IS -11 :
VAL ft.cooked           IS -12 :
VAL ft.raw              IS -13 :
}}}
```

D.6   FILERHDR

These constants are provided for communications between an **EXE** and the user filer, using the channels **from.user.filer** and **to.user.filer**. These channels provide an interface between a running **EXE** and the folded file store.

To declare these constants include the following line in a compilation unit:

```
#USE filerhdr

{{{  user filer interface constants (named filestore)
--VAL file.store.is.named IS FALSE:
-- unnamed filestores not supported by TDS3
VAL file.store.is.named IS TRUE: -- for named filestores
VAL fi.unnamed                   IS 1 :
VAL fi.named                     IS 0 :

{{{  filing system interface array sizes
VAL max.files IS 4 :
VAL max.record.size IS 512 :  -- also in userhdr, uservals
VAL max.string.size IS 256 :  -- also in userhdr, uservals
VAL max.file.id.size IS 256 :
VAL max.fold.depth IS 50 :
VAL max.file.depth IS 20 :
}}}
{{{  fold attribute values 27-1-88
-- Values for type
VAL ft.opstext IS 0 :
VAL ft.opsdata IS 1 :
VAL ft.opscode IS 2 :
VAL ft.foldset IS 3 :
VAL ft.voidset IS 4 :

-- Values for content
VAL fc.comment.text IS 0 :
VAL fc.source.text  IS 1 :
VAL fc.code.data    IS 2 :
VAL fc.occam1.sc    IS 3 :
VAL fc.desc.data    IS 4 :
VAL fc.debug.data   IS 5 :
VAL fc.occam1.prog  IS 6 :
VAL fc.occam1.util  IS 7 :
VAL fc.occam1.exe   IS 8 :
VAL fc.link.data    IS 9 :

VAL fc.occam2.sc    IS 10 :
VAL fc.occam2.prog  IS 11 :
VAL fc.occam2.util  IS 12 :
VAL fc.occam2.exe   IS 13 :
VAL fc.occam2.lib   IS 14 :

VAL fc.imp.proc     IS 15 :
VAL fc.imp.sc       IS 16 :
VAL fc.imp.lib      IS 17 :

VAL fc.c.proc       IS 18 :
VAL fc.c.sc         IS 19 :
VAL fc.c.lib        IS 20 :

VAL fc.pascal.proc  IS 21 :
```

```
VAL fc.pascal.sc      IS 22 :
VAL fc.pascal.lib     IS 23 :

VAL fc.fortran.proc IS 24 :
VAL fc.fortran.sc     IS 25 :
VAL fc.fortran.lib    IS 26 :

VAL fc.check.data     IS 27 :
VAL fc.image.data     IS 28 :
VAL fc.map.data       IS 29 :

VAL fc.config.info  IS 31 :
VAL fc.analyse.info IS 32 :
}}}
--    User filer protocol
{{{  attribute indexes
VAL attr.size          IS 3 :
VAL attr.fold.type     IS 0 :
VAL attr.fold.content IS 1 :
VAL attr.fold.indent   IS 2 :
}}}
{{{  uf. commands to user filer control
VAL uf.open.data.read    IS BYTE 1 :
VAL uf.open.data.write   IS BYTE 2 :
VAL uf.open.fold.read    IS BYTE 3 :
VAL uf.open.fold.write   IS BYTE 4 :
VAL uf.open.text.read    IS BYTE 5 :
VAL uf.open.text.write   IS BYTE 6 :

VAL uf.number.of.folds   IS BYTE 7 :
VAL uf.test.filed        IS BYTE 8 :
VAL uf.read.fold.string  IS BYTE 9 :
VAL uf.read.fold.attr    IS BYTE 10 :
VAL uf.read.file.id      IS BYTE 11 :

VAL uf.write.fold.string IS BYTE 12 :
VAL uf.make.fold.set     IS BYTE 13 :
VAL uf.unmake.fold.set   IS BYTE 14 :

VAL uf.create.fold       IS BYTE 15 :
VAL uf.delete.fold       IS BYTE 16 :
VAL uf.make.filed        IS BYTE 17 :
VAL uf.derive.file       IS BYTE 18 :
VAL uf.unfile            IS BYTE 19 :
VAL uf.delete.contents   IS BYTE 20 :
VAL uf.attach.file       IS BYTE 33 :
}}}
{{{  fsd. tags from file stream and from user filer control
VAL fsd.record        IS BYTE 60 :
VAL fsd.number        IS BYTE 61 :
VAL fsd.fold          IS BYTE 62 :
VAL fsd.filed         IS BYTE 63 :
VAL fsd.endfold       IS BYTE 64 :
VAL fsd.endstream     IS BYTE 65 :
VAL fsd.attr          IS BYTE 66 :
VAL fsd.file.id       IS BYTE 67 :
VAL fsd.error         IS BYTE 68 :
VAL fsd.result        IS BYTE 69 :
VAL fsd.number.of.folds IS BYTE 70 :
VAL fsd.startstream IS BYTE 71 :
```

```
VAL fsd.endfiled    IS BYTE 72 :
}}}
{{{   fsc. tags to file stream
VAL fsc.read           IS BYTE 40 :
VAL fsc.enter.fold     IS BYTE 41 :
VAL fsc.exit.fold      IS BYTE 42 :
VAL fsc.repeat.fold    IS BYTE 43 :
VAL fsc.read.attr      IS BYTE 44 :
VAL fsc.read.enc.attr  IS BYTE 45 :
VAL fsc.read.file.id   IS BYTE 46 :
VAL fsc.close          IS BYTE 47 :
VAL fsc.write          IS BYTE 48 :

-- these not used by the user filer:
VAL fsc.seek           IS BYTE 49 :
VAL fsc.truncate       IS BYTE 50 :
VAL fsc.file.length    IS BYTE 51 :
}}}
{{{   filer error numbers
VAL fi.ok                     IS 0 :
VAL fi.eof                    IS 1 :

{{{   -302 to -319
VAL fi.name.too.long          IS -302 :
VAL fi.too.many.locks         IS -303 :
VAL fi.cannot.lock.write.top  IS -304 :
VAL fi.too.many.releases      IS -305 :
VAL fi.too.many.suspends      IS -306 :
VAL fi.too.many.resumes       IS -307 :
VAL fi.illegal.chrc           IS -308 :
VAL fi.cannot.change.chrc     IS -309 :
VAL fi.filer.error            IS -310 :
VAL fi.cannot.copy            IS -311 :
VAL fi.not.all.data.written   IS -312 :
VAL fi.illegal.extension      IS -313 :
VAL fi.illegal.attr           IS -314 :
VAL fi.cannot.seek            IS -315 :
VAL fi.not.all.data.read      IS -316 :
VAL fi.unknown.record.length  IS -317 :
VAL fi.illegal.record.length  IS -318 :
VAL fi.cannot.create          IS -319 :
}}}
{{{   -1101 to -1143
VAL fi.file.not.open          IS -1101 :
VAL fi.file.already.open      IS -1102 :
VAL fi.file.does.not.exist    IS -1103 :
VAL fi.invalid.file.number    IS -1104 :
VAL fi.invalid.operation      IS -1105 :
VAL fi.wrong.page             IS -1106 :
VAL fi.bad.page               IS -1107 :
VAL fi.no.random.access       IS -1108 :
VAL fi.invalid.page.number    IS -1109 :
-- file not currently open, but is not flagged as closed on disk:
VAL fi.file.not.closed        IS -1110 :

-- internal errors
VAL fi.bad.max.claimed.page   IS -1111 :
VAL fi.page.out.of.range      IS -1112 :
VAL fi.file.map.error         IS -1113 :
VAL fi.device.error           IS -1120 :
```

```
VAL fi.volume.write.protected    IS -1121 :
VAL fi.volume.full               IS -1122 :
VAL fi.incorrect.volume.number   IS -1123 :
                                 -- no volume mounted on device
VAL fi.volume.must.be.dismounted IS -1124 :
                                 -- a volume is already mounted
VAL fi.volume.not.mounted        IS -1125 :
                                 -- the volume asked for not mounted
VAL fi.invalid.volume.number     IS -1126 : -- volume.number = no.volume
VAL fi.invalid.device.number     IS -1127 :
VAL fi.volume.already.mounted    IS -1128 :
VAL fi.bad.format                IS -1129 :
VAL fi.too.many.files            IS -1130 :
VAL fi.fold.too.deep             IS -1131 :
VAL fi.file.too.deep             IS -1132 :
VAL fi.unmatched.release         IS -1133 :
VAL fi.unmatched.resume          IS -1134 :
VAL fi.invalid.lock              IS -1135 :
VAL fi.invalid.chrc.number       IS -1136 :
VAL fi.invalid.command           IS -1140 :
VAL fi.filehandler.full          IS -1141 :
VAL fi.foldmanager.full          IS -1142 :
VAL fi.file.too.big              IS -1143 :
}}}
{{{   -1201 to -1210
VAL fi.invalid.command           IS -1201 :
VAL fi.fold.not.empty            IS -1202 :
VAL fi.operation.failed          IS -1203 :
VAL fi.no.such.fold              IS -1204 :
VAL fi.fold.in.use               IS -1205 :
VAL fi.not.filed                 IS -1206 :
VAL fi.invalid.fold              IS -1207 :
VAL fi.failed.to.open.fold       IS -1208 :
VAL fi.file.already.exists       IS -1209 :
VAL fi.not.on.a.valid.fold       IS -1210 :
}}}

}}}
```

E Error numbers

These error numbers may sometimes be presented to the user in error messages when things have gone wrong. They may also appear across the software interface as status results. Some numbers are generated only if an internal consistency check in a system process fails. These latter errors are marked '(*)' and should never occur.

The system is built out of a number of processes, structured hierarchically. Each process may detect an error, and is designed to pass on errors that occur at lower levels.

If a process detects an error but has not received an appropriate number from another process it may report an error **result = 0**.

The server may return result codes whose values are given in appendix D.3 or error codes defined by the C library with which it was compiled. The undefined result code 129 may be returned in the absence of further information.

E.1   File server errors

These errors may be generated by the TDS.

−**204** Write protect error; a write has been attempted on a disk that is marked for write protect

−**302** Name too long; a file name was received which exceeded the maximum length (63 characters).

−**303** Too many locks; an attempt was made to lock a file when the number of files currently locked on this channel is already the maximum allowed (twenty). (*)

−**304** Cannot lock-write top level; a lock-write command was issued with a null parent identifier, and a null file identifier. This is interpreted to be the top level file, which is read-only.

−**305** Too many releases; a release command was issued when there was no file locked on the file channel. (*)

−**306** Too many suspends; an attempt was made to suspend a file when the number of files currently suspended on this file channel is already the maximum allowed (twenty). (*)

−**307** Too many resumes; a resume command has been issued when there is no file suspended on the file channel.(*)

−**308** Illegal characteristic; an attempt was made to access an undefined file characteristic. (*)

−**309** Cannot change characteristic; an attempt was made to write a read-only characteristic. (*)

−**310** Filer error; this is an internal error and should be reported. (*)

−**311** Cannot copy; the server has not been able to create a new file name for the copy command.

−**312** Not all data written; some of the characters in a write command were not sent to the device. In the case of a disk file, this indicates that the volume is full. For a device, it indicates that some characters were not recognised (e.g. unprintable characters for a printer).

−**313** Illegal filename extension; a command supplying a filename did not have a legal TDS filename extension.

−**314** Illegal attribute; one or both of the attributes passed to the file server was not a TDS attribute suitable for making a file. The most common reason for this is attempting to file a foldset. Foldsets may not be filed folds; their components must be filed folds.

−**315** Cannot seek; could not seek to the desired record in variable length record file mode.

−**316** Not all data read; for alien file modes, a complete record was not read in.

−**317** Unknown record length; a variable record length file could not be read as the record length is unknown.

−**318** Illegal record length; an incorrect record length was specified in an alien file mode.

−**319** Cannot create; the file server failed to create a unique name for a filed fold.

E.2   DOS errors

These errors are generated by the DOS operating system.

−**401** DOS error ; invalid DOS function number. This is an internal error and should be reported.
For an explanation of this and other DOS errors (those in the range 400–499) see the DOS manual.

The DOS error number is given by making the TDS error number positive and subtracting 400 (e.g. −409 = DOS error 9). (∗)

−**405** DOS error; Access denied.

−**406** DOS error; Invalid file handle. This is an internal error and should be reported. (∗)

−**407** DOS error; Memory control blocks destroyed.

−**408** DOS error; Insufficient memory.

−**409** DOS error; Invalid memory block address.

−**410** DOS error; Invalid environment.

−**411** DOS error; Invalid format.

−**412** DOS error; Invalid access code.

−**413** DOS error; Invalid data.

−**416** DOS error; Remove current directory.

−**417** DOS error; Not same device.

E.3   TDS internal errors

−**501** Fold with invalid type encountered.

−**502** Fold with invalid contents encountered.

E.4   Filer errors

These errors may be generated by the TDS, in response to operations on the filer channels.

−**1103** File does not exist; the file that contains the data of a a filed fold cannot be found.

−**1104** Invalid file identifier; a file identifier has been used which is not valid.

−**1105** Invalid operation; a read, write, seek or close command has been issued on a file that is not open or has been opened with the wrong access type, e.g. a write to a file opened for reading.

−**1126** Unknown unit.

−**1127** Invalid drive name.

−**1129** Bad format; a file's contents do not correspond to TDS format.

−**1130** Too many files; a multi file operation (e.g. copy a fold and all it's contents) finds the fold it is operating on overflows it's 'nested file' stack.

−**1131** Fold too deep; a fold has been encountered at too great a fold depth.

−**1132** File too deep; a file has been encountered at too great a filed fold depth.

−**1140** Invalid command; a command outside the range of valid commands has been issued.

−**1141** Too many open files.

−**1142** Fold manager full; the fold manager has run out of storage space.

−**1143** File too big; a file is too big to be read in and stored in the fold structure.

E.5   File streamer errors

These errors are generated by the file streamer, either because some operation has failed or the file streamer has been driven incorrectly.

−**1201** Invalid command; a command has been issued that is not permitted for a particular file or a command has been issued which is outside the range of valid commands for this process. E.g. a write stream command when a file is opened for reading.

−**1202** Fold not empty; a command has been issued which requires an empty fold has been applied to a non-empty fold.

−**1203** Operation has failed; caused by filing system error when trying to obey a command. E.g. failing to delete a filed fold.

−**1204** Fold does not exist; either fold specified in a command does not exist or the top fold of a bundle has been opened, rendering those folds inside it inaccessable.

−**1205** Fold is already in use; fold is already opened for reading or writing and a new attempt to open it has been made.

−**1206** Fold is not filed; a fold must be filed before it can be opened for reading or writing. Also may be returned in reponse to is fold filed test, if fold is not filed.

−**1207** Invalid fold; a command has been issued which is not permitted on this fold.

−**1208** Failed to open fold; filing error opening a filed fold for reading or writing.

−**1209** File already exists; copy command has been issued for a filed fold on a system where the copying of filed folds is not permitted.

−**1210** Only generated by library procedures. The program has been called with the cursor not on a line appropriate for the current operation.

F Fold attributes

F.1 Fold attributes in the TDS

The three attributes of a fold are the fold type, the fold contents, and the fold indent. These attributes are communicated in an array of three integers, indexed by the values **attr.fold.type**, **attr.fold.content** and **attr.fold.indent** defined in appendix D.6. The meaning of each attribute is described in the following sections.

F.1.1 Fold type

The foldtype attribute tells the editor what operations it may carry out on the fold.

Fold types defined are:

ft.opstext Indicates that the fold can be opened and displayed by the system editor.

ft.opsdata Indicates that the fold contains non-ASCII data, or data unsuitable for display by the editor.

ft.opscode Indicates that the fold contains executable code which may be loaded and run by the system; this is non-ASCII data and is hence unsuitable for display by the editor.

ft.foldset Indicates a compilation fold that has been compiled and the source has not subsequently been changed.

ft.voidset Indicates a compilation fold that has not been compiled or in which the source has been changed since compilation.

F.1.2 Fold contents

The contents type attribute indicates what the fold contains, for the benefit of any utilities or programs which may have to operate on the fold.

Fold contents values defined include:

fc.comment.text Comment text can be edited, but is ignored by all compilers. May be created by the editor or as output from various TDS tools.

fc.source.text occam source text can be edited and input to a compiler.

fc.code.data A data fold containing the unlinked output from a compilation.

fc.desc.data A data fold containing descriptor information from a compilation.

fc.debug.data A data fold produced by a compiler containing information required to help diagnosis of run time errors.

fc.link.data A data fold produced by the compiler containing information needed by the linker.

fc.occam2.sc An occam 2 SC compilation fold containing separately compilable procedures. May be a foldset or a voidset or a corresponding **CODE** fold.

fc.occam2.prog An occam 2 **PROGRAM** compilation fold containing a program for a transputer network. May be a foldset or a voidset or the corresponding **CODE** fold.

fc.occam2.util An occam 2 **UTIL** compilation fold containing a utility package. May be a foldset or a voidset or the corresponding **CODE** fold.

fc.occam2.exe An occam 2 **EXE** compilation fold containing a procedure executable within the TDS. May be a foldset or a voidset or the corresponding **CODE** fold.

fc.occam2.lib A **LIB** library fold containing a library identifier and a sequence of text folds, containing constant or protocol definitions, or separate compilation units, containing library procedures. May be a foldset or a voidset.

Other values have been allocated and are either used for obsolete purposes or are reserved for future use in connection with support for other programming languages, etc.

F.1.3   Fold indent

The fold indent attribute is a number indicating the indentation of this fold, relative to the enclosing fold.

Text lines contained within a fold are stored relative to the indentation of their enclosing fold. Before displaying a line, the editor calculates the indentation of the line from the sum of the indent attributes of its enclosing folds. It then inserts that number of spaces before the text of the line. When reading a file in data stream mode, the user filer performs a similar task.

By writing bad or misleading attribute values into the fold structure it is possible to create a fold structure that crashes the system. Therefore system conventions should be followed.

Attribute values not listed here are reserved by INMOS for possible future expansion.

F.2   Attribute constant values

These are defined in the tables **Values for type** and **Values for content** in appendix D.6.

F.3   Attributes of common fold types

This table shows what attributes are used for the fold types commonly found in the TDS fold structure.

Type of fold	fold.type	fold.contents	Key word(s)	Filename ext
occam source	opstext	source text		.tsr
comment	opstext	comment text	COMMENT	.tcm
executable program	opscode	occam2 exe	CODE EXE	.cex
utility package	opscode	occam2 util	CODE UTIL	.cut
network program	opscode	occam2 prog	CODE PROGRAM	.cpr
linked separate comp	opscode	occam2 sc	CODE SC	.csc
object code	opsdata	code data	code	.dcd
descriptor	opsdata	desc data	descriptor	.dds
linkage information	opsdata	link data	link	.dlk
debug information	opsdata	debug data	debug	.ddb
configuration information	opstext	config info	CONFIG INFO	.tci
separate compilation unit	foldset	occam2 sc	SC	
executable program bundle	foldset	occam2 exe	EXE	
utility package bundle	foldset	occam2 util	UTIL	
network program bundle	foldset	occam2 program	PROGRAM	
library fold	foldset	occam2 library	LIB	
library version	opsdata	source text	Library version	

In the above table the type and contents attributes are descrbed by their names, see appendix D.6. Those fold types shown with type *foldset* have type attribute **ft.foldset** or **ft.voidset** according as they have been compiled (and linked, etc) or not since the contents were most recently changed.

Those fold types for which no filename extension is given cannot be filed.

A source text fold in a *foldset* must be filed if it is to be checked or compiled.

The *opsdata* folds can only be created inside a *foldset*, by applying the compiler to it.

G File formats

This appendix discusses exactly how folded files and code files are mapped onto disk files.

G.1 Structure of folded files

In this section the structure of folded files is described. It is not necessary to understand this in order to use the TDS. It is of use if the user wishes to inspect the contents of folded files other than through the TDS user filer interface.

The stored version of a folded file consists of a sequence of bytes. This can be viewed as a list structure containing numbers and records. A record is a sequence of bytes, of any length up to 512. The stored elements, which are data elements (principally lines of text), fold elements and filed fold elements, are then constructed as lists from these elements.

File elements

A file is made up of a sequence of elements. The following types of elements may occur:

- A *number.element*. This element includes a (non-negative) value.

- A *record.element*. This element includes a record length value, followed by a sequence of bytes

- *Startlist.element* and *endlist.element*. These elements are used to group a sequence of elements together.

- *Startfold.element* and *endfold.element*. These elements are used to bracket the elements of a fold.

- *Startfiled.element* and *endfiled.element*. These elements are used to bracket the elements of a filed fold.

Element encoding

The representation used for elements is designed to be totally independent of the word length or order of significance of bytes within words on the computer on which it is implemented.

Elements are identified by their first byte, the tag byte. The tag byte is made up of two bit fields, the 6-bit data field occupying the 6 least significant bits of the byte, and a 2-bit tag field.

tag.byte   =   *tag.field data.field*

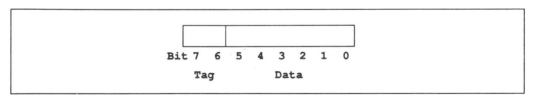

Figure G.1 Element tag byte

The following tag bytes identify elements:

record.byte   A *tag.byte* with tag 0 (byte value **0 – #3F**)
number.byte   A *tag.byte* with tag 1 (byte value **#40-#7F**)
function.byte A *tag.byte* with tag 2 (byte value **#80-#BF**)

The data field is used to give the (non-negative) value of a number element or a function element and the length of the record element. A record element is concluded by the number of bytes as given by the value of its tag byte.

As 6-bits is insufficient to represent the full range of values we will be required to store in files, there is a fourth tag byte, the prefix byte:

prefix.byte   A *tag.byte* with tag 3 (byte value #C0 - #FF)

The data field of a prefix byte is used to extend the data field of the succeeding tag bytes, the data field of the earlier prefix bytes being for the more significant bit positions of the resulting value. A full algorithm is given below.

The elements of a file can be defined as follows:

number.element	=	{ *prefix.byte* } *number.byte*
record.element	=	{ *prefix.byte* } *record.byte* { *byte* }
function.element	=	{ *prefix.byte* } *function.byte*

The following function values are defined:

startlist.element	A *function* whose data value is 0. (#80)
endlist.element	A *function* whose data value is 1. (#81)
startfold.element	A *function* whose data value is 2. (#82)
endfold.element	A *function* whose data value is 3. (#83)·
startfiled.element	A *function* whose data value is 4. (#84)
endfiled.element	A *function* whose data value is 5. (#85)

```
VAL data.field        IS #3F :
VAL data.field.size   IS 6 :
VAL tag.field         IS #C0 :
VAL record.byte       IS 0 << data.field.size :
VAL number.byte       IS 1 << data.field.size :
VAL function.byte     IS 2 << data.field.size :
VAL prefix.byte       IS 3 << data.field.size :

-- PROC read.tag reads tag bytes until a non prefix byte is found.
-- INT tag is the non prefix tag encountered.
-- INT value is value built up from the data fields of the prefix byte
--     and the terminating tag byte.
-- NOTE that if a record byte tag is encountered, we need to read
--      the data bytes of the record before looking for the next
--      element.

PROC read.tag(INT tag, value)
  INT b, data :
  SEQ
    value := 0
    read.byte(b) -- read byte takes an INT parameter
    tag := (b /\ tag.field)
    data := (b /\ data.field)
    WHILE tag = prefix.byte
      SEQ
        value := (value + data) << data.field.size
        read.byte(b)
        tag := (b /\ tag.field)
        data := (b /\ data.field)
    value := value + data
```

Structure of a file

A file is made up of a sequence of elements, bracketted by *startlist* and *endlist*.

> *file   =   startlist.element { element } endlist.element*

where an element may be one of a data element, a fold element, or a filed fold element.

> *element   =   data.element*
> *| fold.element*
> *| filed.element*

A data element

A data element is stored as a record, or as a number.

> *data.element   =   record.element*
> *| number.element*

A fold element

A fold element is stored as a pair: a list of attributes (of which the first is the header string) and the bracketted contents.

> *fold.element   =   startfold*
> *startlist header { attribute } endlist*
> *startlist { element } endlist*
> *endfold*

A filed fold element

A filed fold is stored as a pair: the attribute list and a file pointer.

> *filed.element   =   startfiled.element*
> *startlist.element header.element*
> *{ attribute.element } endlist.element*
> *file.pointer.element*
> *endfiled.element*

The *file.pointer.element* is a *record.element* giving the file id of the file where the contents are to be found.

Note that this means that the header and attributes of a filed fold are contained in the enclosing file, not in the file with the contents.

> *file.pointer.element   =   record.element*
> *header.element   =   record.element*
> *attribute.element   =   number.element*

G.2   DOS files produced by the TDS

TDS format files and host text files produced by the TDS file server are flat, binary DOS files. They contain no structural information. The server enforces 512-byte block boundaries upon reads from block files, by maintaining its own idea of position within a block.

Host text format files are written as:

text.file   =   { *record* **cr lf** }
record   =   { *ch* }

where *record* is the byte slice parameter to the **tkf.write** command. No control-Z is appended to the end of a text file as this will generate a 'not all data written' error.

Host text format files are read as:

{{ *record* } **cr** [**lf**] } [**ctrl-Z**]

(i.e. a line feed after carriage return and a control-Z after the end of the file are optional.) **cr**, **cr lf**, and **cr lf ctrl-Z** combinations are stripped from the record before sending it to the transputer.

G.3   CODE PROGRAM files

A CODE PROGRAM file created by EXTRACT is a folded file whose contents is a sequence of records. If such a file is exported to DOS by means of WRITE HOST, the contents of these records are written contiguously into a host binary bootable file.

The contents of these records are such that if they are communicated sequentially into a transputer network through its boot link, as by the LOAD NETWORK operation, then the whole network will receive its code and start executing. This implies that all necessary bootstrapping and loading code is included in the CODE PROGRAM file. The conventions for this code are described in Technical Note 34.

A particular kind of bootable file is made by the **addboot** tool from a compiled standard hosted procedure. This is not a folded file. It starts with a primary and a secondary loader and then a block of header information and the rest of the code. The primary loader is restricted to less than 256 bytes and is preceded by its length in a byte. The secondary loader starts with an integer, whose size is the word length of the processor type, which is its length in bytes. The contents of the header block are defined in the description of a CODE SC below.

Type	Value	Unit
BYTE	Primary loader code size	bytes
[]BYTE	Primary loader code block	
INT32	Secondary loader code size	bytes
or **INT16**	(for 16 bit transputers)	
[]BYTE	Secondary loader code block	

G.4   CODE SC, CODE EXE and CODE UTIL files

A CODE SC may be created explicitly by applying EXTRACT to a compiled SC fold, or will be created automatically if such a fold is within a configured PROGRAM fold. A CODE EXE or CODE UTIL will be generated automatically when an EXE or UTIL fold is compiled.

Any of these forms of code file is suitable for loading and execution by a controlling program. The TDS itself is an example of such a controlling program. As created such a code file is a folded file consisting of a sequence of data records. Optionally the record brackets may be removed from such a file by using

WRITE HOST . The file format so produced is then identical to loadable code files produced by INMOS toolset products. The records consist of a header block, a size block and code records.

The components of this structure and their meanings are:

Type	Value	Unit
INT32	Interface descriptor size	bytes
[]BYTE	Interface descriptor	
INT32	Compiler id size	bytes
[]BYTE	Compiler id	
INT32	Target processor type	
INT32	File format version	
INT32	Program scalar workspace requirement	words
INT32	Program vector workspace requirement	words
INT32	non-occam stack requirement	words
INT32	Program entry point offset	bytes
INT32	Program code size	bytes
[]BYTE	Program code block	

When writing a program to interpret this format it important to note that in general the integers will not be word-aligned.

The interface descriptor defines the language implementation version. For TDS3 programs it is **occam 2 product compiler (10th March 1988)** This is unchanged from the previous TDS. In INMOS toolset products this string is empty.

The compiler id is a string defining exactly which compiler version was used to compile this code.

The target processor type is derived from the compilation parameters:

Possible values are:

2	T2 series
4	T414
8	T8 series
9	T425 series
10	TA transputer class
11	TB transputer class
12	TC transputer class

Version number is an integer defining the presence or absence of the vector workspace and non-occam stack space fields in this header.

	Vector Space	Stack Space
<10	No	No
10	Yes	No
≥11	Yes	Yes

In TDS3 CODE SCs are generated with version 10, EXEs and UTILs with version 12. In TDS2 EXEs and UTILs had version 11. This distinction is used by the TDS to call old EXEs and UTILs with the appropriate old parameter list.

Scalar workspace is the number of words required for the program's run-time stack.

Vector space is the number of words required for separate vector space for arrays so declared.

Non-occam stack size is the number of words required by some compilers for an additional run-time stack. The TDS cannot generate code files with this requirement.

In TDS format CODE files the next two integers (8 bytes) are held as a separate record.

The entry point offset is the offset from the base of the code of the first instruction within the code that is to be executed.

The code size is the total size of the block of code which follows. This will normally be rounded up by the compiler which generated it to a whole number of words on the target processor. The code in the file may not be word aligned as it is preceded by strings of arbitrary length. Care must be taken when loading code from a code file of this kind, to ensure that it is word aligned in memory. This is because it may contain word aligned constant tables.

G.5   Other compiler outputs

The structures of other compiler outputs are not documented. However, examples of programs which convert descriptor folds and debug information folds into readable form are included as examples with the software.

H Transputer instruction support

This appendix contains the list of transputer instructions supported by the **RESTRICTED** code insertion facility, and indicates the mnemonics for each instruction. These instructions are available when the compiler is targetted to a **T2**, a **T4**, a **T425**, or a **T8**, unless otherwise indicated. Instructions supported only by the **T8** are given in the final section. For the full instruction set the reader is referred to the book 'Transputer instruction set: a compiler writer's guide'.

H.1   Direct functions

ADC	Add constant
CJ	Conditional jump
EQC	Equals constant
J	Jump relative
LDC	Load constant
LDL	Load local
LDLP	Load local pointer
LDNL	Load non local
LDNLP	Load non local pointer
STL	Store local
STNL	Store non local

H.2   Short indirect functions

ADD	Add
BSUB	Byte subscript
DIFF	Difference
GT	Greater than
LB	Load byte
PROD	Product
REV	Reverse
SUB	Subtract
WSUB	Word subscript

H.3   Long indirect functions

AND	And
BCNT	Byte count
CCNT1	Check count from 1
CFLERR	Check single length floating point infinity or NaN (**T4** only)
CSNGL	Check single
CSUB0	Check subscript from 0
CWORD	Check word
DIV	Divide
FMUL	Fractional multiply (**T4** and **T8** only)
LADD	Long add
LDDEVID	Load device identity
LDIFF	Long difference
LDINF	Load single length infinity (**T4** only)
LDIV	Long divide
LDPI	Load pointer to instruction
LDPRI	Load current priority
LDTIMER	Load timer

LMUL	Long multiply
LSHL	Long shift left
LSHR	Long shift right
LSUB	Long subtract
LSUM	Long sum
MINT	Minimum integer
MOVE	Move message
MUL	Multiply
NORM	Normalise
NOT	Not
OR	Or
REM	Remainder
ROUNDSN	Round single length floating point number (**T4** only)
SB	Store byte
SETERR	Set error
SHL	Shift left
SHR	Shift right
STTIMER	Store timer
SUM	Sum
TESTERR	Test error false and clear
TESTHALTERR	Test halt-on-error
TESTPRANAL	Test processor analysing
UNPACKSN	Unpack single length floating point number (**T4** only)
WCNT	Word count
XDBLE	Extend to double
XOR	Exclusive or
XWORD	Extend to word

H.4   Additional instructions for IMS T425 and IMS T800

The following IMS T425 and IMS T800 instructions are supported by the **RESTRICTED** code insertion facility.

BITCNT	Count bits set in word
BITREVNBITS	Reverse n bits in word (where $1 \leq n \leq 32$)
BITREVWORD	Reverse all bits in word
CRCBYTE	Calculate CRC on byte
CRCWORD	Calculate CRC on word
DUP	Duplicate top of stack
MOVE2DALL	Two-dimensional block copy
MOVE2DINIT	Initialise data for two-dimensional block move
MOVE2DNONZERO	Two-dimensional block copy non zero bytes
MOVE2DZERO	Two-dimensional block copy zero bytes
WSUBDB	Form double word subscript

H.5   Additional instructions for IMS T800 only

The following IMS T800 instructions are supported by the **RESTRICTED** code insertion facility.

FPADD	Floating point add
FPB32TOR64	Bit32 to real64
FPCHKERR	Check floating error
FPDIV	Floating point divide
FPDUP	Floating duplicate

FPEQ	Floating point equality
FPGT	Floating point greater than
FPI32TOR32	Int32 to real32
FPI32TOR64	Int32 to real64
FPINT	Round to floating integer
FPLDNLADDDB	Floating load non local and add double
FPLDNLADDSN	Floating load non local and add single
FPLDNLDB	Floating load non local double
FPLDNLDBI	Floating load non local indexed double
FPLDNLMULDB	Floating load non local and multiply double
FPLDNLMULSN	Floating load non local and multiply single
FPLDNLSN	Floating load non local single
FPLDNLSNI	Floating load non local indexed single
FPLDZERODB	Load zero double
FPLDZEROSN	Load zero single
FPMUL	Floating point multiply
FPNAN	Floating point NaN
FPNOTFINITE	Floating point finite
FPORDERED	Floating point ordorability
FPREMFIRST	Floating point remainder first step
FPREMSTEP	Floating point remainder iteration step
FPREV	Floating reverse
FPRTOI32	Real to int32
FPSTNLDB	Floating store non local double
FPSTNLI32	Store non local int32
FPSTNLSN	Floating store non local single
FPSUB	Floating point subtract
FPTESTERR	Test floating error false and clear
FPUABS	Floating point absolute
FPUCHKI32	Check in range of type int32
FPUCHKI64	Check in range of type int64
FPUCLRERR	Clear floating point error
FPUDIVBY2	Divide by 2.0
FPUEXPDEC32	Divide by 2**32
FPUEXPINC32	Multiply by 2**32
FPUMULBY2	Multiply by 2.0
FPUNOROUND	Real64 to real32 without rounding
FPUR32TOR64	Real32 to real64
FPUR64ROR32	Real64 to real32
FPURM	Set rounding mode to round minus
FPURN	Set rounding mode to 'nearest'
FPURP	Set rounding mode to round positive
FPURZ	Set rounding mode to round zero
FPUSETERR	Set floating point error
FPUSQRTFIRST	Floating point square root first step
FPUSQRTLAST	Floating point square root end
FPUSQRTSTEP	Floating point square root step
LDMEMSTARTVAL	Load value of MEMSTART (not **T8**)
POP	Pop operand stack (not **T8**)

I Bibliography

This appendix contains a list of some transputer-related publications which may be of interest to the reader. The References section details publications referred to in this manual, other than the standard INMOS documents detailed below.

I.1 INMOS publications

D Pountain and D May
A tutorial introduction to occam programming
Blackwell Scientific 1987

INMOS
occam 2 Reference Manual
Prentice Hall 1988

INMOS
occam
Keigaku Shuppan Publishing Company 1984
(In Japanese)

INMOS
Transputer reference manual
Prentice Hall 1988

INMOS
Transputer instruction set: a compiler writer's guide
Prentice Hall 1988

INMOS
Transputer technical notes
Prentice Hall 1989
Contains technical notes 0, 1, 2, 5, 9, 10, 17, 18, 19, 24, 27, 29, 46 and 49.

INMOS
Communicating process architecture
Prentice Hall 1988
Contains technical notes 6, 7, 20, 21, 22, 23, 32, 36, 37, 47 and 51.

INMOS
Digital signal processing
Prentice Hall 1989

INMOS
The graphics databook
INMOS 1989

INMOS
The transputer development and iq systems databook
INMOS 1989

INMOS
The transputer applications notebook — systems and performance
INMOS 1989
Contains technical notes 0, 1, 2, 5, 9, 10, 17, 18, 19, 24, 27, 29, 33, 34, 46, 49 and 58.

I.2   INMOS technical notes

R Shepherd
> *Extraordinary use of transputer links*
> Technical note 1
> 72 TCH 001

P Moore
> *IMS B010 NEC add-in board*
> Technical note 8
> 72 TCH 008

S Ghee
> *IMS B004 IBM PC add-in board*
> Technical note 11
> 72 TCH 011

M Poole
> *occam program development using the IMS D700D transputer development system*
> Technical note 16
> 72 TCH 016

P Atkin
> *Performance maximisation*
> Technical note 17
> 72 TCH 017

D May and R Shepherd
> *The transputer implementation of occam*
> Technical note 21
> 72 TCH 021

N Miller
> *Exploring Multiple Transputer Arrays*
> Technical note 24
> 72 TCH 024

G Harriman
> *Notes on graphics support and performance improvements on the IMS T800*
> Technical note 26
> 72 TCH 026

R Shepherd and P Thompson
> *Lies, damned lies, and benchmarks*
> Technical note 27
> 72 TCH 027

M Poole
> *occam input and output procedures for the TDS*
> Technical note 28
> 72 TCH 028

L Pegrum
> *Configuring occam programs*
> Technical note 31
> 72 TCH 031

R Shepherd
> *Security aspects of occam 2*
> Technical note 32
> 72 TCH 032

J M Wilson
> *Analysing transputer networks*
> Technical note 33
> 72 TCH 033

J M Wilson
> *Loading transputer networks*
> Technical note 34
> 72 TCH 034

S Redfern
> *Implementing data structures and recursion in occam*
> Technical note 38
> 72 TCH 038

T Watson
> *Module motherboard architecture*
> Technical note 49
> 72 TCH 049

M Poole
> *Example programs in the TDS*
> Technical note 56
> 72 TCH 056

I.3   References

W J Cody and W M Waite
> *Software Manual for the Elementary Functions*
> Prentice Hall 1980

D E Knuth
> *The Art of Computer Programming*
> 2nd edition, Volume 2: Seminumerical Algorithms
> Addison-Wesley 1981

IEEE
> *IEEE Standard for Binary Floating-Point Arithmetic*
> ANSI-IEEE Std 754-1985

J Glossary

Alias check Ensure all elements are identified by a single name within a given scope.

Alien file server See *Host file server*.

Alien language Sometimes used to refer to programs written in a language other than occam (such as C or FORTRAN), when these programs are embedded as processes in an occam program.

Analyse Assert a signal to a transputer to tell it to halt at the next descheduling point, and allow the state of the processor to be read. In the context of 'analysing a network', analyse all processors in the network. One of the system control functions on transputer boards.

Attach Make a filed fold which refers to a file already existing in the filing system. The opposite of *Detach*.

Autoload Load the standard working set of utilities and user programs in the Autoload fold using the AUTOLOAD key.

Backtrace Using the debugger, move from a position within a procedure or function body to the call of that procedure or function.

Bootstrap A transputer program, loaded from a ROM or over a link after the transputer has been reset or analysed, which initialises the processor and loads a program for execution (which may be another loader).

Bootable file A file whose contents is a sequence of bytes which may be sent down a transputer link with a reset transputer, which will then bootstrap itself, and optionally other transputers attached to it by other links, to a state where a complete network program is loaded and running.

Configuration The association of components of an occam program with a set of physical resources. Usually used in this manual to refer to the specific case of allocating occam processes to processors in a network, and channels to links in the network. The term is used, depending on the context, to describe the act of deciding on these allocations for a program, the occam code which describes such a set of allocations, or the act of applying the COMPILE function to the occam description.

Configurer A program which will place and execute processes on a specified configuration.

Core dump Memory dump.

Deadlock A state in which two or more concurrent processes can no longer proceed due to a communication interdependency.

Detach Make a filed fold into an empty fold without deleting the underlying file in the filing system. The opposite of *Attach*.

Element A syntactic structure (a name, subscripted name or segment) which selects variables, channels, timers or arrays.

Error modes The error mode of a compilation determines what happens when a program error (such as an array bounds violation) occurs. A program may be compiled in one of three error modes: HALT, STOP, or REDUCED.

Error signal In the transputer, an external signal used to indicate that an error has occurred in a running program. One of the system control functions in transputer boards, in which error signals are OR-ed together to indicate an error has occurred in one of the transputers in the network.

Extract Synonym for *Link*, used in the context of **SC** and **PROGRAM** folds.

Filed fold A fold whose contents are stored in a separate file in the filing system.

Fold The basic unit of data storage within the TDS programming environment. When used for the storage of text, it consists of a sequence of text lines which may be displayed on the screen (when the fold is open), or hidden away and replaced by a single line (when the fold is closed). When used for the storage of data, the fold contains a sequence of data records; it appears within the editor as a single line. A text fold may contain nested folds within it, and these may be data folds.

Fold bundle A fold which contains a sequence of nested folds. When running a user program with the cursor on the fold bundle, the user filer interface allows separate access to each of the nested folds in the bundle.

Foldset A special kind of fold bundle used for compilation folds. It has the property that its fold type attribute is either **foldset** or **voidset**. If the fold type attribute value is set to **foldset** (which happens after a program has been compiled) then if any changes are made to the contents of the fold the attribute value is set to **voidset**. The latter is also known as an 'uncompiled foldset'.

Folded file stream In the context of the user filer interface, the sequence of communications required to read or write a file, including all the fold information stored with it.

Free variables The variables which are referred to in a procedure, but declared outside of it.

Hard channels Channels which are mapped onto links between processors in a transputer network (used in contrast to *Soft channels*).

Host Depending on the context, this term is used either to describe the computer which is running the server to the TDS, and providing the filing system and terminal I/O, or (when used in contrast to *target system*) is used to describe the combination of the computer and the transputer which is running the TDS.

Host file server A file server which provides access to the filing system and terminal I/O of a host operating system, which may be used when running standalone programs. Sometimes known, for historical reasons, as the 'alien file server'. A different file server is used to support the TDS.

Library A collection of text folds (containing constant and **PROTOCOL** definitions), and **SC** folds (containing procedure and function definitions) which may be shared between parts of a program or between different programs.

Library logical names Names which may be looked up in a fold in the Toolkit fold to provide host file names for libraries for particular targets and error modes.

Link In the context of transputer hardware, a noun referring to a serial communication link between processors. Used as a verb, in the context of program compilation, to collect together all the code for a compilation unit (which may involve making copies of code stored in subsidiary compilation units or in libraries which are used), and put the collected code into a single file.

Linker The program (part of the compiler) which links a compilation unit.

Loader Depending on the context, refers to the part of the TDS which loads a transputer network or, more commonly, refers to a small program which is loaded into a transputer, and which may then distribute code to other transputers, and load a larger program on top of itself.

Locate Change the current cursor position in the fold structure to the line which caused an error.

Network A set of transputers connected together using links, as a connected graph (i.e. in such a way that there is a path, via links and other transputers, from one transputer to every other transputer in the set).

Peek and poke Read and write locations in a transputer's memory, by communication over a link, while the transputer is waiting for a bootstrap.

Preamble Part of a transputer loader program; this part initialises the state of the processor.

Priority In the transputer, the priority level at which the currently executing process is being run. The IMS T800, IMS T414 and the IMS T212 all support two levels of priority, known as 'high priority' and 'low priority'.

Protocol The pattern of communications between two processes, often including communications on more than one channel. When appearing as **PROTOCOL**, refers to the occam meaning of the term (see the occam 2 Reference Manual).

Reset Transputer system initialisation control signal.

Root processor (or Root transputer) The processor in a transputer network which is immediately connected to the host, and through which the network is loaded or analysed.

Separate compilation A self-contained part of a program may be separately compiled, so that only those parts of a program which have changed since the last compilation need to be recompiled.

Server A program running in a host computer attached to a transputer network which provides access to the filing system and terminal I/O of the host computer. The server is normally used to boot up the network as well.

Soft channels Channels declared and used within a process running on a single transputer. (used in contrast to *Hard channels*).

Standalone program A program running outside of the TDS, and without being connected to the TDS; usually booted onto a transputer network from a host computer, and supported, while it is executing, by a server running on the host.

Subsystem In a transputer board architecture, the combination of the Reset, Analyse and Error signals which allows the board to control another board on its subsystem port.

Target processors Processor types or classes determining the details of the code generated in a compilation.

Toolkit fold A special fold maintained by the TDS which is always accessible using the ENTER TOOLKIT key. It is used to hold pointers to tools and utilities, the library logical names fold and parameter folds.

Tools fold A fold within the toolkit fold containing pointers to tools.

Transputer classes Sets of transputer types for which identical target code may be compiled.

Usage check Check a program to ensure that it obeys the occam rules preventing the sharing of variables between parallel processes, and the rules about the use of channels as unidirectional point-to-point connections.

User filer A component of the TDS which may be communicated with by a program running within the TDS, to allow the program to read and write data within the fold structure.

Valid fold set A fold set is marked valid by the system when it is known that the data contained in the fold set is consistent with the current version of the source file.

Validate In the context of 'validating a library' check the components placed into the library fold, and give the library a new version number.

Vector space The data space required for the storage of vectors (arrays) within an occam program.

Worm A program that will distribute itself through a network of transputers (perhaps with an unknown topology) and allow all the processors in the network to be loaded, tested or analysed.

Workspace The data space required by an occam process; when used in contrast to *Vector space*, it means the data space required for scalars within the process.

Index